T0281442

Water Resources Development and Management

Cecilia Tortajada
Editor

Increasing Resilience to Climate Variability and Change

The Roles of Infrastructure and Governance in the Context of Adaptation

Editor
Cecilia Tortajada
Institute of Water Policy, Lee Kuan Yew
School of Public Policy
National University of Singapore
Singapore
Singapore

ISSN 1614-810X ISSN 2198-316X (electronic)
Water Resources Development and Management
ISBN 978-981-10-9476-7 ISBN 978-981-10-1914-2 (eBook)
DOI 10.1007/978-981-10-1914-2

Printed on acid-free paper

This Springer imprint is published by Springer Nature
The registered company is Springer Science+Business Media Singapore Pte Ltd.

Foreword by Benedito P.F. Braga

In the spirit of cooperation that embodies the World Water Council, collaborative thinking is central to shaping approaches to climate resilience. While humanity experiences increasing demographic and socioeconomic stresses, recent episodes of extreme climate around the world bring additional complexities in finding solutions to reduce these stresses. Water is one of the most impacted resources, but water also provides solutions to these challenges. It is key in mitigating and adapting to uncertainty, both now and in the future.

Present climate variability poses important social and economic impacts in the vulnerable, less affluent counties in the world. IPCC reports that extreme hydrological events, such as droughts and floods, are likely to be more intense and frequent in the future. At the same time, increasing demands for multiple water uses (irrigation, hydropower, domestic needs, industrial use, or ecosystems) place additional pressure on already-limited water resources, intensifying competition for water within and between states and raising the importance of how we manage this precious resource. Increased uncertainty can also reduce the potential for investments, particularly in countries with the greatest needs.

In this book, the World Water Council and the Government of Mexico, whose generous support enabled this work to be carried out, drew on influential and inspiring experts worldwide to examine the role of water infrastructure, management practices, and governance in generating increased resilience under increasing climatic uncertainty. The findings of studies in the following pages from across multiple continents suggest inspiring best-bet options for building and managing water systems that increase resilience to climate change.

I am confident that this book will be an important contribution to shaping the water security and sustainability agendas in the coming years. In alignment with the role of the World Water Council, this endeavour seeks to encourage continued dialog between the scientific community and decision-makers for better water management under uncertainty. Importantly, it also moves us another step forward towards improving understanding between water and climate change professionals—because after all, climate is water.

Benedito P.F. Braga
President of the World Water Council

Foreword by Roberto Ramirez de la Parra

Climate change is a reality affecting all productive, economic, social, and environmental sectors. However, its effects are mainly reflected in the hydrological cycle behavior and temperature anomalies, which result in substantial changes in the average rainfall patterns and in the frequency and intensity of extraordinary hydro-meteorological phenomena (droughts and floods).

In particular, precipitation variation and the consequential surface runoffs have caused once-efficient and safe hydraulic infrastructure to become inefficient and hazardous. In some cases, this has resulted in a lack of water discharge capacity or conveyance.

To address this problem, society at large has established new public policies and programs, including the implementation of mitigation and adaptation to climate change actions, especially of governance actions undertaken by the society jointly with the authorities to reduce these effects.

The experiences described in this book will provide the reader with a better overview of the current climate challenge and of the adaptation actions using successful case studies that could be replicated in other regions of the world.

This book is not just a compilation of good intentions. It is a testimony of the ongoing global efforts to reduce the vulnerability of the population and infrastructure faced with an increasingly evident and extreme climate change.

Roberto Ramírez de la Parra
Director General of the National Water Commission of Mexico
(CONAGUA)

Foreword by Dogan Altinbilek

Uncertainty about the potential impacts of climate change and variability on human and natural systems globally requires that the government, private sector, academia, and nongovernmental organizations continue with their efforts to further generate and disseminate knowledge on this topic.

The World Water Council has joined the international community in its search for a deeper understanding of these themes. In collaboration with the National Water Commission of Mexico (CONAGUA) and the National Association of Water and Sanitation Utilities of Mexico (ANEAS), the Council has supported in-depth studies in several regions, basins, and projects in different parts of the world. The lessons learnt are many. Specific conditions aside, the studies confirm the relevance of water infrastructure in the overall resilience and adaptation of systems (economic, social, and environmental). A word of caution is that fundamental infrastructure that is built to store and regulate water under normal years may not be as efficient under climate change conditions, where extreme events are expected to increase.

As Chair of the Steering Committee, I trust that case studies and analyses in the book will be an important reference for policymakers and new infrastructure developments. I extend my most sincere thanks to all who contributed to this project.

Dogan Altinbilek
Vice President of the World Water Council
Chair of the Steering Committee

Foreword by Claudia Coria

Climate has no borders, nor can it differentiate between rich and poor countries, social conditions. It does not know how efficient the institutions responsible for resource management can be. What is a fact is that today the world is facing more complex problems related to water resources management. Therefore, this publication presents the most representative cases of actions and experiences of different countries on water infrastructure management and operation. We are certain it can be used as a benchmark for solving similar problems around the world, helping to build resilience to climate change.

This book contributes to the achievement of Sustainable Development Goal (SDG) 6: "*Ensure availability and sustainable management of water and sanitation for all.*" In particular, it contributes to the target of expanding international cooperation for capacity-building activities and programs on water and sanitation, including water harvesting and storage.

Claudia Coria
International Affairs Manager
National Water Commission of Mexico
Member of the Steering Committee

Foreword by Jerome Delli Priscoli

Climate, water and security debates are raising public anxiety about potential changes and its impacts. Most of the projected impacts have to do with water such as droughts floods, sea level rise, tsunamis and more. While anxieties around these events are rising, relatively little investment is going into helping the people who we know will be affected and to increase their capacity to adapt to climate change impacts. This gap is likely to become a significant ethical as well as scientific public policy issue. Water resources investments and management are critical means for people and social systems to adapt to such projected climate changes and to increase resilience of social systems in the face of their potential impacts. They have been so to social stability and to civilizations throughout our history and are even more so today. The collaboration between the World Water Council, National Water Commission of Mexico and the National Association of Water and Sanitation Utilities of Mexico has produced an important volume which illuminates pathways to address our need to increase our resilience worldwide.

Jerome Delli Priscoli
Chair Global Water Partnership Technical Advisory Committee (TAC)
Member of the Steering Committee

Foreword by Liu Zhiguang

I am both proud and delighted to see the publication of this book. As the world is witnessing the increasing impacts of climate change on water resources, it is high time for countries around the world to share their experience and perspective on effective adaptation strategies. The adjustment and optimization of water infrastructure management will undoubtedly will be an important component of these strategies. I applaud the authors and supporting organizations for their valuable contribution to this important endeavor jointly launched by World Water Council and the National Water Commission of Mexico. I also hope the case studies presented in this book will contribute to the ongoing global efforts in addressing the impacts of climate change.

Liu Zhiguang
Consul, Ministry of Water Resources, China
Member of the Steering Committee

Foreword by Roberto Olivares

Achieving water security is a priority objective for governments around the world because, in recent years, the uncertainty caused by climate change and variability has alerted the international community. Water management demands adaptation strategies to face the challenge of ensuring that water resources meet the needs of the biosocial systems. In this regard, this study conducted by the World Water Council (WWC), the National Water Commission (CONAGUA), and the National Association of Water and Sanitation Utilities of Mexico (ANEAS) provides an in-depth analysis of relevant related factors such as governance, policy, technology, innovation, and infrastructure, identifying drivers for change and new management models. We hope this publication will become a reference to maximize the adaptive strategies worldwide.

Roberto Olivares
Director General of ANEAS
Member of the Steering Committee

Preface

Some of the main concerns globally at present include climate variability and change, their potential implications, and related uncertainty. Extreme events have become development challenges of increasing importance from all points of view—economic, social, and environmental—for countries worldwide. Floods that displace thousands of people and droughts that cause reservoirs to decline to record limits affect the economies of entire countries, constrain livelihoods, and affect the environment—and they seem to have become too-frequent events, not only in developing countries but also in developed ones. There is also the impression that economic, social, and ecological impacts have become more serious. More important, there does not seem to be an end in sight.

Decision makers, water users, researchers, and the general public all wonder what the future will look like. How will the environment change? How much will we have to adapt to the changing conditions and what this will mean regarding quality of life? These concerns have resulted in increasing attention to adaptation strategies. These strategies are expected to potentially allow growing populations to maintain, and also improve, living standards and make the world as resilient as realistically possible.

Regarding water resources, in order to provide water in the required quantity and quality for all sectors (domestic, livestock, food, energy, the environment), adaptation strategies have to ensure that resources are conserved and supply is maintained (and even increased under situations of stress). This requires further studies on whether past events are analogues for the future and to what extent, and if lessons from previous droughts and floods can be applied to future events.

Given that reservoirs are some of the most important buffers against droughts and that one of their most important roles is flood protection, the question arises as to whether construction of new reservoirs should be encouraged or whether small projects should be developed instead. Because new construction may not always be possible for economic, social, environmental, or dogmatic reasons, a feasible alternative is to look into their reoperation, which may be more effective under the present, and perhaps also future, conditions. A limitation could be, however, that reoperation of reservoirs requires management, governance, and financial considerations that are not always easy to realize.

With the objective to advance the understanding of infrastructure and governance aspects in the context of resilience and climate change adaptation as well as the policy implications for water resources management, the World Water Council and the National Water Commission of Mexico (CONAGUA) in collaboration with the National Association of Water and Sanitation Utilities of Mexico (ANEAS) have supported a series of case studies to study these issues in depth. They are included in this book.

The case studies discuss floods and droughts events in projects, basins, and regions. The geographical focus includes the Arid Americas (United States and Mexico), Australia, Brazil, China, Egypt, France, Nepal, Mexico, Pakistan, Turkey, and South Africa. The case studies discuss the relevance of infrastructure in adaptation strategies; how infrastructure has been planned, operated, and managed to date; how operation is changing; how it should change more to respond to changing climatic, economic, social, and environmental situations; and what the constraints are for any change. Governance aspects (policies, institutions, and decision-making) and technical and knowledge limitations are a sizeable part of the analyses. These comprise how decision-making has improved and should improve more, considering the needs and wants of economic sectors, societies, natural environments, users, riparian countries, etc.

The studies explain that extreme events are not new and that civilizations have adapted to them throughout history. What is different at present is the extent to which human activities have overexploited natural resources, including water. As a consequence, resources are now scarce, polluted, misallocated, mismanaged, and misgoverned, which exacerbates the impacts of both droughts and floods. A mix of policy options, managerial, governance, technology, and also behavioral alternatives will have to be tried based on multisectoral needs and resource availability under local conditions so that resilience becomes an achievable aim under specific scenarios.

In this context, water infrastructure is essential to build resilience contributing to adaptation to climate variability and change. However, for it to be effective, it needs to be planned and managed within a governance framework that considers long-term perspectives and multisector and multilevel actor needs and perspectives.

Improved monitoring will be more essential to identify shock events and understand to what degree physical and natural events can adapt to them and how. Furthermore, it is necessary to plan for strategies that are able to contribute toward building their resilience. Progress in knowledge, science, and technology will allow a more comprehensive understanding of future climate events. In the end, however, it will be decision-making that will predetermine the course of events.

Climate variability and change are forcing the implementation of policies and practices that aim at reducing water demand; improving management, development, and governance; and building systems that are more resilient. Sustainable water management will be improved in particular, but it will be the overall socioeconomic development of the countries and their populations that will gain in the long run.

Singapore

Cecilia Tortajada
Senior Research Fellow

Contents

Chapter 1
Water, Governance, and Infrastructure for Enhancing Climate Resilience

Cecilia Tortajada

Abstract Water management, climate change, and climate variability—along with their numerous interlinkages and the extent of related hydrologic, economic, social, environmental, and political impacts over time and space—have become issues of increasing global concern. Despite their importance for freshwater systems and the fundamental services they provide, numerous factors and uncertainties prevent us from forecasting their likely future multidimensional and multisectoral impacts with any accuracy. These factors include, but are not necessarily limited to, a lack of scientific understanding of their complexity and interrelationships, as well as an absence of reliable data that would otherwise allow us to understand them more reliably. Therefore, policy alternatives, management and development decisions, and investment choices on any adaptation strategy are challenging tasks under the best of circumstances. Given the uncertainties related to climate change and variability, nonclimatic factors have thus become more relevant. Governance—or decision-making by multiple actors with dissimilar interests grouped under formal and informal institutions—is one of the most important. This chapter discusses several case studies that analyze the roles that infrastructure and governance play in the context of adaptation in increasing resilience to climate variability and change in different regions, basins, and projects. Water infrastructure is fundamental to build resilience and adapt to climate variability and change. However, to be effective, it needs to be part of a governance framework that considers multisector needs and multilevel actors in the longer term.

Keywords Resilience · Climate change · Infrastructure · Governance · Adaptation

C. Tortajada (✉)
Lee Kuan Yew School of Public Policy, Institute of Water Policy, National University of Singapore, 469C Bukit Timah Road, Singapore 259772, Singapore
e-mail: cecilia.tortajada@nus.edu.sg

C. Tortajada (ed.), *Increasing Resilience to Climate Variability and Change*, Water Resources Development and Management,
DOI 10.1007/978-981-10-1914-2_1

1.1 Introduction

According to the Intergovernmental Panel on Climate Change (IPCC 2007), resilience is the ability of a social or ecological system to absorb disturbances while retaining the same basic structure, ways of functioning, capacity for self-organization, and capacity to adapt to stress and change. The concept of resilience is often linked to climate change; however, the related issues are much broader. In fact, social and ecological systems have always had capacity to adapt to stress and change. They have been essential to the progress of societies throughout the history of humankind. The numerous local and global changes in human and natural environments, global use of and competition for scarce natural resources, growing populations, globalization, information and communication revolutions, increasing social expectations, and rapid progress in scientific and technological developments have affected—either positively or otherwise—the resilience of populations and their environments all over the world.

Adaptation, in the context of climate change, has been described as "initiatives and measures to reduce the vulnerability of natural and human systems against actual or expected climate change effects" (IPCC 2007). *Mitigation*, also in the context of climate change, has been described as "technological change and substitution that reduce resource inputs and emissions per unit of output. Although several social, economic and technological policies would produce an emission reduction, with respect to climate change, mitigation means implementing policies to reduce greenhouse gas emissions and enhance sinks" (IPCC 2007).

Freshwater systems and their future development depend on climate-related factors such as precipitation, temperature, and evaporative demand, as well as on nonclimatic factors. Given the uncertainties related to climate change and variability and the lack of data to predict them with certainty within given timeframes, nonclimatic factors have become more relevant. Examples include management, governance and policy issues, land use considerations, infrastructure development (reservoirs, groundwater storage, and recovery), technology and innovations, and diversification of water resources through water reuse and desalination.

Reservoirs have become an integral part of basic infrastructure by offering indispensable benefits, such as domestic, livestock, industrial, and irrigation water supplies, as well as hydropower, flood control, drought mitigation, navigation, fish farming, and recreation. The construction of new reservoirs has often been controversial during recent decades because social and environmental impacts have not always received due consideration (Scudder 2012). However, the limited and skewed distribution of water over time and space has created difficulty in meeting the increasing demands of uses and users at national, regional, and global levels—some of them new. Thus, the world has realized that more reservoirs, their improved operation, and broader policy, management, development, and governance alternatives are necessary to promote development and meet basic human needs. Global dynamics in terms of water, energy (including electricity trade), food, and climate securities have recast the importance of the roles of reservoirs in national development. In many cases, this has triggered massive investments in the construction and modernization of multiple projects in both developed and developing countries, such as Turkey, China, India, Laos PDR, Bhutan, and the United States (Tortajada 2014; Tortajada et al. 2014).

The World Water Council and the National Water Commission of Mexico (CONAGUA), in collaboration with the National Association of Water and Sanitation Utilities of Mexico, have supported a series of case studies on the roles that infrastructure and governance play in the context of adaptation to increasing resilience to climate variability and change in different parts of the world. These case studies aimed to improve understanding of the policy, management, governance, and development of water resources in changing environments.

These case studies focus on regions, river basins, and projects, including the arid Americas (with examples from the United States and Mexico), the Murray–Darling River basin in Australia, the Koshi River basin in Nepal, the Indus River basin in Pakistan, the Yellow River basin in China, and the Durance–Verdon River basin in France. In terms of reservoirs, the case studies examine Sobradinho in Brazil, high-risk dams in Mexico, Seyhan Dam in Turkey, High Aswan Dam in Egypt, and Sterkfontein Dam in South Africa. Some of the studies are transboundary, such as Arizona–Sonora and the Indus, Koshi, and Nile River basins. These cases have very different geographical and hydrological characteristics; they also differ in the related governance frameworks.

In all of the studies, the authors—who are from the government, private sector, and academia—analyze the numerous challenges faced by policymakers and society at large at the basin, national, and regional levels. These issues are related to governance (policies, institutions, and decision-making), infrastructure development, technical and knowledge constraints and limitations, and the economic, social, and environmental impacts of climate change and variability. Decision-making has changed but still needs to improve in the face of climatic uncertainty, knowledge and data gaps, aging infrastructure, increasing demands for water resources that tend to be scarcer and more polluted for an increasing number of uses and users, and numerous unresolved water management-related concerns. Many problems that need to be addressed more immediately seem to have lost urgency because of climate concerns; in reality, all concerns should be addressed concurrently.

The case studies demonstrate that decision-making has become quite complex. Traditional decisions on infrastructure development are based on historical trends that lack the understanding of long-term uncertainty in climate variability and change. There are concomitant issues concerning erosion and sedimentation, ecosystem degradation, social impacts, sharing benefits with local communities, and financial and institutional implementation modalities. However, as climate-related conditions change, knowledge and understanding of their social, economic, environmental, and political effects need to improve and become more comprehensive—a goal that is still in progress globally. More robust policy, institutional, development, and management frameworks need to be developed and implemented in collaboration with users and affected sectors of society.

The authors of the case studies argue that overall decision-making should become more efficient. In many cases evolve from monolithic and unidirectional structures to more responsive and inclusive ones, and concerned with the interests, contributions, and needs of several sectors, including society generally.

1.2 Infrastructure in a Changing Environment

Infrastructure is an indispensable component of economic growth, poverty reduction, and climate change adaptation. Infrastructure has distinct potential to help overcome growth constraints, respond to urbanization pressures, improve social and environmental conditions, encourage competitiveness and productivity, underpin improvements in quality of life and social inclusion, and enlarge and speed up communication and mobility. Thus, infrastructure should be an ongoing priority for all public and private sectors in society (Bhattacharya et al. 2012). Developing infrastructure in isolation of development policies, efficient management, and consideration of social needs and environmental concerns has only limited value.

In relation to water resources management, it is assumed that the stationarity on which water resources have been managed historically is no longer valid and that water systems have to be optimized in different ways. It is argued that the extent of alteration to the means and extremes of precipitation, evapotranspiration, and rates of river discharge due to anthropogenic effects make it necessary to identify new nonstationary probabilistic models of relevant environmental variables. In turn, these variables will provide the relevant information necessary when renewing and building water infrastructure (Milly et al. 2008. For a discussion on Milly et al. 2008 and nonstationarity, see WMO no date).

Most dams and reservoirs were constructed when climate change was not a consideration. Despite this, they have been able to respond to potential impacts of climate change and cope with extreme events by means of operation or modernization in some cases (Lima and Abreu 2016; Muller 2016) but not all cases (Alcocer-Yamanaka and Murillo-Fernandez 2016; Pittock 2016; Selek et al. 2016).

In the uncertain face of climate change, governments are undecided on the most appropriate water management practices and development-related decisions to help their countries adapt positively while continuing to grow economically and socially. Rather than focusing on models alone, the case studies in this book suggest that more attention should be given to adaptation mechanisms that consider water infrastructure as a means to build resilience by augmenting and regulating water resources and thus availability. More attention is also required for more efficient water management; poor practices may not only constrain development under normal conditions but also exacerbate extreme events.

Because extreme events are likely to increase in the future, it is essential that countries prepare for them. Although infrastructure is critical, it is not enough. Policies, institutions (both formal and informal), laws, regulations, management practices, pricing instruments, and participation models that encourage the efficient management of water resources and water-related services are fundamental to build resilience (Molden et al. 2014). Given the changing local, regional, and global economic situation, social expectations, and environmental considerations, it is essential to assess the extent to which infrastructure can actually help build resilience, how it can be improved, if its operation needs to be changed, or, in extreme cases, if it needs to be put out of operation or even decommissioned. This assessment will be achieved only within a broader framework of development, with governance as an essential component.

Policymakers are aware of the importance of strategies for adaptation to climate change, particularly with respect to water resources and ecosystem management. They are also aware that adaptation measures are needed to deal with more or less precipitation. Nevertheless, the devil is in the details. After all, it may be that societies will have to learn how to live with less water and more widespread landscape transformation in the non-too-distant future (Overpeck and Udall 2010). Adaptations are needed either to more or less precipitation. These adaptations include policy measures that are more comprehensive, the involvement of formal and informal institutions at different levels, demand management strategies that focus on the basins rather than the cities (even when the basins are transboundary), conservation practices, technological innovations, market-based solutions, infrastructure adaptation, and, most importantly, the engagement of users and the public in general (AghaKouchak et al. 2015).

The use of all means available and the development of new alternatives for the purposes of adaptation to climate change—but more immediately, to local, regional, and global changes—are in the political agendas of governments all over the world. Climate change challenges have joined the long list of challenges that governments and populations face on daily basis and have made them re-evaluate their water and development policies. These challenges include the provision of services to larger urban populations and to impoverished rural communities; a race to achieve economic growth and compete globally for the provision of products and services; increasing demand for natural resources that are more limited, polluted, scarce, and overexploited and for which competition has become global in scope; and very complex decision-making processes that frequently involve formal and informal institutions concurrently from more than one sector, basin, region, and (in many cases) country. As mentioned by Scott and Lutz-Ley (2016), extreme events seriously limit management responses to increase the supply of water resources. Therefore, it is important to improve the effectiveness of current management practices, which in many cases leave much to be desired.

1.3 Governance to Build Resilience

Governance usually refers to decision-making that is more inclusive and considers multiple partners, including formal and informal institutions at different levels and in several sectors. It provides several actors with the opportunity and responsibility to contribute local experiences and traditional knowledge, which have the potential to render great benefits. Governance is expected to result in better management of water resources under normal conditions and especially during extreme events, which will result in greater resilience (Pittock 2016).

The comparative assessment of Arizona and Sonora in this book focuses on factors affecting water security in the two states of two different countries, United States and Mexico. These factors include hydroclimatic and water-demand uncertainties, the capacity for institutional learning, the level of flexibility of policies for infrastructure that also address water governance, and the existence of science–policy linkages to underscore reservoir infrastructure and governance mechanisms to improve resilience. One of the most important arguments that Scott and Lutz-Ley (2016) make is that the provision of water requires the implementation of policy instruments (such as pricing and water rights for water allocation or reallocation), institutional arrangements or rearrangements, new rules or rules as necessary, financing mechanisms, social learning, and, of course, built infrastructure for it to be effective.

Governance can be an important element when the development of infrastructure and reallocation of water resources are necessary, as any change in flow allocation that will benefit any use and users will affect all of them—more so in the so-called closed basins or vulnerable basins (Falkenmark and Molden 2008; Pittock 2016). Among these basins, the Murray–Darling is an example where institutional and policy reforms as well as governance practices have evolved by responding to the extreme climatic variability faced through the years. The numerous reforms have focused, for example, on groundwater extraction and water accounting to purchase and manage water entitlements for environmental flows (Pittock 2016). Even though opinions vary on the effectiveness of reform implementation, environmental flows have been an essential component of the development and policy agendas in the basin.

A basin where infrastructure is a priority but where institutional and policy reforms for them still have to become more effective is the Koshi (also known as Kosi) River basin in Nepal. Although water infrastructure is urgently needed for water supply, energy generation, and flood control (even when this has been debated), it will render more positive results if the infrastructure is developed within a framework of responsive governance, both nationally and at the transboundary level (Wahid et al. 2016). As the authors discuss, it is not only the need to develop infrastructure under conditions of climate uncertainty but also development challenges that should encourage more comprehensive decision-making. To be more effective, institutional arrangements, policy options, and the costs and benefits of irrigation and hydropower development projects should be based on cumulative effects economically, socially, and environmentally, as well as at multiple levels rather than individually (as is done at present). Planning and implementation should also benefit from governance practices that consider more successful models of stakeholder engagement that include responsibility, accountability, and transparency aspects. For infrastructure development to benefit the country in the long term, planning and implementation should consider the potential impacts of climate change and, more broadly, the economy of the country, the benefits to society, and the impacts to the natural environment on which development of the country depends.

The study of the Indus River in Pakistan shows how extensively flood disasters can hinder not only the development of the basin but of the country. Mendoza and Khero (2016) argue that the management of water resources and infrastructure development, as well as its proper operation prior and during floods, should enable the river system to absorb adverse impacts. They also argue that this has proven to be a real challenge because of the multiplicity of actors involved in individual planning, decision-making, and management of the river, both nationally and at the transboundary level.

The case study explains that the 2010 floods were the worst in the country's history, causing damages of approximately USD 10 billion. They affected all provinces, inundated 38,600 km^2, and resulted in the deaths of some 1600 people. The reasons for the flood included not only the intensity of rainfall but also an inadequate early warning system and reduced capacity of the available waterways. Despite the large number of reservoirs in the basin, the floods in 2010 (and also 2014) indicate not only that storage is needed—its operation needs to be more efficient. In the specific case of the Indus River basin, provinces have to coordinate water management options rather than deciding on them individually. Systems are interdependent and management practices taken in different places in the basin have impacts in other places. Water infrastructure is a requirement to build resilience in the Indus River basin. However, also necessary are comprehensive policies, institutions that are able to set systems that are cost effective, multisector and multilevel governance structures, data collection and sharing of information, monitoring of multihazards, risk assessment evaluations, technology for flood early warning systems, communication of risk information to the vulnerable population, and the building of response capabilities at community and national levels (Mendoza and Khero 2016).

A very large river basin in China—the Yellow River basin—has seen the benefits of infrastructural development as well as improvements in the institutional structure and policy—and decision-making. A total of 30 large dams and hydropower facilities have been built on the main stream of the river for water supply for domestic, agricultural, and industrial uses, as well as power generation, drought management, and flood control. This extensive infrastructure development has had both positive and negative impacts and has resulted in significant changes of flow regime and water allocation. For example, in the 1990s and until 2003, the river would run dry before it reached the sea. A more comprehensive development framework was implemented by the Yellow River Conservancy Commission.

The infrastructure was reoperated with a focus on water scarcity, water quality, environment flow and ecological restoration, flood discharge, and optimization of flow and sediment transportation (Sun and Fu 2016). The end result has been the river flowing into the sea.

Sun and Fu (2016) argue that, despite the enormous benefits of the infrastructure, there are still numerous challenges related to the management of water resources at the river basin level, for which both structural and nonstructural measures are under implementation. Governance is one of the most important nonstructural measures. It includes changes in institutional arrangements as well as consideration of laws, regulations, policies, and economic approaches; management at the river basin level; and a policy framework that considers declining water availability, water quality degradation, riverine wetlands, geomorphology, sedimentation, and water conservation.

The last study on a basin focuses on the Durance–Verdon River basin in the south of France (Branche 2016). In this basin, water resources are under increasing pressure from considerable abstractions for domestic, irrigation, industrial, hydropower, and recreational uses. Given the importance of climate variability and change, population growth, and socioeconomic development, Électricité de France (EDF) developed three strategies for climate change adaptation: the R2D2 2050 program, value creation methodology, and water-savings agreements. The R2D2 2050 program tries to evaluate possible impacts of climate change on the quantity and quality of water resources, biodiversity, and changing demands and uses in the river basin in 2050. Model-based scenarios for future water demand were developed in collaboration with different stakeholders. The value creation methodology has the objective to identify the socioeconomic benefits of hydropower. The water-saving agreements are voluntary economic instruments that require EDF to pay back to users a part of the savings if specific targets are reached.

The three initiatives have been positive but have proven to be complex. For example, the scenarios developed under R2D2 2050 were very useful for collecting physical, biological, and socioeconomic data; understanding the impacts of hydropower development in the basin; and further realizing the complexity of the potential economic, social, and environmental impacts of climate change. Even when they were not able to consider water management in its totality, potential research topics were identified for further study. Through the value creation methodology, it was possible to establish a relationship between the financial and economic contributions of hydropower-related services, tax payments to the region, governance models, and thus the involvement of a broad base of stakeholders. Finally, the water-savings methodology is useful for understanding how complex decision-making can be when there are conflicting interests. A main finding of the study has been that more infrastructure does not necessarily contribute to adaptation or resilience to climate change impacts. What contributes more effectively is appropriate governance frameworks of the water storage available. This reinforces the arguments of other case studies in this book in the sense that infrastructure is necessary but so are governance practices.

Moving from river basins into projects, the first case study in the book is that of Sobradinho Reservoir in the Sao Francisco drainage basin. The basin covers 7.5 % of Brazil, with 60 % of it located in the semi-arid region of the country, which is subject to long periods of drought due to low precipitation and high evapotranspiration. Sobradinho—a multipurpose reservoir for human, livestock, irrigation, hydropower, fish farming, navigation, recreation, tourism, and environmental purposes—also plays an important role in flood control. Opinions vary on the effectiveness of the traditional reservoir's operation as allocation had favored one sector over the others. According to Lima and Abreu (2016), the main user of water of the reservoir was originally the hydropower sector; however, this has changed with time in order to meet multipurpose uses.

The case study discusses that the years 2014, 2015, and part of 2016 experienced the lowest average annual unregulated flows historically. This reduced the volume of water stored in the reservoir, affecting the related productive and nonproductive activities and thus the thousands of jobs and quality of life of the population. A series of mitigation measures were implemented to prevent further reduction of the water stored and minimum outflows were discussed, agreed upon, and implemented. The users in the Sao Francisco River basin are aware of the situations and of the possible restrictions. Lima and Abreu (2016) highlight the importance of good governance practices in the Sobradinho Reservoir during drought conditions, when the demands of many sectors have to be addressed concurrently. This is important because effective decision-making strategies need to be established; extreme events will likely be more frequent and more severe in the future.

Considering the possible impacts of more frequent and more severe extreme events, the Mexico study analyzes high-security dams (Alcocer-Yamanaka and Murillo-Fernandez 2016). The study assesses the mitigation measures that are needed to operate the dams under conditions of change in climate, the resulting and probable floods, and possible impacts on populations living nearby and downstream, including property and the environment. Measures are both corrective and preventive, including management and structural considerations, operating policies, dam safety laws, regulations and standards that define conservation, and monitoring and supervision procedures.

The study concludes that changes in meteorological, hydrological, and structural conditions, in addition to negligence, have resulted in approximately 115 dams becoming unsafe in the country. As in many other cases, land use changes (in this specific case, from agriculture and forested areas to mainly agricultural and urban areas) have resulted in increasing runoff and thus floodwater. Dams in urban areas require limiting storage during the rainy season to avoid any flooding downstream. The fact that some of them are for recreational purposes also creates a problem because they are kept full most of the year, resulting in lower flow capacity.

According to the study, the owners of the dams are aware of their conditions and possible alternatives (which may include taking the dams permanently out of service) and the final decisions on how to proceed depends on them. As Alcocer-Yamanaka and Murillo-Fernandez (2016) discuss, knowledge and understanding of hydroclimatological variables, evolution and forecasting, the current conditions of storage infrastructure, and flood control are essential for safe operation of the dams. Effective participation and collaboration of the water users and those in charge of the operation of infrastructure are necessary because they have practical implications in terms of safety.

The Turkish case study focuses on the Seyhan Dam in the Seyhan River basin. Seyhan Dam provides water for domestic, irrigation (mainly cotton), and industrial uses, as well as energy generation and flood control. It also has environmental benefits for the Adana Plain. Predictions for the Seyhan River basin in the future vary, but they estimate that precipitation will decrease (in one case up to 25–35 %), with changes in seasonality resulting in earlier snowmelt over the mountains. Less available water is expected to put additional stress on local populations, wildlife, and groundwater resources for irrigation, as well as increase the risk of pollution and salt water intrusion along the coastal regions, sometimes up to 10 m inland (Selek et al. 2016).

Given the importance of the Seyhan River basin for agricultural production locally and nationally and the expected lower precipitation in the future, Selek et al. (2016) argue that new water projects may have to be constructed: more efficient irrigation will be necessary but not be enough. Therefore, various adaptation measures for resilience building will have to be implemented in all sectors. These will include policy, legal, regulatory, institutional, governance, and coordination aspects, in addition to water conservation initiatives, the building of capacities at the basin-scale for water users and rural communities, and managerial and technological development and transfer. Resilience has to become more robust in all sectors in order to adapt to changing conditions—be they climatologic, economic, social, or environmental.

According to Selek et al. (2016), reservoirs that store and regulate river flow are essential to manage the impacts of climate variability on water resources. However, efficient institutions, management, land use planning, and conservation measures (not only storage) will determine the equitable allocation of available water resources to all users under stress conditions.

High Aswan Dam and Lake Nasser in Egypt are the focus of the case study by Biswas (2016). The author explains that Egypt depends entirely on a single source of water and that the needs and expectations of the population have been sustained for millennia by the waters of the Nile River. Water is a cross-cutting issue. In the case of Egypt, it is even more essential for water, food, energy, environment, and climate security, not to mention national security. Even so, poor management of water resources is still the rule rather than the exception, which has the potential to constrain development and exacerbate the impacts of extreme events. Water storage is as essential in Egypt as it is in the upstream countries in the Nile River basin. In the case of Egypt, Lake Nasser is critical because it stores and regulates the only source of water the country has. Therefore, additional storage options, more efficient water management that considers institutional rearrangements, and collaboration with upstream countries should be pursued. Otherwise, the task of delivering water derived from only one source for a growing population and for the increasing uses of water by the water, food, and energy sectors will become increasingly complex.

The final study presented in the book focuses on South Africa. Muller (2016) analyzes how the Sterkfontein Dam has contributed to the economic and social development as well as the water security of a large region in the country. The author explains how, in a variable river system, the operation of the dam has helped to manage the impacts of extreme events, such as droughts. As explained in the case study, as important as the dam and its operation have been, so have been the institutional arrangements that have enabled the most appropriate decisions to be made and implemented. During the 2016 drought, for example, the operation of the dam was so efficient that the users and sectors supplied from this system have not yet experienced water restrictions at the time of writing.

1.4 Water Infrastructure Within a Governance Framework: An Evolving Paradigm

The construction of dams, reservoirs, and other water infrastructure has traditionally been proposed as the most effective way to address water scarcity, meet increased water demand for several uses, and protect the infrastructure itself in the case of breaching of levees, such as barrages and bridges. Taking into consideration the potential impacts of climate variability and change, water infrastructure now has the objective of building resilience to more frequent and more severe events within a framework of uncertainty.

As discussed in all case studies in this book, water scarcity may not only be the result of climate change, but also of policy, management, development, and governance-related long-term decisions. Resilience (economic, social, environmental, and often also political) depends on multiple variables. For example, impoverished and vulnerable populations either in urban or rural areas are less able to face any type of pressure—be it economic, social, or environmental—because they lack basic services and infrastructure. Therefore, they are often simply unable to cope with extreme events (Scott and Lutz-Ley 2016).

In many places, planning for extreme events is still limited to reactive measures rather than preventive ones, to protect populations prior to the events and, in fewer cases, to support them after the events have occurred. Infrastructure is thus essential to provide the whole array of services that a population needs and demands. Comprehensive, well-thought-out plans are needed to make society resilient or resistant to extreme events in the long term. Poor management of resources and poor decision-making can limit very seriously any type of support public or social organizations can provide.

As suggested in all studies, one way or another, the benefits and costs of reservoirs must be considered at the river basin level rather than at the project level to be able to provide a broader view of the situation. This together with planning, management, financial, and governance strategies, as well as scientific advances and information systems, are likely to result in more comprehensive planning, efficient management, wider social and ecological processes, and more effective contributions and commitments from the different parties involved. An assessment of benefits and costs is fundamental not only in monetary terms (Scott and Lutz-Ley 2016) but also from an overall resilience viewpoint. This may be a challenging situation, but it will have to become a reality given the serious development and climate-related challenges that the world is currently facing. Resilience needs to be built at all levels.

References

AghaKouchak A, Feldman D, Hoerling M et al (2015) Recognize anthropogenic drought. Nature 524:409–411

Alcocer-Yamanaka V, Murillo-Fernandez R (2016) Adaptation and mitigation measures for high-risk dams, considering changes in their climate and basin. In: Tortajada C (ed) Increasing resilience to climate variability and change: the roles of infrastructure and governance in the context of adaptation. Springer, Berlin

Bhattacharya A, Romani M, Stern N (2012) Infrastructure for development: meeting the challenge. Centre for Climate Change Economics and Policy, London

Biswas AK (2016) Lake Nasser: alleviating impacts of climate fluctuations and change. In: Tortajada C (ed) Increasing resilience to climate variability and change: the roles of infrastructure and governance in the context of adaptation. Springer, Berlin

Branche E (2016) The Durance—Verdon river basin in France: the role of infrastructures and governance for adaptation to climate change. In: Tortajada C (ed) Increasing resilience to climate variability and change: the roles of infrastructure and governance in the context of adaptation. Springer, Berlin

Falkenmark M, Molden D (guest eds) (2008) Special Issue: Closed basins highlighting a blind spot. Int J Water Resour D 24(2):201–318

IPCC (Intergovernmental Panel on Climate Change) (2007) Fourth assessment report: climate change 2007. https://www.ipcc.ch/publications_and_data/ar4/syr/en/annexessglossary-a-d.html . Accessed 22 May 2016

Lima AAB, Abreu F (2016) Sobradinho reservoir—Brazil case study. In: Tortajada C (ed) Increasing resilience to climate variability and change: the roles of infrastructure and governance in the context of adaptation. Springer, Berlin

Mendoza G, Khero Z (2016) Building Pakistan's resilience to flood disasters in the Indus River basin. In: Tortajada C (ed) Increasing resilience to climate variability and change: the roles of infrastructure and governance in the context of adaptation. Springer, Berlin

Milly PCD, Betancourt J, Falkenmark M et al (2008) Stationarity is dead: whither water management? Science 319:573–574

Molden D, Vaidya RA, Shrestha AB et al (2014) Water infrastructure for the Hindu Kush Himalayas. Int J Water Resour D 30(1):60–77

Muller M (2016) Greater security with less water: Sterkfontein Dam's contribution to systemic resilience. In: Tortajada C (ed) Increasing resilience to climate variability and change: the roles of infrastructure and governance in the context of adaptation. Springer, Berlin

Overpeck J, Udall B (2010) Dry times ahead. Science 38:1642–1643. doi:10.1126/science.ll86591

Pittock J (2016) The Murray–Darling basin: climate change, infrastructure and water. In: Tortajada C (ed) Increasing resilience to climate variability and change: the roles of infrastructure and governance in the context of adaptation. Springer, Berlin

Scott CA, Lutz-Ley AN (2016) Enhancing water governance for climate resilience: Arizona, USA—Sonora, Mexico comparative assessment of the role of reservoirs in adaptive management for water security. In: Tortajada C (ed) Increasing resilience to climate variability and change: the roles of infrastructure and governance in the context of adaptation. Springer, Berlin

Scudder T (2012) The future of large dams: dealing with social, environmental, institutions and political costs. Earthscan, London

Selek B, Demirel-Yazici D, Aksu H, Özdemir AD (2016) Seyhan Dam, Turkey, and climate change adaptation strategies. In: Tortajada C (ed) Increasing resilience to climate variability and change: the roles of infrastructure and governance in the context of adaptation. Springer, Berlin

Sun Y, Fu X (2016) Yellow River: re-operation of infrastructure system to increase resilience to climate variability and changes. In: Tortajada C (ed) Increasing resilience to climate variability and change: the roles of infrastructure and governance in the context of adaptation. Springer, Berlin

Tortajada C (2014) Dams: an essential component of development. J Hydrol Eng 20(1):A4014005. doi:10.1061/(ASCE)HE.1943-5584.0000919

Tortajada C, Altinbilek D, Biswas AK (eds) (2014) Impacts of large dams. Springer, Berlin

Wahid SM, Mukherji A, Shrestha A (2016) Climate change adaptation, water infrastructure development, and responsive governance in the Himalayas: the case study of Nepal's Koshi River basin. In: Tortajada C (ed) Increasing resilience to climate variability and change: the roles of infrastructure and governance in the context of adaptation. Springer, Berlin

WMO (World Meteorological Organisation) (no date) A note on stationarity and nonstationarity. http://www.wmo.int/pages/prog/hwrp/chy/chy14/documents/ms/Stationarity_and_Nonstationarity.pdf. Accessed 22 May 2016

Chapter 2
Enhancing Water Governance for Climate Resilience: Arizona, USA—Sonora, Mexico Comparative Assessment of the Role of Reservoirs in Adaptive Management for Water Security

Christopher A. Scott and America N. Lutz-Ley

Abstract Climate variability and change exert disproportionate impacts on the water sector because water is a crosscutting resource for food production, energy generation, economic development, poverty alleviation, and ecosystem processes. Flexible surface water and groundwater storage together with adaptive water governance are increasingly recognized and deployed to strengthen climate resilience, specifically by buffering drought and flood extremes, bridging interannual variability, and providing for multiple uses of water, including environmental flows. Adaptation can be further enhanced by the following: (1) accounting for hydro-climatic and water-demand uncertainties; (2) strengthening institutional learning in relation to reservoirs (reoperations as well as mechanisms to address growing civil-society critiques of "hard-path dependence"); (3) increasing flexibility of policies for infrastructure (including readaptation to past cycles of infrastructure development); and (4) building on science-policy dialogues that link infrastructure and governance. An array of complementary adaptation tools will buttress climate resilience. Some emerging techniques include underground storage, distributed basin-wide enhancement of water retention, efficient water use (with limits on the expansion of new demands on saved water), and wastewater reclamation and reuse (with their own emerging storage and recovery techniques). Each of these techniques is directly linked to reservoirs in practical and operational terms. Conjunctive surface-water and groundwater storage must be further developed through infrastructure, institutional, and policy approaches including groundwater banking, trading and credit schemes, water swaps (substitutions and exchanges), and a robust approach to targeted water storage for climate resilience.

Keywords Reservoirs · Infrastructure · Groundwater banking · Water security · Adaptation · Readaptation · Hard path · Soft path · Arizona · Sonora

---

C.A. Scott (✉) · A.N. Lutz-Ley
University of Arizona, Tucson, AZ, USA
e-mail: cascott@email.arizona.edu

C. Tortajada (ed.), *Increasing Resilience to Climate Variability and Change*,
Water Resources Development and Management,
DOI 10.1007/978-981-10-1914-2_2

2.1 Introduction

Climate change exerts significant and increasing impacts on human societies and ecosystems globally. Rising atmospheric concentrations of greenhouse gases, particularly carbon dioxide, have resulted in warming temperatures and more variable precipitation accompanied by shifts in land cover, species distribution, and altered hydrological processes. For decades, the societal implications of climate variability and change have been recognized across a range of economic sectors, from agriculture to energy to urban development. The monitoring of environmental conditions and modelling tools have greatly aided the understanding of the spatial variability of climate impacts and their temporal trends (Overpeck et al. 2013). However, future uncertainty—particularly with regard to the distribution, timing, and intensity of precipitation (including its partitioning as rain or snow)—make freshwater availability an especially acute dimension of climate change. Additionally, climate variability resulting from El Niño Southern Oscillation (ENSO) and other decadal timescale variations in global circulation processes drives extreme events, including the unpredictable and often rapid sequencing of droughts and floods. As a result, the dynamics and uncertainty of terrestrial hydrological processes underlying the spatial and temporal distribution of water resources are central impacts of climate change.

Coupled with natural processes of global environmental change are human drivers, especially rapid demographic growth and accelerating economic development (Leichenko and O'Brien 2008). Together, natural and human drivers contribute to the integrated process of global change (Vitousek 1994). The exploitation of natural resources, mining, energy development, land conversion for agriculture and livestock production, soil-nutrient depletion, urban growth, industrialization, pollution (land, air, and water), and increasing carbon emissions from these and other human activities serve to intensify global change. Freshwater availability and quality are at the epicenter of global change, given the crosscutting importance of water for food (production using surface water and groundwater irrigation and food safety and handling), energy (electricity generation, biofuels, and environmental impacts of fossil and renewable energy sources), economic growth (essential provision of water and sanitation services for poverty alleviation, commercial and industrial development), and ecosystem processes (in-stream flows, wetlands, and biodiversity maintenance). Global change has resulted in systemic imbalances in water resource quantity and quality (Vörösmarty et al. 2013), particularly in arid and semi-arid regions (Scott et al. 2013). Expanding human demands for water in the context of erratic surface water availability are increasing the need for water storage to buffer extremes (Palmer et al. 2008), while at the same time shifting dependence to groundwater (Taylor et al. 2013).

The current understandings of climate and global change adaptation in the water sector are based on an underlying pressure-state-response (PSR) conceptual model. As extreme events, especially drought-induced water scarcity, exert pressure (P) and limit water-use activities (S), management responses (R) include efforts to increase supply and improve the effectiveness of current practices in the medium term. Water demand management is most often viewed as a strategy to alleviate short-term scarcity under the prevailing assumption that earlier conditions of supply will equilibrate. There is growing recognition, however, that interactive and bidirectional influences of mutual feedbacks in social-ecological systems can lead to unanticipated outcomes (Walker and Salt 2013; Moser and Pike 2015). Global change is producing a new normal; in other words, extreme conditions of the past are here to stay and indeed may well recur with increasing frequency. This raises the urgency of understanding global change adaptation in water management, including its spillover effects in other sectors.

Adaptation in the prevailing context of water scarcity produces a set of responses in individual sectors: for example, agricultural managers alter the timing and distribution of irrigation supplies, electrical power operators gear up for heightened electricity use, and urban water utilities seek to augment supply. Clearly, better integrated, cross-sectoral approaches to environmental and water governance (Tortajada 2010) are recognized, yet institutional disarticulation limits effective anticipatory planning and coordinated responses during and after times of crisis. Thus, short-term adaptation can be linear in that it posits a tangible set of manageable outcomes resulting from adaptive response measures. Such approaches are also based on static assumptions: that earlier conditions will recur with known frequency and that outcomes are predictable. These assumptions have too often led to chronic shortages in water availability and deteriorating quality for human and ecosystem purposes. This invariably entails iterative adaptation measures to respond to earlier cycles of management and, indeed, mismanagement—a process we refer to as readaptation, as expanded upon below.

The need for readaptation in the water sector responds to path dependency; that is, the prevailing courses of action and options that are available to managers are limited by the current stock of infrastructure, investments (both financial and institutional investments), and professional norms (Lach et al. 2004). Two principal, often contrasting, sets of options have been termed hard path and soft path strategies (Gleick 2003; Wolff and Gleick 2002, who acknowledge the use of hard/soft terminology from earlier thinking along similar lines in energy sector policy by Lovins (1976, 1979). Hard path options are primarily based on infrastructural or physical management (e.g., dams, reservoirs, aqueducts, and interbasin transfers), whereas soft path options are those based on social and organizational strategies (e.g., water pricing, public awareness, and changes in people's behavior; Brooks and Brandes 2011). New approaches based in engineering resilience (LRF 2015; Hollnagel et al. 2006) identify opportunities to enhance infrastructure robustness while ensuring safety and reliability of infrastructural assets, such as storage reservoirs and the services they provide—for example, water supply (Scott et al. 2012) and flood control. In most situations, particularly where water stress is pronounced and cumulative, hard and soft path measures coexist and may be judiciously combined. In other contexts, where infrastructural stock is low or the reduction of losses from disasters like flooding is an overriding imperative, hard path options have tended to receive emphasis.

The heightened variability of global change processes, the context-specific nature of impacts and responses, and above all, the uncertainty inherent in spatial and temporal trends in natural and human drivers raise the need for new understandings of water security. We define water security here following Scott et al. (2013) as "the sustainable availability of adequate quantities and qualities of water for resilient societies and ecosystems in the face of uncertain global change." We take water security to be more of a goal than a static condition or fixed end-state. This is particularly evident given the dynamic and evolving set of mutual interactions among water, energy, and food security. By pursuing flexible adaptive water governance, the resilience of societies and ecosystems to the water sector dimensions of global change can be significantly strengthened.

2.2 A Comparison of Arizona, USA and Sonora, Mexico

This study undertakes a comparative assessment of storage reservoir infrastructure for adaptive management and water security together with other emerging techniques to enhance climate resilience outcomes in Arizona, USA and Sonora, Mexico (see Fig. 2.1). The transboundary focus (Wilder et al. 2010; Browning-Aiken et al. 2004) is complemented by a discussion of broader relevance for the arid Americas (Varady et al. 2013), which addresses resilience to climate change and variability considering infrastructure, soft path measures, and governance.

2.2.1 The Arizona Case

Arizona's modern water history is inextricably linked to the development of water resources of the Colorado River and its tributaries. In 1911, Roosevelt Dam on the Salt River was completed with federal funding under the National Reclamation Act of 1902. After the 1922 Colorado River Compact that apportioned water to the seven states comprising the U.S. portion of the river basin, negotiations, political deals, and legal battles resulted in Arizona's current annual water allocation from the Colorado River of 3450 million cubic meters (MCM). However, the state was initially unable to utilize this allocation; agricultural lands and population centers of south-central Arizona were hundreds of miles away and at significantly higher elevations. The development of large dams on the Colorado River began with Hoover Dam, formally dedicated in 1935, which impounded Lake Mead with a maximum storage capacity of 35,700 MCM. This was later complemented by Glen Canyon Dam upstream, which created Lake Powell (Dean 1997) with a maximum storage capacity of 30,000 MCM. Lakes Powell and Mead are each capable of storing 2 years of average flow of the Colorado River, not considering climate change impacts.

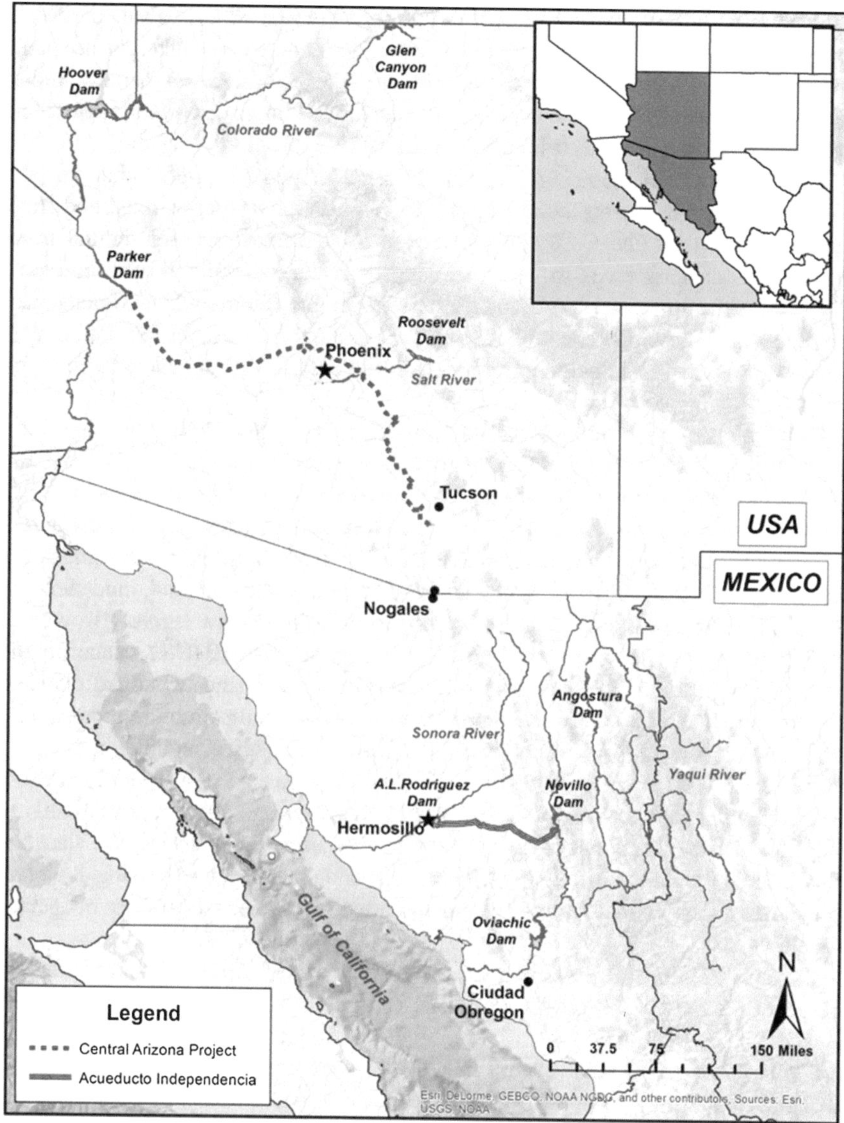

Fig. 2.1 Arizona, USA and Sonora, Mexico study locations

In 1947, the U.S. Bureau of Reclamation proposed to the Secretary of the Interior the construction of the Central Arizona Project (CAP). CAP is a 540-km aqueduct to convey water from Lake Havasu (capacity 798 MCM) behind Parker Dam on the Colorado River to urban and agricultural users in the central and southern parts of the state. CAP comprises pumped lift totaling almost 880 m over the length of the aqueduct, moving water up multiple individual pipeline systems with gravity flow in open-air, lined canals until the next lift (Eden et al. 2011). Construction of CAP began in 1973 and lasted 20 years, when water was delivered at Pima Mine Road just south of the city of Tucson. CAP is Arizona's largest supplier of water (1970 MCM annually) and its largest single consumer of electrical power (2.8 million megawatt-hours annually) (CAP 2010).

Development and financing of the CAP were imbued with hard- and soft-path considerations. The Carter administration in Washington was unwilling to approve federal funding for the CAP (or other large infrastructure projects of the time) unless Arizona took steps to better manage its water resources (Glennon 1995). These considerations led to Arizona's passage of the Groundwater Management Act in 1990 and establishment of active management areas around the state's main urban centers as well as irrigation non-expansion areas in water-scarce agricultural areas.

Climate change is having significant impact in the Colorado River Basin (Kates et al. 2012). During the late 1990s, through a combination of high inflows and alternative sources of water to at least partially meet off-river demand for irrigation and urban supply, the reservoirs were relatively full. However, persistently low inflows and high reservoir evaporation resulting from drought conditions in the 2000s through to the present coupled with growing demand and diversions of Colorado River water have reduced storage to near-record low levels.

As Arizona's population grew to 6.4 million people in 2010 (US Census Bureau 2010), in part due to assured water supply from CAP to the two main cities of Phoenix and Tucson, water management and associated infrastructure increasingly prioritized urban supply (Megdal 2012). Current strategies include interannual carryover of CAP water, particularly in the face of prolonged drought and interstate water-sharing of the Colorado River, and long-term storage, recovery, and banking of groundwater (Megdal et al. 2014). The readaptation challenges in the state are centered on flexibility in planning, demand management, and retooling the state water administration that was partially dismantled in 2008, ostensibly for budgetary reasons.

Roosevelt Dam on the Salt River, a Colorado River tributary within Arizona, presents an illustrative example of climate resilience opportunities and limitations. Constructed in 1911 to a height of 85 m, it was raised in 1996 at a cost of USD 430 million to its present level of 109 m with a maximum storage capacity of 2040 MCM. The dam, water supply, and hydroelectricity generation are managed by the Salt River Project (SRP), a community-based, not-for-profit public power utility in the greater Phoenix metropolitan area that is also the largest supplier of water to municipal, urban, and agricultural water users. SRP is an award-winning utility that reports highest customer satisfaction.

The river basin is subject to significant interannual variability (Wentz and Gober 2007). High rainfall in the watershed resulted in record inflows in February 1980, although flood operations of the reservoir proved capable of averting major damage downstream in the cities of Mesa and Phoenix. Complementing the reservoir's flood control functions is the Flood Control District of Maricopa County, which provides up-to-date hydroclimatic information, flood-hazard maps, and details on flood insurance. SRP is partnering with researchers at the University of Arizona to better incorporate downscaled climate model information into reservoir operations.

Downstream of Roosevelt Dam, conversion of agricultural water rights to urban supply (Gooch et al. 2007) in urban areas with low population densities supplied by the Salt River Project mean that on a per-area basis water allocations are considerably high. Thus, Roosevelt Dam, SRP, and complementary functions provided by the flood control district in the lower Salt River basin have ensured water security through active and engaged institutional learning. The effects of expanding demand in the context of climate change and variable flows on the Salt River, however, require planning and control of urban growth that extend beyond SRP's mandate and are under the purview of zoning regulations and assured water supply rules.

2.2.2 The Sonora Case

Sonora's water administration is qualitatively different from that in Arizona, because formal institutions are centralized at the federal level in Mexico. The main organization managing water resources is the National Water Commission (CONAGUA), which is under the Ministry of Environment and Natural Resources (SEMARNAT). The modern water history in Mexico is centered on the creation of CONAGUA in 1989 and the enactment of the Law of National Waters in 1992 (Aboites 2009). In 1998, CONAGUA divided the Mexican territory into 13 hydrologic-administrative regions to facilitate water management and also designated river basin organizations including the Northwest Basin Organization (OCNO) in Sonora and tributary rivers in Chihuahua State (Diario Oficial de la Federación 1998). Although these actions were intended to decentralize water management, in reality the process was only partial because CONAGUA still retains control over water allocation, concession titles, regulations, and policy-making (Scott and Banister 2008).

Sonora also has its own state-level Water Commission (CEA Sonora) in charge of implementing local policies, data generation, and infrastructure development within the institutional boundaries set by the federal law. Approximately 90 % of all the water allocated in Sonora is for the agricultural sector, with 5 % for urban public supply (CEA Sonora 2008). Sonora relies heavily on groundwater, as does the entire country; it is the largest groundwater user in Latin America (Scott and Banister 2008). The state's 2.7 million population (INEGI 2010) is served by five river basin systems: Sonoyta, Concepción, Sonora-San Miguel, Yaqui-Mátape, and Mayo (CEA Sonora 2008). The Yaqui-Mátape basin—the largest in terms of area, availability of water resources, and irrigated area—is transboundary (with the United States) and interstate (with Chihuahua). In the lower basin lies the second largest city in the state and the second largest agricultural district in Mexico, which is the major wheat producer in the country. The Sonora-San Miguel basin includes the capital city of Hermosillo, the majority of the state's population, and the largest copper mine in Mexico.

In the heyday of infrastructure development, the Ministry of Hydraulic Resources (SRH), a predecessor of CONAGUA, built most of the 31 dams existing in Sonora. Three of the largest ones are located in the Yaqui basin, with volume stored of more than 6500 MCM. Of these, El Novillo Dam (also referred to as Plutarco Elias Calles), was built in 1964 in the mid-basin and is managed by the Federal Electricity Commission (CFE) for hydroelectricity generation. El Oviáchic (Álvaro Obregón) Dam, the largest in the basin, is a multipurpose reservoir like El Novillo; it controls water flows, generates electricity, serves eco-tourism enterprises (such as sport fishing), and, most importantly, is the primary source of water for Irrigation District 041, Río Yaqui, and for Ciudad Obregón, with its public water supply system connected to the district's network. Two other dams—Abelardo L. Rodriguez and El Molinito—were constructed in the Sonora River basin to regulate water flows and supply Hermosillo. Together, these two store almost 350 MCM (CONAGUA 2015).

Because of its geographic location, Sonora is subject to significant interannual climate variability with a strong influence of El Niño Southern Oscillation (Robles-Morua et al. 2015). It has a bimodal precipitation regime dominated by the North American Monsoon, with frequent occurrence of Pacific cyclones making landfall. In the last two decades, Sonora has had very low precipitation accompanied by record high temperatures. The Abelardo L. Rodriguez reservoir has been in a state of near-total depletion since the end of the 1990s. The Hermosillo water utility and state authorities have been looking for different solutions to meet urban water demand, in many cases relying on poor-quality, unreliable groundwater sources such as peri-urban wells, or purchasing agricultural water rights from ejidos surrounding the city, which generates social and political opposition (Pineda-Pablos et al. 2012). The recent history in Hermosillo and the Sonora River Basin (SRB) reflects the vicissitudes of achieving water security in a context of prolonged drought, intersectoral and intrasectoral competition, and less than transparent decision-making processes.

The strategies all have the same underlying logic—appropriating and transferring water from nearby sources to Hermosillo city through ad hoc combinations of soft and hard path measures, sometimes involving socioeconomic and institutional measures (such as purchasing agricultural water rights) and at other times construction of pipelines, new wells, or other infrastructure. In 2004, the recently municipalized urban water utility, Agua de Hermosillo, bought water rights from Ejido Las Malvinas, north of Hermosillo city, within the San Miguel watershed (SMW). This action generated strong opposition from CONAGUA and local residents of the area at the time. During fieldwork conducted in mid-2015, small-scale farmers and ranchers in the ejidos of the lower SMW believed that Hermosillo's water extractions were one of the reasons for the depletion of the San Miguel Aquifer, although Hermosillo's rights account for less than 15 % of the aquifer's annual recharge of 52 MCM (Pineda-Pablos et al. 2009), while large-scale agriculture has become an important player in the region. In 2006, the municipal utility bought rights for 17.4 MCM of groundwater from the Costa de Hermosillo Valley at Los Bagotes by negotiating the purchase with organized ejido and private farmers. In 2008, the State Government built a 28.3-km pipeline connecting El Molinito to a water treatment plant close to the Abelardo L. Rodríguez Dam. Although this was conceived as an augmentation strategy for providing Hermosillo with water during drought, in real terms the transfer was redundant because both dams were already part of the city's supply (Scott and Pineda-Pablos 2011). Soft path strategies during the 2000s included rationing and rotational water supply, leak detection, and educational programs in schools, communities, and the business sector under the general banner of water culture. These programs were based on discursive contents, not differentiated for each social group, more than on capacity building. Their effectiveness has not systematically been assessed (Lutz-Ley 2013). Hermosillo managed to cover urban demand during the drought in 2014; however, long-term supply challenges remain.

The 2006–2012 state administration developed the program known as Sonora SI (Sonora Sistema Integral), an ambitious plan of water infrastructure that included new and refurbished past projects, with investments ranging from improving irrigation channels to building dams for urban supply in different parts of the state (Robles-Morua et al. 2015; Prichard and Scott 2013). The centerpiece of these projects is the Acueducto Independencia, an interbasin transfer scheme originally projected to convey 75 MCM of water per year from El Novillo Dam in the Yaqui basin via a 150-km aqueduct to Hermosillo in the lower Sonora basin. Rights to the water to be transferred via the aqueduct were purchased from agricultural users in communities upstream of El Novillo. The construction of the aqueduct was accompanied by allegations of corruption, lack of transparency, and highly conflictive relationships between the water stakeholders of the Yaqui Valley (including urban, agricultural, and indigenous users) on the one hand, and the population of Hermosillo city and the state government on the other (Scott and Pineda-Pablos 2011; Lutz-Ley et al. 2011). The aqueduct is currently an important source of water for Hermosillo, especially after pollution in the Sonora basin in 2014 resulted from a major spill of the copper mine tailing ponds upstream in Cananea. Heavy metal contamination reached the Molinito Dam and is suspected to have also affected some of Hermosillo's wells in that area (El Imparcial 2015).

Several important issues emerge from the development and implementation of these strategies. First, it is clear that hard path and soft path strategies are intermingled and it is difficult to draw clear lines between their respective outcomes. The building of infrastructure for water transfers has led to institutional reorientation in the form of innovative responses, new rules, incipient water markets, and social learning. Evidence of this is the past two decades characterized not only by severe drought, but also by intense institutional learning and governance changes in Sonora.

Realistic options for the reallocation of water rights as required by water transfers are themselves dependent on the stock of infrastructure in place. In other words, the capacity and flexibility to actually move water require built infrastructure. In Sonora, this capacity is less well developed and, therefore, the policy mechanisms for reallocation, such as market and pricing, are less well articulated than in Arizona. It is clear that different levels of development and the evolution of governmental institutions affect the perceptions of decision-makers. Diverse actors across multiple levels of government and civil society have shifted roles and power positions in terms of water decision-making, characterized in general by the retreat of federal authorities to give room to more empowered local intervention, although CONAGUA fully conserves its regulatory capacity. Although not formalized in new legislation, this is a major governance change in comparison to management by CONAGUA and prior federal agencies relying on highly bureaucratic, centralized, and top-down approaches.

Second, the shifting power positions are also accompanied by innovations in infrastructure funding. Insufficient monetary resources have been one of the main reasons for the delay or cancellation of several water projects in Sonora (Pineda-Pablos et al. 2012; Scott and Pineda-Pablos 2011). The financial burden has progressively moved from the governmental bodies—particularly at the federal level—to more distributed financing schemes involving private and users' participation to a greater degree. For example, the Sonora SI plan included a major proportion of capital from build-operate-transfer (BOT) contracts, federal and state funds, irrigation districts, and Hermosillo city (Scott and Pineda-Pablos 2011). On the other hand, financing from local sources has not necessarily been matched by a commensurate degree of local self-determination in water or climate governance, because water legislation still remains in the hands of federal authorities. For example, an important component for starting the Sonora SI plan advocacy was directly related to the declaration of drought emergency in the state. This declaration must be approved by the Mexican Ministry of Interior (SEGOB) and is the key for accessing funds from the federal Fund for Natural Disasters (FONDEN). Drought declaration also played a major role in the social construction of water scarcity in Sonora (Lutz-Ley et al. 2011), which would ultimately serve to legitimate the decisions made by the Sonora SI plan.

On the other hand, the strong El Niño that started in 2014 after several years of drought evidenced the combined effects of both hydroclimatic stress and governance solutions to water scarcity. These outcomes are associated with: (1) the priority of managing drought instead of a broader planning framework that considers the full range of extreme events in the region (i.e., flooding and cyclones; Farfán et al. 2013); and (2) the imbalance of water allocation decision-making over spatial and power differences among water sectors. Urban and large-scale irrigation actors have been the most powerful negotiators when it comes to infrastructure (Scott and Pineda-Pablos 2011). This has promoted a large gap in the physical infrastructure available to deal with drought and flooding in rural settings in contrast to cities. While drought conditions have affected all reservoirs in the state, leading them to historic low levels of storage, the extraordinary rainfall in late 2014 and early 2015 entailed water releases from El Molinito and partial filling of the chronically depleted Abelardo L. Rodríguez Dam. In the SMW towns, located less than 100 km north of Hermosillo, the river washed away many hectares of cropland, according to the report of farmers interviewed in the summer of 2015. The rural communities depending on agricultural livelihoods are vulnerable to climate extremes because they lack the infrastructure to withstand drought for crops, livestock, and human domestic consumption. They are served by aging water pipelines, inadequate or inexistent storage facilities, and lack of wastewater treatment for effective wastewater reuse. Prevention of flooding risks is also a major concern in a basin where reservoirs and flow control are concentrated only in its lower section. In the words of some rural inhabitants, they are at their capacity limits when it comes to drought, not prepared at all when flooding or other extreme climate events hit the region. Cities such as Hermosillo do not have effective strategies to deal with flooding, other than reactive measures to protect or support the population after these events. However, certainly in terms of flood control, the dams existing close to cities have prevented major disasters so far. Tropical cyclones making landfall in Sonora (and other Pacific coast states in Mexico) pose very widespread flood risk from intense precipitation occurring over the entire coastal plain (Farfán et al. 2013). Cyclone events are difficult if not impossible to manage with infrastructure alone.

Today, Sonora faces important water challenges related to dual climatic and human impacts. Persistent gaps in wastewater treatment, partial enforcement of rules and legislation, and the effects of policies that have prioritized hard path measures all raise important (re)adaptation challenges in the state. In this sense, the potential of reservoirs to increase resilience to climate change is critically assessed in the context of new institutional arrangements for water transfers in urban–rural settings. These arrangements need to be referenced to particular contexts, especially based on the strong opposition that many water works have raised in Sonora in the recent past. The implications are not the elimination of reservoirs from the water security portfolio but the careful analysis of cost-benefit ratios of investments in ecosocial terms—that is, analyses that factor in social and ecological impacts. Climate change is projected to increase the frequency of extreme events in the region, requiring urban and rural populations to deal with continued and more severe droughts as well as with more extreme flooding. Robles-Morua et al. (2015) assessed the impacts of the SONORA SI proposed Sinoquipe and Las Chivas reservoirs (27 and 41 MCM, respectively) under the context of future climate change in the SRB and found that, in contrast to projections for the Southwest United States, more precipitation and water inflow can be expected for the next 30 years in Sonora, together with a shift in rainfall seasonality. However, the proposed reservoirs would add little to the existing dams in terms of flood control. The role of reservoirs thus must be viewed in full river basin terms, with soft path strategies that enhance partial resilience offered by infrastructure-based solutions.

2.2.3 *Arizona–Sonora Comparative Assessment of the Role of Reservoirs in Enhancing Water Security*

The concept of water security provides an analytical framework for the effectiveness of reservoirs as a tool for adaptive management. Water security entails the dual character of water resources as productive—when water makes possible the existence of irrigated agriculture—and destructive—when an extreme event produces fatal flooding (Scott et al. 2013). To be water secure means to have the amount and quality of water required for the specific uses that are intended. In addition, the criterion of resilience of societies and ecosystems means that water distribution and management should generate flexible and equilibrated dynamics that do not decrease the capacity of one social group or ecosystem to adapt when this same capacity is improved somewhere else in the system.

Reservoirs must be viewed in broader social-ecological resilience terms. The academic literature tends to stress the adverse social and environmental impacts of dams and related infrastructure when they are viewed as part of a management approach that prioritizes water storage as an isolated or unique supply-oriented objective. In this view, reservoir management is based on simple economic cost-benefit analysis, fragmentary visions of the social-natural landscape, and static climatic and environmental conditions. The policy outcomes under this type of management have been historically characterized by environmental degradation, increased social inequality between winners and losers, inflexible management in the long run, and, in numerous cases, social protests related to water and land appropriation for the construction of storage reservoir (World Commission on Dams 2000).

By contrast, multipurpose surface reservoirs as well as additional water storage and supply augmentation techniques, including groundwater recharge-recovery and banking, water reuse, and rainwater and storm water harvesting, can enhance resilience. In order to do so, however, adaptive management must be based in social and ecological processes, incorporate institutional learning, and assess benefits and costs not solely in monetary terms but in the ability to increase flexibility and minimize risks. Typically, the timeframes for such analysis should be centuries, even though infrastructure may have a useful life of decades. Crucial to enhancing the role of reservoirs in adaptive management are appropriate siting, design, construction, operation, and decommissioning.

In terms of the comparison of the role of reservoirs in achieving water security in Arizona and Sonora, dams to store water have helped in physically managing the two main impacts of climate variability in the region: water shortage resulting from drought and heightened demand, on the one hand, and flooding and the loss of life, property, and certain ecosystem functions and services on the other. However, dependence on single hard path strategies has translated into the neglect, even rejection, of alternative options. For instance, the lack of a comprehensive portfolio of supply and demand management policies in the case of Sonora led to chronic water insecurity for Hermosillo city after the main supply reservoir was depleted in the late 1990s. Similar conditions were endured for the larger Yaqui river reservoirs that reached record low storage levels after the drought of 1995–2003, causing the loss of many cultivated hectares in the Yaqui Valley Irrigation District. In 2003, the district planned 40 % more land area for irrigation than was feasible given the drought conditions. The economic output of the district dropped to less than 40 percent of the average real output during the preceding decade (McCullough and

Matson 2011). Although the district has a diverse array of technological and scientific sources of information (McCullough and Matson 2011), path dependency on dams prevented further development of watershed-based complementary strategies to face the increased variability of water supply.

In Arizona, the reservoirs built on the Colorado River (much larger than both the Sonora and Yaqui River reservoirs), together with the CAP aqueduct, have buffered water scarcity. However, long-term drought and heightened water demand have challenged the capacity of this complex system to provide water and energy in the quantity, timing, and quality it was designed for. Arizona has developed combined hard and soft path strategies to support the river storage system, particularly in urban areas (i.e., the active management areas [AMAs]). However, climate change and variability coupled with rapid demographic and economic growth continue to exert pressure on the Colorado River system. In contrast to Sonora, in Arizona, a better organized civil society and academia have pushed the consideration of ecological flows in water planning, which may contribute to enhanced adaptive management in the long term. Nevertheless, in both cases, hard and soft path policies still need further modification (readaptation) particularly to better integrate multiple sectors, especially as integrative approaches like the food-energy-water nexus gain momentum. Also in both states, a predominance of the urban and large-scale agricultural sectors in defining water policy directions (and historically, also determining the construction of dams) has been detrimental to environmental security, in general, and to water security of rural communities in particular (Lutz-Ley 2014).

To strengthen water security and adaptive outcomes in the future, reservoir planning needs to do the following:

(1) Better account for hydroclimatic and water demand uncertainties over longer timeframes;
(2) Strengthen institutional learning in relation to reservoirs (reoperations as well as mechanisms to address growing civil society critiques of hard path dependence);
(3) Increase the flexibility of policies for infrastructure that better address governance through further inclusion of all stakeholders involved, not only those directly benefitted or financially involved; and
(4) Build decision-making on science–policy linkages to underscore reservoir infrastructure coupled to governance mechanisms, which would need improved information systems and regionalized climate models (if possible, to the level of basins).

Based on the overview provided above, the principal characteristics of water security management in Arizona and Sonora are synthesized in Table 2.1.

Following this, an empirical study of stakeholder perceptions of water security and the role of infrastructure comparing the two states is presented.

Table 2.1 Comparative overview of factors contributing to water security

Global-change resilience factors contributing to water security	Sonora	Arizona
Hydroclimatic variability:	Medium-high:	High:
– Drought occurrence	– High drought risk	– Extreme drought risk
– Flooding, severity	– Frequent, severe floods	– Moderate flooding
Water storage, conveyance, infrastructure:	Hard path:	Post-hard path:
– Reservoirs	– Under development	– Overbuilt, no new plans
– Interbasin transfers	– Early stages	– Post-CAP soft measures
– Groundwater storage, recovery, banking	– Not currently practiced	– Well developed
– Water reuse	– Informal, undeveloped	– Well developed
Adaptive water management:	Nascent stages:	Moderately advanced:
– Water allocation frameworks	– Flexible administered	– Rigid, rights-based
– Risk assessment/insurance	– Emerging	– Developed
– Public awareness	– Moderate	– Moderate, politicized
– Demand management/pricing	– Nascent	– Active in urban areas
– Integrated water-energy-food security	– Nascent	– Nascent
– Institutional learning	– Active	– Moderate
Institutional arrangements:	Centralized:	Polycentric:
– Water rights	– Regulated, unenforced	– Legal rights, enforced
– Agency mandates, financing	– Clear, inadequate $	– Overlapping, declining $
– Stakeholder coordination	– Opaque	– Transparent
– Civil society participation	– Low/non-existent	– Medium/structured
– Science-policy dialogues	– Nascent dialogues	– Active, results unclear
Water security:	Moderate-low:	Moderate-high:
– Societal	– Moderate	– High
– Ecological	– Low	– Low

2.3 Empirical Study Methods

2.3.1 Goals and Questions

The purpose of the empirical study of stakeholder perceptions is to undertake a comparative assessment of the factors affecting water security in Arizona and Sonora considering: (1) hydroclimatic and water demand uncertainties; (2) the capacity for institutional learning in relation to reservoirs in the broader context of hard path and soft path strategies; (3) the level of flexibility of policies for infrastructure that also address water governance; and (4) the existence of science–policy linkages to underscore reservoir infrastructure coupled with governance mechanisms to improve resilience. Through application of surveys to selected water sector stakeholders in both Sonora and Arizona, we aim to answer the following questions:

- What are the most important factors influencing water security in Arizona and Sonora?
- How do stakeholders in both states perceive the effectiveness of hard path strategies to improve water security in comparison to soft path strategies?
- Which are the discrepancies between the investment in the water sector perceived by stakeholders in both states and the preferences for future funding strategies considering water security?
- How can hard path and soft path strategies be compared in terms of their outcomes in both states?
- Which are the institutional strategies that stakeholders in Sonora and Arizona consider to be important for improving governance and water security?

2.3.2 *Participants*

Recruitment of 105 potential respondents (50 in Arizona and 55 in Sonora) was done through e-mail invitations that provided a brief explanation of the study, the assurance of confidentiality and protection of data, and the link to the online survey. The potential respondents were purposively selected from governmental, academic, and civil society groups representing water and environmental issues or management. A total of 50 completed surveys (48 %) were received online—34 from Arizona (68 % response rate) and 16 from Sonora (29 % response rate). An additional eight surveys were applied in person by the researchers through a snowball technique. These participants include governmental officers from the water utility in Hermosillo, the state's water agency (CEA Sonora), and the federal water agency (CONAGUA). Of the total of 24 participants in Sonora, 50 % were from the academic sector, 33 % from the government, and 17 % from civil society and private sector organizations. An advantage of the in-person surveys was the elicitation of detailed comments and perspectives on water security from the participants. Overall, the combined response rate was 55 %.

2.3.3 *Instruments*

A questionnaire survey was developed, related to ongoing water security research by the University of Arizona's Udall Center for Studies in Public Policy, which has existing approvals for human subjects research via the Institutional Review Board. The survey was administered online through Survey Monkey in both English and Spanish and consisted of descriptive data (organizational affiliation and state of respondent) and seven questions using a combination of Likert-type ranking text boxes and checklist formats. In Survey Monkey, the random sequencing of potential responses for each question was used in order to minimize the bias that might result from preset sequencing. While 50 of the surveys were completed online, eight were applied in person in Spanish. These were printed versions of the same survey that was available online.

2.3.4 Analysis Procedures

Preliminary analysis was done using the data processing tools provided online by Survey Monkey; secondary analysis was conducted in Microsoft Excel. Survey Monkey generates descriptive statistics of responses, the average score of ranking and Likert-type questions, and total summaries for checklist questions. Those preliminary results were further processed to obtain final results and graphical representations of participants' responses. We accompany these with a discussion of factors likely to explain the observed results for individual questions and components of the survey.

2.4 Results

2.4.1 Factors Affecting Water Security (Fig. 2.2)

In both Sonora and Arizona, inadequate management of water was ranked most frequently as the most important factor affecting water security. Arizonans gave

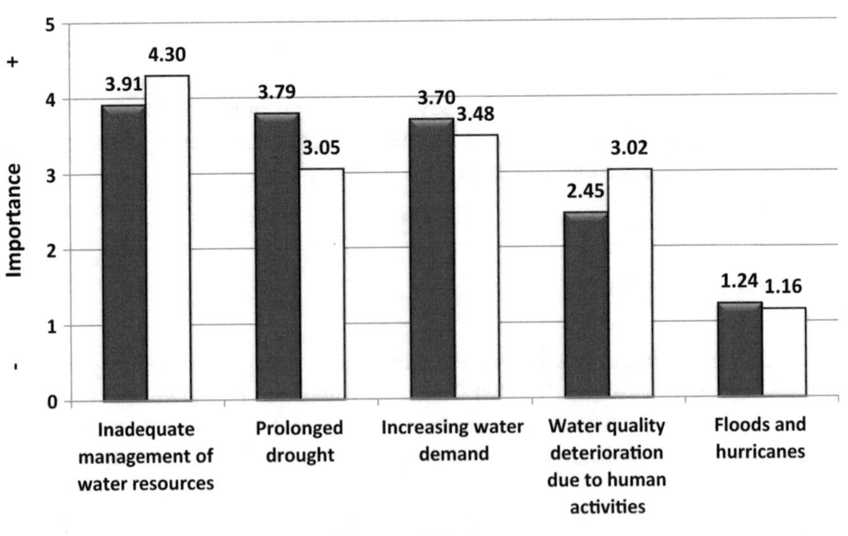

Fig. 2.2 Factors affecting water security

higher importance than Sonorans to prolonged drought as the next most important factor affecting water security, while Sonorans considered increasing water demand and water quality deterioration due to human activities as the next most important factors after inadequate management of water.

These findings are partly related to the term prolonged drought being considered in Arizona as shorthand for "climate change," which carries political implications particularly with conservative lawmakers and decision-makers. In Sonora, by contrast, ongoing water-quality impacts of mining (with the recent 2014 tailings-dam spill) and urban sanitation and wastewater challenges are considered to exert important influences on the responses. Water quality in Sonora was a recurrent topic during the surveys conducted in person with governmental officials, who mentioned that quality rather than the quantity was one of the major motivations for construction of the Acueducto Independencia. Because the Abelardo L. Rodríguez reservoir dried in the late 1990s, Hermosillo city has been relying on deep wells in the city's periphery. In the view of the local utility officials, overpumping promoted high concentrations of certain dangerous elements (particularly arsenic) in several wells of the supply system during the last years. The levels in some cases were very close to the limits established by the Mexican Official Norm (NOM) regulating the quality of water for human consumption. However, one official said these finding had not been revealed to the general public in order to prevent further conflicts and demands.

2.4.2 *Effectiveness of Strategies for Enhancing Water Security* (Fig. 2.3)

Participants in Arizona ranked price mechanisms as the most effective way to enhance water security. Participants in Sonora ranked policies for efficiency higher than Arizonans in terms of achieving improved water security. In general terms, Sonoran participants considered soft path strategies to be more effective for enhancing water security than did participants in Arizona.

2.4.3 *Perceived and Desired Investment in Water Management Strategies* (Fig. 2.4)

Participants were asked their perceptions on the percentage of total investment in different types of strategies in Arizona and Sonora over the last 10 years, as well as the percentage they would like to see invested in each of these in the coming decade.

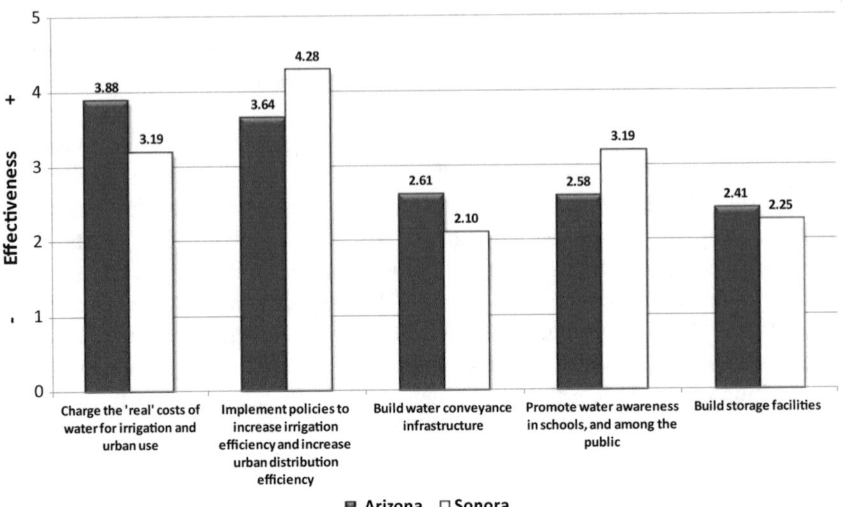

Fig. 2.3 Effectiveness of strategies for enhancing water security

Participants in Sonora think that an average of 45 % of financial resources during the last 10 years have been invested in building new aqueducts and interbasin transfer projects, followed by 17 % in increased groundwater pumping (considering costs of electricity and equipment). This arises from the recently commissioned, often controversial, Acueducto Independencia that was built to transfer up to 75 MCM from the mid-Río Yaqui basin to Hermosillo city, in the lower Río Sonora basin. Groundwater pumping and aquifer drawdown are pervasive practices in both states. Clearly, each of these strategies incurs costs to different sectors in society; i.e. aqueducts and transfer projects are publicly funded while pumping is largely private. In the coming decade, Sonoran respondents would like to see at least 22 % of investments in the water sector going to wastewater treatment, 19 % to improved drainage for stormwater recharge, and 14 % to awareness and conservation programs.

Participants in Arizona think that approximately half of the investments during the last 10 years in their state have gone to three main strategies: water awareness and conservation programs (18 %), increased wastewater treatment (16 %), and increased groundwater pumping (16 %). Arizona respondents would like to maintain the current principal strategies into the coming decade, but decrease groundwater pumping and improve water markets and policies connecting water and energy (20 % of the investment on average).

What is notable in both states, at present and especially in the future, is the relatively low priority accorded to increased water storage in dams (or new dams), aqueducts, or interbasin transfer projects.

2.4.4 *Strategies to Enhance Governance for Water Security* (Fig. 2.5)

In terms of the strategies for improving governance of water security, respondents in both Arizona and Sonora consider evaluation and dissemination of water management outcomes as the most effective way for enhancing governance of water security.

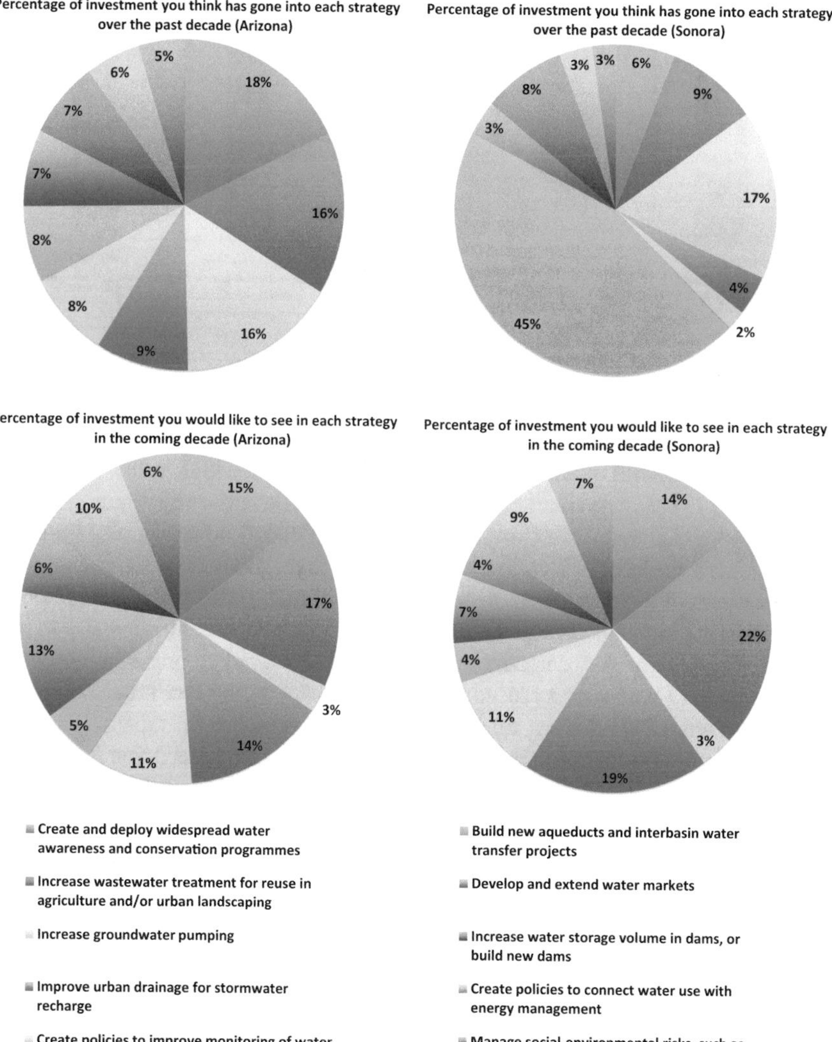

Fig. 2.4 Perceived and desired investment in water management strategies (Arizona, *left*; Sonora, *right*)

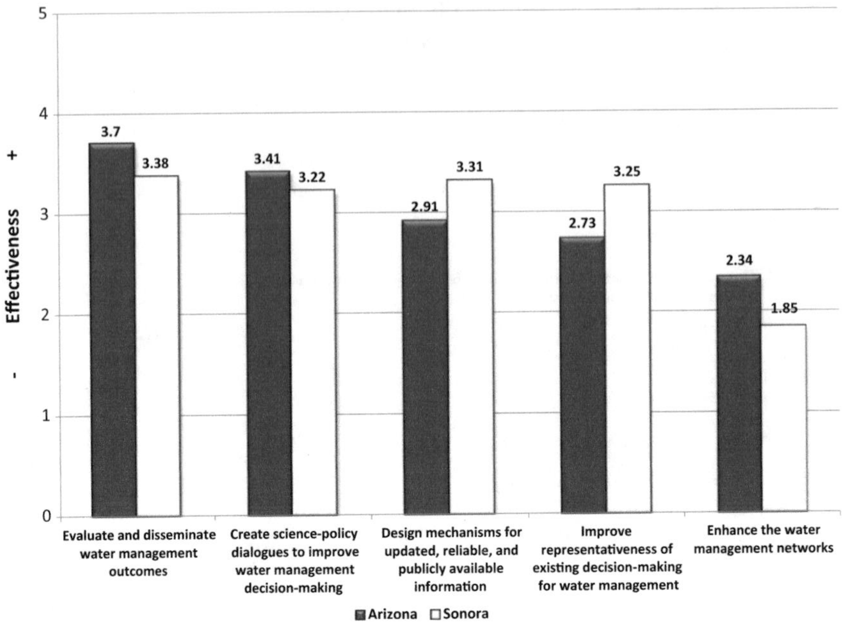

Fig. 2.5 Strategies to enhance governance for water security

In addition, both Arizona and Sonora participants place high value on updated, reliable, and publicly available information on variables affecting water resources. By contrast, enhancing water management organizational networks was ranked in both states as the least effective strategy to improve governance. The context of Sonora as part of a country with recurrent issues of transparency and corruption are reflected in the higher scores given by Sonoran participants to the necessity to have reliable sources of information, as well as improved representativeness in decision-making bodies and procedures. For example, during the in-person surveys with public officials in Sonora, all the respondents agreed that the aqueduct solved a notorious supply problem for Hermosillo city, but they also agreed that the political and social treatment of the issue could have been much better and more inclusive than was done by the state government.

2.4.5 *Outcomes Attributable to Hard Path and Soft Path Strategies* (Fig. 2.6)

This section of the study requires some clarification. The intent was to reverse the line of questioning from the first part of the survey in which survey respondents were asked which strategies contributed to water security. Here, respondents were

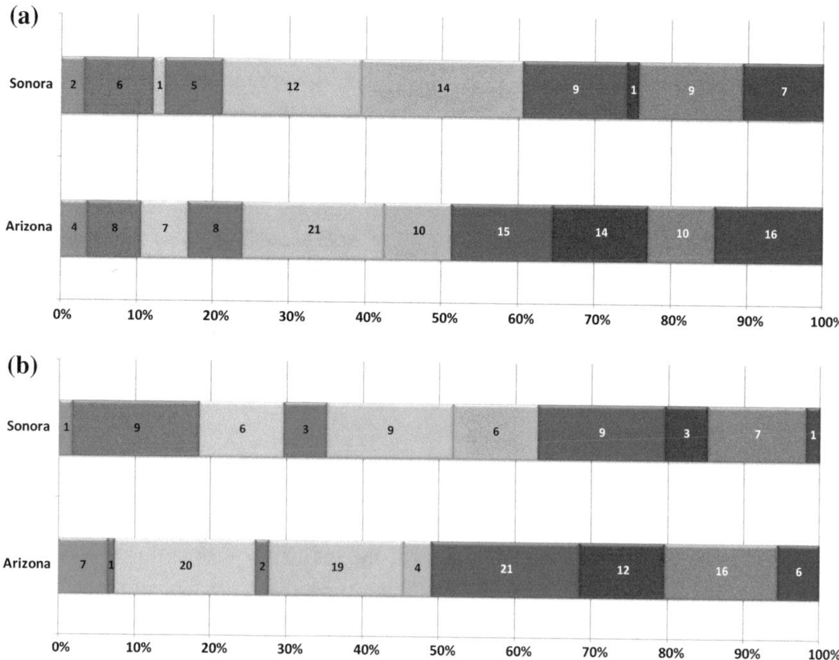

Fig. 2.6 Outcomes attributable to **a** hard path and **b** soft path strategies

asked to check the principal outcomes of hard path (infrastructure-led) strategies compared to the same set of outcomes they considered to result from soft path (policy-led) strategies.

In Sonora, the two most frequently indicated outcomes of water infrastructure are less inclusion and diversity of stakeholders participating in water management (selected 14 times) and a better match between water demand and availability (12). In turn, the outcomes of water policies (soft path strategies) more frequently mentioned were the reduced access of vulnerable groups to good water quality due to increased cost (selected 9 times), better match between water demand and availability (9), and better preparation to face future socioeconomic and environmental impacts (9). There is understandable concern that pricing water as a means to reduce demand will result in disadvantaged populations losing access. In fact, one official from the water utility in Hermosillo mentioned high cost as one of the main constraints for pursuing alternative sources of water, such as a desalination plant on the Gulf of California coast. Nevertheless, the principal representative from the federal agency CONAGUA indicated that a desalination plant was still economically feasible if big agricultural producers helped to offset the costs for the city.

In Arizona, too, a better match between water demand and availability (21 responses selected) is the most frequently mentioned outcome of water infrastructure, and better preparation to face future socioeconomic and environmental impacts (21) is the most mentioned outcome of water policies. Another frequently mentioned outcome of water policies is improved participation of diverse stakeholders in water management (20). This reflects social learning and participatory approaches to water management.

2.5 Discussion

In general terms, the results are indicative of the major impacts assigned to human management as compared to natural or climate-related factors that influence water security. In particular, reservoirs are viewed as one of multiple adaptation strategies. It is interesting to note that floods and hurricanes did not receive a high score, especially in Sonora, which has borne severe impacts from these events in the near past and is expected to face further extreme events in the future. This coincides with the broader managerial approach in Sonora that emphasizes adaptation to drought, instead of a framework that includes the full range of potential hydroclimatic events in the region under climate change scenarios. Underestimation of the full range of events could also have been associated with the higher priority accorded to soft path measures in Sonora compared with Arizona. While scarcity of water can be addressed through a combination of supply and demand management policies (and Sonoran respondents actually think that a big proportion of water investments during the last decade were directed to hard path strategies, and that educational programs are necessary), riverine flooding may require more than behavioral or institutional (soft path) solutions—that is, improved operation and management of existing reservoirs and associated floodworks. As commented, however, cyclone landfall and coastal-zone flooding are less amenable to infrastructural solutions.

The recent experience of Sonora in building the Acueducto Independencia, and the implementation of Sonora SI in general, are reflected in the responses of participants, who gave lower scores to infrastructural solutions than in Arizona in terms of enhancing water security. "Traditional" or "monumental" infrastructural solutions in Sonora have also been associated with lower public participation, often contributing to important social conflicts, making large infrastructure undesirable in public opinion. However, on the other hand, "greener" and more flexible water infrastructure is deemed necessary (for example, drainage for storm water recharge and wastewater treatment plants). This also reflects the gap—both real and perceived—in localized water infrastructure to solve specific problems: for example, the lack of wastewater treatment that still exists in large Sonoran cities such as Hermosillo as well as in rural communities. Additionally, hard and soft measures are needed to promote sustainable groundwater yield and recharge to minimize the effects of the overpumping that has depleted the Costa de Hermosillo aquifer and threatens other aquifers surrounding the capital city as well as the agricultural valleys of northwest Sonora.

2.6 Conclusions

Infrastructure development has played a critical role in water supply, often fueling growth in demand, in both Arizona and Sonora. Water security in both states is significantly influenced by hydroclimatic drivers, chiefly drought, while growing populations and expanding economic activity (much of it water intensive) drive water demand and therefore the risk of water insecurity. While it can be argued that Arizona embarked on the hard path of dams and reservoir construction earlier than Sonora and therefore has entered a post-hard path stage sooner, the relations that each state has with its federal government (and the role of federal financing, and therefore incentives) are crucially important determinants of the options available to stakeholders. Federal–state institutional design in both Mexico and the United States also influence the locus of adaptive response—that is, which organizations and stakeholders are engaged in assessing water (in) security and subsequently in proposing and modifying adaptation plans for implementation. Arizona's polycentric institutional arrangements are in contrast with Sonora's reliance on centralized federal decision-making.

In addition to surface water storage in reservoirs, infrastructure-based adaptive responses include flexible systems of groundwater storage and recovery as well as wastewater treatment and reuse. In the urban sphere in Arizona particularly, rainwater harvesting from rooftops, parking lots, and streets forms an increasingly important set of water augmentation approaches. All of these interventions must be accompanied by flexible governance arrangements to facilitate priority setting, maintenance and upkeep, and iterative processes of adaptation.

The findings of the empirical survey, set in historical context, point towards an emerging integrated view of water security resulting from infrastructure (especially that which is already built and in place) combined with more flexible and adaptive water management and policy. It appears evident, particularly with tight budgets and growing concern over social equity, that greater emphasis will be placed on soft path measures in the future than on hard path measures.

Particularly in river basins where water resources are overallocated, reliance on increasing reservoir storage as the principal adaptive governance tool in the context of climate change must increasingly be complemented by policy measures that respond to hydroclimatic and water demand uncertainties. Additional adaptation tools include basin-wide planning of water storage (especially to avoid low-elevation, high-evaporation storage), efficiency in irrigation and urban water distribution with regulations and enforcement on expansion of new demands for saved water, wastewater reclamation and reuse including effluent storage and recovery techniques, and conjunctive surface and underground storage programs that innovate with groundwater banking and flexible water allocation (e.g., substitutions and exchanges). Social learning will be enhanced through information deployment and uptake, tailored modelling studies that include user-generated data, and collaborative adaptation planning. Finally, management and policy for infrastructure that accounts for multiparty cost-shared financing and civil society participation in decentralized decision-making will enhance water governance for climate resilience.

Acknowledgments The authors gratefully acknowledge support from the World Water Council as well as the Inter-American Institute for Global Change Research (project CRN3056, which is supported by U.S. National Science Foundation grant GEO-1128040).

References

Aboites L (2009) La decadencia de agua de la nación. Estudio sobre desigualdad social y cambio político en México. Segunda mitad del siglo XX. El Colegio de Mexico, Mexico

Brooks DB, Brandes OM (2011) Why a water soft path, why now and what then? Int J Water Res Dev 27(2):315–344

Browning-Aiken A, Richter H, Goodrich D, Strain B, Varady R (2004) Upper San pedro basin: fostering collaborative binational watershed management. Int J Water Res Dev 20(3):353–367

CEA Sonora (2008) Estadísticas del Agua en el Estado de Sonora, Edicion 2008 (Water Statistics in Sonora, 2008th edn. CEA Sonora, Hermosillo, Sonora

Central Arizona Project (CAP) (2010) The Navajo Generating Station White Paper. http://www.cap-az.com/includes/media/docs/Navajo-Generating-Station-White-Paper–2-.pdf. Accessed Nov 15 2015

CONAGUA (2015) Sistema de Seguridad en Presas (Dams Security System). http://201.116.60.136/inventario/hinicio.aspx. Accessed Nov 15 2015

Dean R (1997) Dam building still had some magic then: stewart udall, the central arizona project, and the evolution of the pacific southwest water plan, 1963–1968. Pac Hist Rev 66:81–98

Diario Oficial de la Federación (Mexican Federal Bulletin) (1998) DOF: 18/05/1998. http://www.dof.gob.mx/nota_detalle.php?codigo=4878553&fecha=18/05/1998. Accessed Nov 23 2015

Eden S, Scott CA, Lamberton ML, Megdal SB (2011) Energy-water interdependencies and the central arizona project. In: Kenney D, Wilkinson R (eds) The water-energy nexus in the american west. Edward Elgar, Cheltenham, UK, pp 109–122

El Imparcial (2015) Desfogan otra vez presa 'El Molinito'. El Imparcial newspaper. http://www.elimparcial.com/EdicionEnlinea/Notas/Sonora/04092015/1004775-Desfogan-otra-vez-presa-El-Molinito.html. Accessed Nov 16 2015

Farfán LM, Alfaro EJ, Cavazos T (2013) Characteristics of tropical cyclones making landfall on the pacific coast of Mexico: 1970–2010. Atmósfera 26(2):163–182

Gleick PH (2003) Global freshwater resources: soft-path solutions for the 21st century. Science 302(5650):1524–1528

Glennon RJ (1995) Coattails of the past: using and financing the central arizona project. Ariz St LJ 27:677

Gooch RS, Cherrington PA, Reinink Y (2007) Salt river project experience in conversion from agriculture to urban water use. Irrigat Drain Syst 21(2):145–157

Hollnagel E, Woods DD, Leveson N (eds) (2006) Resilience engineering: concepts and precepts. Ashgate Publishing, Farnham, UK

INEGI (2010) México en cifras (Mexico in figures). http://www3.inegi.org.mx/sistemas/mexicocifras/default.aspx?e=26. Accessed Nov 15 2015

Kates RW, Travis WR, Wilbanks TJ (2012) Transformational adaptation when incremental adaptations to climate change are insufficient. P Natl Acad Sci 109(19):7156–7161

Lach D, Ingram H, Rayner S (2004) Maintaining the status quo: how institutional norms and practices create conservative water organizations. Tex Law Rev 83:2027–2053

Leichenko RM, O'Brien KL (2008) Environmental change and globalization: double exposures. Oxford University Press, New York

Lloyd's Register Foundation (LRF) (2015) Foresight review of resilience engineering: designing for the expected and unexpected. Report Series: No. 2015.2. LRF, London

Lovins AB (1976) Energy strategy: the road not taken. Foreign Aff 55(1):65–96

Lovins AB (1979) Energy controversy: soft path questions and answers. Friends of the Earth, San Francisco, CA

Lutz-Ley AN, Pablos NP, Adams AS (2011) La política hidráulica del Gobierno del Estado de Sonora y el Programa Sonora SI en el año 2010. In: Mendez E, Covarrubias A (eds) Estudios sobre Sonora 2010. Instituciones, procesos socio-espaciales, simbólica e imaginario. Universidad de Sonora, Hermosillo, Sonora, pp 27–47

Lutz-Ley AN (2013) La construcción social de grupos objetivo en subprogramas de 'Cultura del Agua' en Hermosillo, Sonora. In: Pineda N (coord.) Modelos para el análisis de políticas públicas. El Colegio de Sonora, Hermosillo, Sonora, pp 129–143

Lutz-Ley AN (2014) Institutional frameworks for managing groundwater in rural Arizona and Sonora. The University of Arizona. Unpublished manuscript awarded 2nd place of the Central Arizona Project's Water Research Award 2014. http://www.cap-az.com/documents/education/America-N-Lutz-Ley-Institutional-Frameworks-for-Managing-Groundwater-in-Rural-Arizona-and-Sonora.pdf. Accessed Dec 18 2015

McCullough E, Matson PA (2011) Evolution of the knowledge system for agricultural development in the yaqui valley. P Natl Acad Sci, Sonora, Mexico. doi:10.1073/pnas.1011602108

Megdal SB (2012) The role of the public and private sectors in water provision in Arizona. USA. Water Int 37(2):156–168

Megdal SB et al (2014) Water banks: using managed aquifer recharge to meet water policy objectives. Water 6:1500–1514

Moser S, Pike C (2015) Community engagement on adaptation: meeting a growing capacity need. Urban Climate 14:111–115

Overpeck J, Garfin G, Jardine A et al (2013) Summary for decision makers. In: Garfin G, Jardine A, Merideth R et al (eds) Assessment of climate change in the Southwest United States: a report prepared for the National Climate Assessment. Island Press, Washington, DC, Southwest Climate Alliance, pp 1–20

Palmer MA, Reidy Liermann CA, Nilsson C et al (2008) Climate change and the world's river basins: anticipating management options. Front Ecol Environ 6(2):81–89

Pineda-Pablos N, Scott CA, Wilder M et al (2012) Hermosillo, ciudad sin agua para crecer. Vulnerabilidad hídrica y retos frente al cambio climático. In: Wilder M, Scott CA, Pineda-Pablos N et al (eds) Moving forward from vulnerability to adaptation: climate change, drought, and water demand in the urbanizing Southwestern United States and Northern Mexico. Udall Center for Studies in Public Policy, The University of Arizona, Tucson, pp 125–169

Prichard AH, Scott CA (2013) Interbasin water transfers at the U.S.-Mexico border city of Nogales, Sonora: Implications for aquifers and water security. Int J Water Res Dev 29(2):1–17. doi:10.1080/07900627.2012.755597

Robles-Morua A, Che D, Mayer AS, Vivoni ER (2015) Hydrological assessment of proposed reservoirs in the Sonora River Basin, Mexico, under historical and future climate scenarios. Hydrolog Sci J 60(1):50–66. doi:10.1080/02626667.2013.878462

Scott CA, Bailey CJ, Marra RP et al (2012) Scenario planning to address critical uncertainties for robust and resilient water-wastewater infrastructures under conditions of water scarcity and rapid development. Water 4:848–868. doi:10.3390/w4040848

Scott CA, Meza FJ, Varady RG et al (2013) Water security and adaptive management in the arid Americas. Ann Assoc Am Geogr 103(2):280–289. doi:10.1080/00045608.2013.754660

Scott CA, Pineda-Pablos N (2011) Innovating resource regimes: water, wastewater, and the institutional dynamics of urban hydraulic reach in northwest Mexico. Geoforum 42 (4):439–450

Scott CA, Banister J (2008) The dilemma of water management 'regionalization' in Mexico under centralized resource allocation. Int Water Res Dev 24(1):61–74

Taylor RG, Scanlon B, Döll P et al (2013) Ground water and climate change. Nat Clim Chang 3 (4):322–329

Tortajada C (2010) Water governance: Some critical issues. Int J Water Res Dev 26(2):297–307

United States Census Bureau (2010) United States Census 2010. https://www.census.gov/2010census/. Accessed Nov 23 2015

Varady RG, Scott CA, Wilder M et al (2013) Transboundary adaptive management to reduce climate-change vulnerability in the western U.S.-Mexico border region. Environ Sci Policy 26:102–112. doi:10.1016/j.envsci.2012.07.006

Vitousek PM (1994) Beyond global warming: ecology and global change. Ecology 75 (7):1861–1876

Vörösmarty CJ, Pahl-Wostl C, Bunn SE, Lawford R (2013) Global water, the Anthropocene and the transformation of a science. Curr Opin Environ Sustain 5(6):539–550

Walker B, Salt D (2012) Resilience practice: building capacity to absorb disturbance and maintain function. Island Press, Washington, DC

Wentz EA, Gober P (2007) Determinants of small-area water consumption for the city of Phoenix. Arizona. Water Res Manag 21(11):1849–1863

Wilder M, Scott CA, Pineda Pablos N et al (2010) Adapting across boundaries: climate change, social learning, and resilience in the U.S.-Mexico border region. Ann Assoc Am Geogr 100 (4):917–928. doi:10.1080/00045608.2010.500235

Wolff G, Gleick PH (2002) The soft path for water. In: Gleick PH, Burns WCG, Chalecki EL (eds) The world's water 2002–2003: the biennial report on freshwater resources. Island Press, Washington, DC, pp 1–32

World Commission on Dams (2000) Dams and development: a new framework for decision-making. Earthscan, London

Chapter 3
The Murray–Darling Basin: Climate Change, Infrastructure, and Water

Jamie Pittock

Abstract The extreme climatic variability historically experienced in Australia's Murray-Darling Basin and the institutions developed to respond to it have generated key lessons for managing water infrastructure under climate variability and change. In response to highly variable river flows, surface water storage systems have been constructed in the Murray-Darling Basin that can store nearly three times the average river discharge, and groundwater is increasingly exploited. These interventions have negatively impacted the environmental health of the river system. Severe droughts have exacerbated the environmental impacts, threatened livelihoods, and exposed weaknesses in management. Incremental reforms have been introduced to better manage climatic variability. A number of key water management reforms for climatic variability are described in this chapter, including capping consumptive water extractions; conjunctive management of surface water and groundwater; establishing a water market; rebuilding water storage, distribution, and irrigation infrastructure to reduce water losses; and reallocating water to environmental flows. These have dramatically changed the way that surface water infrastructure is operated. Reforms in 2008 led to the adoption of a new basin plan for implementation from 2019, but there is little allowance for climate change despite projections for changing water availability in the future. The lack of periodic relicensing is a missed opportunity to systematically reoperate water infrastructure. Surface water and groundwater resources are further threatened by other changes, including the exploitation of unconventional gas and carbon sequestration plantations.

Keywords Adaptation · Climate change · Climate variability · Environmental flows · Evaporation · Groundwater · Governance · Infrastructure · Water pricing · Water storage

---

J. Pittock (✉)
Fenner School of Environment and Society, The Australian National University, Canberra, ACT, Australia
e-mail: Jamie.pittock@anu.edu.au

C. Tortajada (ed.), *Increasing Resilience to Climate Variability and Change*, Water Resources Development and Management,
DOI 10.1007/978-981-10-1914-2_3

3.1 Introduction

Lessons from the Murray-Darling River Basin (the Basin) in Australia are illuminating in any consideration of the role of water storage and management of climatic variability and change for three reasons. First, Australian rivers are the most variable in the world, making water storage especially important. The adaptations applied to meet the needs of people while sustaining the environment provide valuable lessons. The interannual variability is so marked that management based on hydroclimatic averages is an ineffective way to manage water. This extreme variability was quickly recognized by the European occupiers in the 19th century. A deliberate decision was therefore made not to adopt either the riparian rights or prior appropriation water entitlement systems, instead favoring shares in the available water resource that vary for each category of water users each year (Garrick 2015; Connell 2007). There was also major investment in surface water storages (as described below).

Second, the rivers of southern Australia are in the mid-latitudes where the impacts of climate change on hydrology are anticipated to be most severe, generating considerable research on response options (Grafton et al. 2012). In response, the Australian Government commissioned projections linking climate change and surface water and groundwater models for the period 2010–2030. The projections of *extreme wet* for 7 % more water, *median* for 12 % less water, and *extreme dry* for 37 % less water illustrate uncertainty (CSIRO 2008). The greater or lower inflows are magnified downstream so that, for instance, the extreme dry projection is for a 69 % reduction in outflows. Greater extremes of flood and drought are expected, together with the uncertain impacts of higher temperatures across a semi-arid to temperate climate subject to severe wildfires, vegetation change, and erosion.

Third, Australia has experimented with a range of reforms in the Basin in attempts to harmonize the competing goals of providing a secure water supply for agriculture and other consumptive uses while maintaining the variable hydrological processes that favor environmental health (Garrick 2015; Grafton et al. 2014). Consequently, management of the basin is producing key examples of what to do and what not to do globally in managing similar water resources under climate change as well as other types of change.

The basin covers an area of more than 1 million km^2 of southeastern Australia and includes the nation's longest rivers (Fig. 3.1). The basin's mid-latitude location results in evapotranspiration of the overwhelming majority of the precipitation (Fig. 3.2; Leblanc et al. 2012). Consequently, small changes in climatic conditions can significantly influence surface water availability, groundwater contributes nearly half of the streamflow, and surface and aquifer storages are significant for managing water supply with great climatic variability.

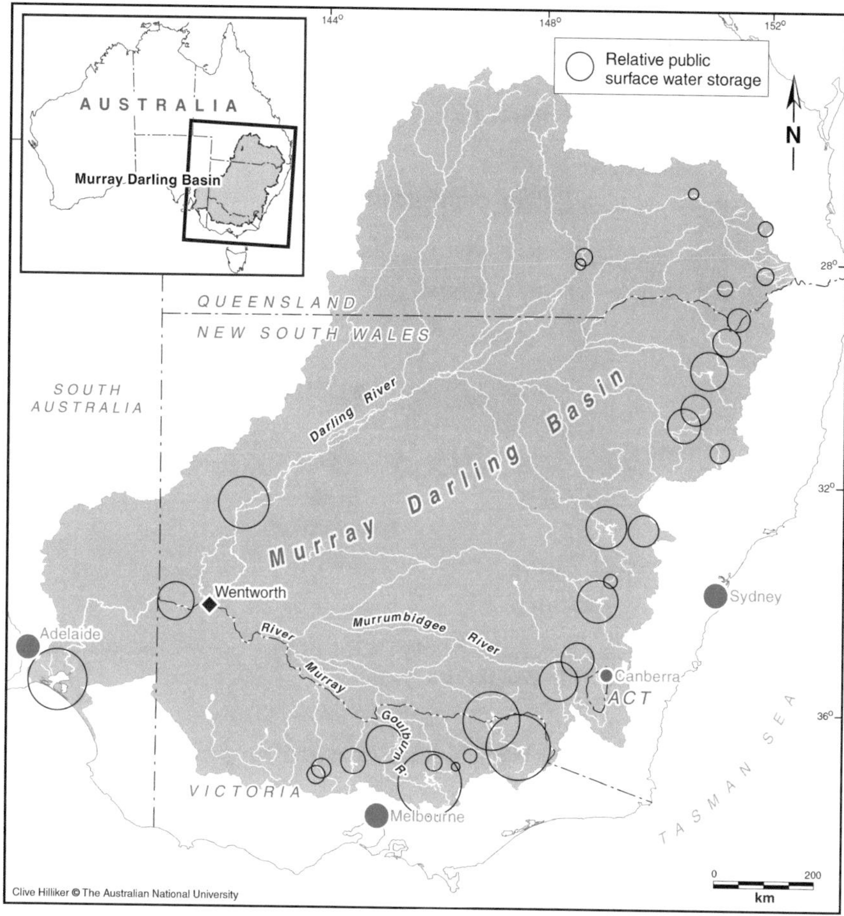

Fig. 3.1 Location of the Murray-Darling Basin, Australia, showing the 30 major public surface water reservoirs as *circles* in proportion to their relative water storage volume (adapted from MDBA 2015). © Clive Hilliker, The Australian National University

The basin supplies water to 2 million residents, as well as a further million people outside the basin. Around 90 % of the water diverted in the basin is used in irrigated agriculture, generating around AUD 5.5 billion in produce per year (MDBA 2010). Around 5 % of the basin's area is wetland, supporting extensive biodiversity, including 16 Ramsar wetland sites of international importance (Pittock et al. 2010).

This assessment of the role of water infrastructure and governance in increasing resilience under climate change is different from others in this volume in assessing the full range of surface water storages across the basin, as well as aquifers. Consequently, this assessment considers individual reservoirs in lesser detail while focusing on the system of water storages in the basin as a whole, in order to discern

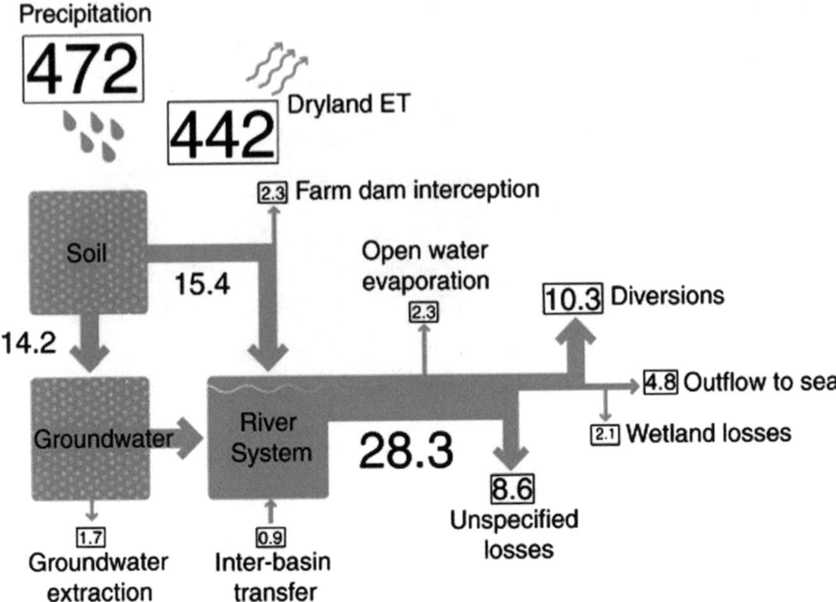

Fig. 3.2 Approximate water budget for the Murray-Darling Basin in mm/year under development conditions prior to the 2012 Basin Plan (Leblanc et al. 2012)

globally relevant lessons in managing a complex system for greater resilience under climate change.

By late last century, storage reservoirs had been built that are capable of storing more than 3 years' average flow of the Murray and Darling rivers (Fig. 3.3). The basin closed, meaning that water had been overallocated to such an extent that the River Murray did not flow to the sea as it had in the past (Falkenmark and Molden 2008; Kingsford et al. 2011). Thus, this assessment differs from many others in this volume as it is undertaken in the context of a "fully developed" river system, where any new water storage or use diverts water from another user. There is no new supply to be gained from the basin's water cycle and the focus is on better management of the existing resource.

While the Murray-Darling river system is subject to extreme droughts and floods, this assessment focuses on the use of infrastructure to manage water scarcity as this has been the primary concern of Australian institutions (Connell 2007; Garrick 2015). Surface water storages in the basin have not been designed to operate for flood control. Although there has been ad hoc and problematic development of flood levees, floodwaters typically inundate the extensive areas of floodplain that occupy around 5 % of the basin's area and have only modest impacts on people and industries. There are better places from which to draw lessons of global relevance regarding flood management infrastructure.

In this context, the approach adopted in this assessment is now outlined.

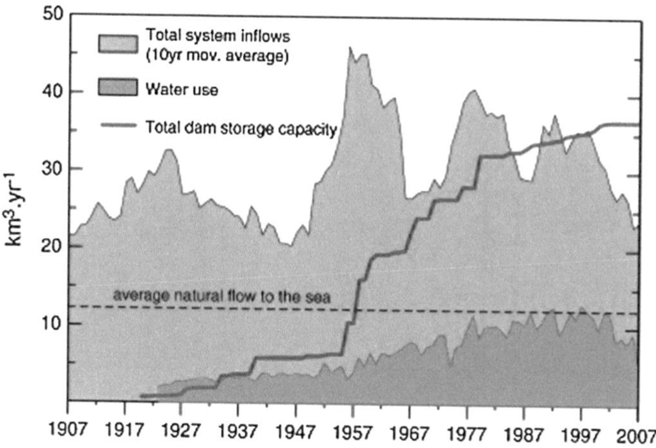

Fig. 3.3 Total dam storage capacity in the Murray-Darling Basin compared to average natural flows to the sea, system inflows, and annual water consumption (Leblanc et al. 2012)

3.2 Assessment Approach

The World Water Council provided an assessment framework for assessing the contribution of infrastructure and its governance for adaptation and resilience to climate variability and change, for application in this volume. A modified version of this has been adopted as described here. The Council assessment framework involves assessing conditions before intervention, describing the structural intervention, and then describing the conditions afterwards, with emphasis on droughts and floods, as well as on socioeconomic resilience following the provision of infrastructure. These before and after conditions are assessed with respect to governance, structures, practices, the positive and negative impacts, and resulting conclusions.

This case study considers the basin as a whole, covering a range of different aspects of infrastructure; this system is so sensitive to climatic and directly human-induced change that focusing on one aspect would mean that important lessons would be missed. Five stages of water infrastructure development in the basin in the past, present, and future have been assessed (Table 3.1) to document the history of water storage and infrastructure, management of climatic variability, and the lessons for climate change adaptation. The categories used to assess the stages of water infrastructure development are characteristics, initiating triggers and limitations; qualitative information on relevant governance, structures, practices, and impacts are listed under these categories.

As groundwater is critical to water storage, ecological health, and future climate change adaptation, this natural infrastructure is included in this assessment. Environmental resilience is also considered as good ecological health is a prerequisite for optimizing socioeconomic resilience in the basin. A short case study is

Table 3.1 Stages of water infrastructure development in the Murray-Darling Basin in the past, present, and future

Stage	Time	Characteristics	Initiating triggers	Limitations
1. Before development	Pre-1880s	Huge variations in flows with historic annual discharge ranging from 1626 to 54,168 GL (Maheshwari et al. 1995) Significant inland fisheries (Pascoe 2014) Management by 34 Aboriginal nations (Weir 2011; MDBA 2010)		River systems would cease flowing in large droughts
2. Public surface water storage development	1880s–1980s	Public investment results in 30 major surface reservoirs with 22,214 GL capacity, mainly in the southern Basin (MDBA 2015) Management by state governments (Connell 2007) Water access entitlements, mainly a share of available resource and tied to land titles (Garrick 2015) Massive irrigation area expansion, peaking at 1,654,000 ha (MDBA 2010) Increasingly significant environmental impacts, such as on salinity and fisheries	Water storage to manage summer low flows and drought for irrigation	Extensive public subsidies for water infrastructure construction and operations (Connell 2007) No devices to reduce environmental impacts, such as fish passage or thermal pollution control
3. Private surface water storage development	1980s–1994	Private investment in surface water reservoirs capable of storing 2600 GL, mainly in the northern Basin (Merz et al. 2010) Extensive re-engineering of river floodplains with canals, levees and weirs (Steinfeld and Kingsford 2013) Cyanobacteria and salinity crises, fish stocks collapse (Bowling and Baker 1996; MDBMC 2001; MDBC 2004) Effective Basin closure (Connell 2007)	Exhaustion of reservoir development opportunities in the southern Basin Development frontier moves to flatter north and west, focused on annual crops like cotton (McHugh 1996)	Development of even more hydrologically variable river tributaries Questionable quality of regulation by state governments (e.g., Foerster 2008) Greater environmental impacts

(continued)

Table 3.1 (continued)

Stage	Time	Characteristics	Initiating triggers	Limitations
4. Surface water cap on extractions and the water market	1994–2008	Surface water extractions capped at 1993–1994 development levels (Connell 2007) Establishment of water market, access entitlements separated from the land (Garrick 2015) Cap on new surface entitlements accelerates groundwater exploitation National competition policy results in significant reductions in state subsidies for water infrastructure and corporatization or privatization of water service providers (NCC n.d.; Hollander and Curran 2001) Water service providers have to justify new expenditure to a public tribunal, including whether water users pay or if the activity should be subsidized for the public good	National microeconomic reform process requires state governments to justify subsidies and separate water service provision from regulatory functions. Basin closure. Environmental crises, especially cyanobacterial blooms, increasing salinity, fish stocks collapse	Increase in water extractions as unused entitlements are activated in the market Inadequate regulation of groundwater and river inflow interception activities e.g. tree plantation establishment (van Dijk et al. 2006; Merz et al. 2010)
5. The Basin plan	2008 onwards	Federal Government intervention from 2008 results in 2012 Basin Plan and sustainable diversion limits to be fully operational from 2019, including inflow interception (Commonwealth of Australia 2012) Financial incentives for greater water use efficiency Conjunctive management of surface water and groundwater systems Reallocation of up to 25 % of surface water entitlements to environmental flows Commonwealth environmental water office established (Connell 2011) Purchased environmental water entitlements can be stored in reservoirs as for irrigation entitlements Effective end of construction of new surface water storages	Severe socioeconomic and environmental impacts of 2000–2010 millennium drought, exacerbated by climate change Closure of the river Murray mouth and desiccation of the lower lakes Death of tens of thousands of hectares of floodplain forest trees (Pittock et al. 2010) Detailed climate change projections available from 2008 (CSIRO 2008) Inadequate pace of reform through decision-making based on consensus of state and federal governments Recognition of the need to better manage groundwater, inflow interception activities and climate change	Water users again subsidized for undertaking water efficiency projects Questionable emphasis on 'environmental works and measures' to conserve biodiversity using less water (Pittock et al. 2012) Focus on water volumes sees less attention on non-volumetric climate change adaptation measures, such as dam reoperation (Pittock and Finlayson 2011) Groundwater extraction entitlements increased (SSCRRAT 2013) Changes in water allocation to manage climate change adaptation postponed to now 2026 (Pittock 2013)

also presented. The assessment concludes by summarizing key learnings from the basin that have international relevance for water infrastructure management under a changing climate.

3.3 Stages of Water Infrastructure Development

A qualitative assessment of five stages of water infrastructure development and management is summarized in Table 3.1. These stages are adapted from those of Connell (2007) and Garrick (2015) to focus on water infrastructure. The key issues emerging from this history and the currently planned management of water resources are discussed here.

Indigenous Australians have occupied the Murray-Darling Basin through climatic changes for at least 50,000 years, from the last glacial period through to the current, more benign, warmer, and wetter climate. Today, 34 major Aboriginal nations maintain an ongoing association with the lands and water of the basin (MDBA 2010). Their past (Table 3.1, stage 1) and present management is based on an intimate knowledge of water resource availability, as illustrated by the expansion and maintenance of small, natural wells in drier parts of the Basin, and complex, in-stream stone fish trap systems. Indigenous livelihoods center on the natural variability of the river system, such as harvesting fish and birds at different stages of the hydrological cycle, which has been harmed by subsequent water resource development. Indigenous knowledge is applicable today—traditional ecological knowledge from the culture and traditions of these Aboriginal nations that informs management of healthier ecological states, which may be the objective of current management; there is also increasing legal recognition of native title and similar rights to natural resources (Weir 2009, 2011).

Following European occupation of Australia, the new settlers quickly realized that the highly variable Murray-Darling river system could not be managed in the ways they were used to in the more benign conditions of Europe. As the irrigated agricultural industry developed from the 1880s, Australian leaders visited California and resolved not to adopt their business-based development and prior appropriation water entitlements system (Garrick 2015; Connell 2007). Instead, a system was adopted of water entitlements as a share of the available resource in any one year, so that shortages and plenty are shared, as well as massive investment in public surface water storages (Table 3.1, stage 2). Thirty major publicly owned storages were constructed in the basin, which were capable of storing 22,214 GL (Fig. 3.1; MDBA 2015). Public investment culminated in the massive Snowy Mountains hydroelectric scheme, transferring water from the Snowy River to the River Murray system (Fig. 3.2) for hydropower generation and irrigated agriculture (Miller 2005).

By the 1980s, major sites for water storage dams in the better watered southern and eastern portions of the basin had been developed and it was possible to store an amount of water equivalent to around 3 years of average river flows (Fig. 3.3);

then, state governments largely withdrew from funding new irrigation water storages. Some 3600 dams and weirs have been built across the Basin (Arthington and Pusey 2003), as well as thousands of kilometers of irrigation and flood levees and canals that alter water flows (Steinfeld and Kingsford 2013). Construction of tens of thousands of farm dams (tanks) across the basin, publicly subsidized as a drought adaptation measure up to the turn of the century, further intercepted stream inflows (Fig. 3.2; Merz et al. 2010). Under the constitution, the Australian state governments exercised primary responsibility for water management.

At this point, private irrigation developers with regulatory support from the New South Wales and Queensland state governments constructed large ring or *turkey's nest* dams capable of storing 2600 GL of water in the flatter, more arid northern and western portions of the basin (Table 3.1, stage 3) to capture the irregular flood pulses (Merz et al. 2010). These storages captured "overland flows" across floodplains, which bizarrely were not regulated as part of the river system or were filled by pumps so powerful that they would make the rivers run backwards. Although the Darling River subbasin where most of these storages were built only provides around 10 % of the basin's outflows on average, these flood pulses were ecologically significant.

A cyanobacterial bloom in 1991 during a drought turned water along almost 1000 km of the Darling River toxic to people and livestock (Bowling and Baker 1996), catalyzing reform. The looming extraction of more water than was available in the basin each year on average, as well as the threat from rising salinity to drinking water supplies in the downstream state of South Australia, prompted the state and federal governments to transform the water infrastructure-focused River Murray Commission into the basin-wide Murray-Darling Basin Commission with a broader natural resources management mandate (Garrick 2015; Connell 2007). A key decision was to cap surface water extractions at 1993–1994 water season levels, obviating incentives to construct large new water storages (Table 3.1, stage 4).

At the same time, the Australian state governments adopted neoliberal, microeconomic reforms directed at increasing economic competition; this included ensuring that state governments systematically separated regulatory functions from service provision (Pittock et al. 2015b). Service provision agencies were corporatized or privatized, with utility fees being regulated to achieve cost recovery and to justify subsidies only where there was a greater public good. Water was just one of a suite of services subject to these reforms, but this had two major implications for water infrastructure. First, investments in water infrastructure have to be undertaken on a cost recovery basis from user fees, unless a public good component is demonstrated, curtailing the more ambitious plans of the hydraulic bureaucracy. Second, the cap on new diversions and competition policy provided new impetus to expanding water market trading, so that in the closed system there are incentives for water to be traded to higher value uses (Grafton and Horne 2014). These reforms represented a significant change in the role of governments from "captain-coach of the irrigation team" to the "umpire" adjudicating amongst competing water stakeholders—a change deeply resented today in irrigation communities.

The crisis grew as the 2000–2010 Millennium Drought emptied storages and exposed the assumptions of hydroclimatic stationarity upon which water management in the basin is based (Pittock 2013; Leblanc et al. 2009). The fragility of the basin's environment was demonstrated when the Murray Mouth closed to the sea, resulting in desiccation and acidification of the lower lakes system and hypersalinity in the lower lakes and Coorong wetlands (Kingsford et al. 2011). Frustrated with the slow pace of reform when consensus was required by the states, in 2008 the federal government asserted greater constitutional authority through a new Water Act (Commonwealth of Australia 2008). Among other reforms, a Commonwealth Government Murray-Darling Basin Authority replaced the consensus-based commission, groundwater extraction was capped, water accounting was nationalized, and a Commonwealth Environmental Water Holder was established to purchase and manage water entitlements for environmental flows.

Creeping loss of water due to changes in land use is also meant to be regulated within the new cap, such as transpiration by plantation forests. The need to reallocate water from agriculture to sustain the services provided by the river system was realized in 2012 with the Basin Plan that sets a surface water sustainable diversion limit 2750 GL/year lower, at 10,873 GL/year from 2019 (Commonwealth of Australia 2012; Table 3.1, stage 5). This illustrates the need for effective, nested water governance that involves setting resource-wide standards at the largest scale while more local governments have the responsibility and opportunity to achieve and build on these standards by applying local knowledge and other resources.

This history, culminating in the management envisaged under the Basin Plan from 2012, raises a number of issues and lessons for water storage and infrastructure management under climate change, which are now discussed in greater depth.

3.4 The Role of Water Infrastructure and Storage

In an already hydrologically variable river basin, climate change increases the need to better manage water storage and infrastructure. From the preceding analysis, a number of key issues of global significance are drawn on and discussed here in turn.

3.4.1 Water for the Environment

Water infrastructure has negative impacts on biodiversity and other ecosystem services. Therefore, the management of dams needs to be periodically assessed to maximize socioeconomic benefits while minimizing environmental and social impacts (WCD 2000; MEA 2005). Environmental flows are needed to maintain basic riverine health (Arthington 2012). The severe impacts of drought and overextraction of water that led to desiccation and floodplain forest death, loss of fisheries, acidification, and salinity in the basin illustrated the need to reallocate more water from agriculture to ecosystem processes to benefit both people and biodiversity (Pittock et al. 2010; MDBA 2010). Reform in the basin demonstrates the value of having a strong and independent office (the Commonwealth Environmental Water Office) to own and manage environmental water, to ensure that environmental requirements are not unduly subjugated to economic interests (Connell 2011). Effective use of environmental water requires its storage in reservoirs for timely release to meet seasonal ecological needs or episodic events, such as the completion of waterbird breeding events, and such storage changes the volumes available for consumptive use.

The legal character of this environmental water is important as currently the water allocation rules favor entitlement-based agricultural users in drier years over rule-based environmental water (Grafton et al. 2014). Although Australia's state governments agreed to give environmental water the same legal security as issued entitlements from 2004, this has not occurred yet (NWC 2011). Consequently, the purchased entitlement water, which may eventually represent around a fifth of the environmental water in a (hypothetical) average year, is important for ecosystem conservation and changes reservoir operations.

Retention of some free-flowing rivers (without reservoirs) is an important measure for conserving biodiversity. It retains the transport of sediment and nutrients, protects natural flow variability that supports freshwater ecosystems, and supports the migration of aquatic wildlife like fish, including to newer habitats with climate change (Lukasiewicz et al. 2013). The protection of the last two large free-flowing rivers in the Basin, the Ovens and Paroo rivers, from water infrastructure development and new water diversions illustrates one approach to conserving freshwater ecosystems while facilitating development in other valleys (Pittock and Finlayson 2011).

The basin demonstrates that full development of water infrastructure is undesirable due to the severe negative impacts on the environment and the people who depend on the services from the river. Reforms in the Basin to preserve some free-flowing rivers and allocate water to the environment highlight two ways in which river management may be optimized.

3.4.2 Water-Efficient Infrastructure Investments

The federal government has allocated over AUD 5 billion for improving the efficiency of water infrastructure to recover water for the environment (Grafton 2015). This investment has been widely criticized because it is up to four times more expensive than the purchase of water entitlements, it subsidizes some in the farming community over others, and it "gold plates" infrastructure in areas that may not be suitable for irrigation under future climates (Productivity Commission 2010;

Adamson and Loch 2014). Under some circumstances, these subsidies can result in reduced net downstream flows (Adamson and Loch 2014; Qureshi et al. 2010).

To reduce the need to reallocate water from agriculture, "environmental works and measures" infrastructure is being constructed "to multiply the environmental benefits achievable from the water available … to enable controlled landscape-scale flooding using environmental water–often in much smaller volumes than would be required without these works" (MDBA 2012a: 56). However, these works have been criticized for benefitting only small areas of wetlands, as well as having negative environmental impacts and high opportunity costs. Pittock et al. (2012) argued that the AUD 235 million allocated to the first tranche of these works may have been more effectively spent on purchasing entitlements for greater environmental flows, which would have increased river flows by 0.9 % on average, or even purchasing wetlands to manage as nature reserves. More importantly, this irrigation and environmental water-efficiency infrastructure is being constructed without any climate change impact assessment and may become redundant and need to be decommissioned under a future climate (Pittock et al. 2012). These perverse impacts illustrate the need to thoroughly evaluate the costs and risks as well as the benefits of investments in water-efficient infrastructure.

3.4.3 Groundwater

In a basin with great hydrological variability and water scarcity, the importance of aquifers for storing water has to be recognized as equally important as the management of reservoirs. It is possible that with good management under climate change, aquifers could be recharged with the periodic high surface flow events (Boening et al. 2012) and called on as a drought reserve.

Groundwater and surface water extractions are conjunctively, systematically regulated in the Basin Plan for the first time. However, the Authority raised the post-2019 groundwater sustainable diversion limit by 948 GL/year to 3334 GL/year above their estimated baseline (historical) diversion limit of 2386 GL/year in the 2012 Basin Plan (Pittock et al. 2015a). Modelling suggests that groundwater use in the basin should be capped in some systems and reduced in others; the cases for increases were modest (CSIRO 2008). There is little evidence of rigorous assessment of the proposed increased levels of groundwater extraction on river flows (MDBA 2012b), which CSIRO (2008) argued could be significant in the long term. Provisions for flooding for groundwater recharge was largely neglected in the Basin Plan (SSCRRAT 2013; WGCS 2012). Under a dry scenario of climate change, most of the priority aquifers in Australia may expect reduced recharge (Barron et al. 2011). The basin highlights the need to conjunctively manage surface water and groundwater resources, prevent overexploitation of aquifers, and maintain their recharge capacity by, for example, conserving floodplains.

3.4.4 Climate Response Measures and Risks to Shared Water

In a closed river basin, changes in land and water management practices may result in significant, cumulative water consumption by increasing the green water component of the water cycle over blue water. The National Water Commission estimated that nationally the water consumption by these "inflow reduction activities" was equivalent to a quarter of all issued water entitlements. This was recognized in the 2004 National Water Initiative and prompted government agencies to assess "risks to shared water resources" from such inflow interception activities as wildfire regrowth, plantation establishment, reuse of irrigation tail water, and on farm dams (tanks) (van Dijk et al. 2006; Sinclair Knight Merz et al. 2010). Fire regrowth, plantation establishment, and farm dams are particular risks to inflows to major reservoirs situated in the higher elevation rim of the basin. It is notable that during the twentieth century, Australian governments were subsidizing farm dam construction as a drought adaptation measure; although a number of Basin states have moved to limit the volume of water captured in these structures, enforcement is questionable. As the Lake Mokoan example (below) suggests, in a warming world shallow, smaller dams may not be an efficient way to store water.

There is a great risk that climate change response policies consume water in the landscape and dewater key storages. For example, modelling of carbon sequestration reforestation of 10 % of the headwater catchments of Lake Burrandong on the overexploited Macquarie River in the Basin projected a 17 % decline in river flows (Herron et al. 2002). Legally, this water would come from the remaining environmental flows that sustain a wetland listed under the Ramsar Convention. The Australian Government has legislated to facilitate carbon farming to sequester greenhouse gases in the landscape; although there are regulations to limit impacts on run-off, they are flawed (Pittock et al. 2013). Similarly, there are risks to water resources from the expansion of unconventional gas extraction as a lower emission energy source, involving aquifer dewatering in parts of the basin (NWC 2010). These examples illustrate the potential for climate change response strategies to negatively impact water resources and storage. Consequently, new kinds of regulation are needed to protect strategic water storages from creeping reductions of inflows through land use and other policy changes.

3.4.5 Climate Change Modelling and Decision-Making

A key question for water managers is when to take action to adapt to climate change because climate projections usually are uncertain. The basin has been subject to extensive climatic and hydrologic modelling (as outlined in the introduction) which projected a range of potential climate change impacts on water availability.

The Commonwealth Scientific and Industrial Research Organisation (CSIRO) did not assign probabilities to these projections, considering that each is possible (CSIRO 2008). The Basin Authority interpreted this in 2010, saying: "While there is uncertainty associated with different predictions of the magnitude of climate change effects by 2030, there is general agreement that surface-water availability across the entire Basin is more likely to decline, with Basin-wide change of 10 % less water predicted" (MDBA 2010: 33). Equating the *median* scenario with *predicted* for planning purposes was not only erroneous, but it did not consider the risk management practice of considering affordable measures to reduce risks arising from less likely but more damaging impacts (Pittock and Finlayson 2011).

An additional reallocation of 3 % of consumptive water to offset climate change impacts was proposed by the Authority in 2010 for the 10-year life of the Basin Plan (Pittock and Finlayson 2011). However, this was abandoned in 2011 because the Authority stated it had "formed the view that there is considerable uncertainty regarding the potential effects of climate change, and that more knowledge is needed to make robust water planning and policy decisions that include some quantified allowance for climate change. Until there was greater certainty MDBA considered that the historical climate record remains the most useful climate benchmark for planning purposes" (MDBA 2012a: 123). Instead, there is a need to start implementing *no-regrets* and robust adaptation measures that offer benefits under a range of different climates. Tools exist to select such measures in the basin (Lukasiewicz et al. 2013) and should be applied, especially the reoperation of water infrastructure and storage.

3.4.6 Reoperation of Infrastructure

Existing water infrastructure in the basin has been built under the premise of hydroclimatic stationarity. Australian engineering standards are now changing with climate change projections (Babister and Ball 2013), leading to programmers to upgrade and reoperate existing infrastructure, such as dam spillway upgrades. While Australian states have dam safety legislation and committees, the narrow mandate of these processes is limiting opportunities to more systematically review the performance of water infrastructure. A process is needed to enable infrastructure upgrading to meet wider economic, social, and environmental objectives, similar to the United States Federal Energy Regulatory Commission's periodic relicensing process for nonfederal government hydropower dams (Russo 2000). For instance, the impacts of both dams and climate change may be reduced by retrofitting fish passages and thermal pollution control devices to many dams in the basin (Pittock and Hartmann 2011). Furthermore, there are a great many redundant structures in the basin that should be decommissioned to increase safety, reduce costs, and restore the river environment. The decommissioning of Lake Mokoan is a key example of the benefits of reoperating infrastructure.

3.4.7 Case Study: Decommissioning Lake Mokoan

Lake Mokoan was constructed as an off-river irrigation, livestock, and domestic farming water supply storage by the Victorian Government in the northeastern part of the state in 1971, in the process inundating the culturally and ecologically significant Winton wetland system. At full supply level, the lake held 365,000 ML but was shallow, covering an area of 3000 ha, leading to evaporation of up to 50,000 ML of the stored water per year—a situation expected to worsen with climate change. From 2003, the basin state and federal governments allocated funds to recover for environmental flows an initial 500 GL per year on average through water-efficiency projects (Water for Rivers 2009). The Victorian Government decided to rely more on other, deeper reservoirs and decommissioned Lake Mokoan in 2009, replacing it instead with much more efficient piped water supply to the farm users, with a total project cost of AUD 108 million (Victorian Auditor General 2010). The Winton wetlands are being restored through a public trust as a focus for conservation of Indigenous heritage and biodiversity, as well as an ecotourism venue.

This case highlights that changing water availability, climate, and public values mean that water storage infrastructure may no longer be fit for purpose. Legislated periodic relicensing of water storage infrastructure is needed to ensure that over periods of decades each structure is reviewed to ensure that it is safe, provides socioeconomic benefits, and minimizes environmental and social impacts (Pittock and Hartmann 2011). In this way, infrastructure can be reoperated as required or decommissioned if it is no longer fit for purpose.

3.5 Conclusion

Australia's Murray-Darling Basin is a key place to assess lessons for managing water infrastructure under climate variability and change because of its extreme climatic variability and the unusual policies applied in this large river basin. Five past and present eras of water management in the basin have been assessed here to document the history of water storage, management of climatic variability, and the lessons for climate change adaptation.

By late last century, storage reservoirs had been built that were capable of storing more than 3 years' average flow of the Murray and Darling rivers, and the basin had closed. Prodded by the environmental and socioeconomic impacts of overextraction of water and severe droughts, Australia's governments have belatedly adapted a range of measures designed to reduce environmental degradation while optimizing socioeconomic benefits. These include allocating shares of the available consumptive water resource; moving towards full-cost recovery in water infrastructure and management; establishing a water market; conjunctive management of surface water and groundwater; reoperation of some water storages; and implementing environmental flows.

Eleven key lessons of global relevance emerge from this assessment of water storage in the Murray-Darling Basin. First, in a closed or closing river basin, additional surface water storage may detract from water security by placing water in sites subject to greater evaporation and by increasing the number of users of the same resource, rather than enhancing adaptation. In such river basins, how the water storages are managed is equally important or more important than the volume of water that they store, as demonstrated by the use of water markets to ensure that smaller volumes of water generate greater socioeconomic return. Second, to effectively maximize the benefits of water storage, surface reservoirs and aquifers need to be managed conjunctively so that groundwater may be used to ameliorate the impacts of surface water users, for the environment and people.

Third, maintaining basic environmental health requires strong environmental flow rules, and these require access to water stored in reservoirs. The fourth lesson is that it also needs a strong and independent management organization to ensure that environmental water is used to maximize ecological outcomes and is not unduly subjugated to economic uses. Fifth, retention of some free-flowing rivers (without reservoirs) is an important measure for conserving biodiversity by retaining flow variability and migratory pathways.

Sixth, considering economic institutions, cap and trade water markets are a key complementary tool to storages, ensuring that available water resources are used to generate higher economic value and employment. In the Murray-Darling Basin, such a system significantly reduced the socioeconomic impacts of climatic variability in the past two decades and has decoupled growth in socioeconomic benefits from water consumption. Seventh, rules that favor full-cost recovery are an important discipline to ensure that proposed new water infrastructure has strong socioeconomic benefits while minimizing environmental impacts, or at least put the onus on governments to be more transparent as to the justification for water infrastructure subsidies.

As the climate changes, the benefits offered by some water storages will decrease or their impacts may increase (for example, with greater evaporation) and may justify decommissioning. Furthermore, the need to re-engineer surface water storages to meet new safety standards (e.g., with bigger spillways) provides cost-effective opportunities to add modern environmental mitigation technologies, including fish ladders, thermal pollution control devices, and larger outlet valves. Hence, the eighth lesson is that periodic relicensing of infrastructure would provide more regular reoperation windows.

Good intentions for adaption are often derailed through a focus on climate change modelling, confusion as to how to manage uncertainty in the resulting projections, and political expediency. A better approach (the ninth lesson) is the assessment of a full suite of adaptation options against a number of climate change scenarios to identify those interventions that may be effective under a range of future conditions, sometimes called robust, no-regret, or low-regret adaptation measures. The tenth lesson is that in addition to direct impacts, climate change response policies will impact on and change the management of surface and groundwater storages requiring ongoing reform of water storage management.

Some climate change mitigation measures threaten to significantly, negatively impact on water resources, such as exploitation of unconventional gas and carbon sequestration plantations.

Finally, the eleventh lesson is that management of water storage is complex and ideally is based on complementary governance at different scales, including establishing key parameters at the larger basin and national scales (e.g., capping water extraction), while offering stakeholders the opportunity and responsibility to apply local knowledge-based management at more local scales (subsidiarity).

The policy experimentation in Australia's Murray-Darling Basin thus provides many lessons for how to better manage water storage and infrastructure to adapt to climate change. Building more water infrastructure does not always increase adaptation or resilience to climate change impacts, but better management and reoperation of existing natural and built infrastructure can deliver environmental and socioeconomic benefits.

References

Adamson D, Loch A (2014) Possible negative feedbacks from 'gold-plating' irrigation infrastructure. Agric Water Manag 145:134–144

Arthington AH (2012) Environmental flows. Saving rivers in the third millennium. University of California Press, Berkeley

Arthington A, Pusey BJ (2003) Flow restoration and protection in Australian rivers. River Research and Applications 19:377–395. doi:10.1002/rra.745

Babister M, Ball J (2013) Statement from the national committee on water engineering. Engineers Australia, Canberra

Barron OV, Crosbie RS, Charles SP, Dawes WR, Ali R, Evans WR, Cresswell R, Pollock D, Hodgson G, Currie D, Mpelasoka F, Pickett T, Aryal S, Donn M, Wurcker B (2011) Climate change impact on groundwater resources in Australia. Waterlines, National Water Commission, Canberra

Boening C, Willis JK, Landerer FW, Nerem RS, Fasullo J (2012) The 2011 La Niña: so strong, the oceans fell. Geophys Res Lett 39(19):L19602. doi:10.1029/2012GL053055, http://onlinelibrary.wiley.com/doi/10.1029/2012GL053055/abstract;jsessionid=35A055A7B721ED6DEC2FC14133C61DD4.f01t03

Bowling L, Baker P (1996) Major cyanobacterial bloom in the Barwon-Darling River, Australia, in 1991, and underlying limnological conditions. Mar Freshw Res 47(4):643–657. doi:10.1071/MF9960643

Commonwealth of Australia (2008) Water Act 2007. Act No. 137 as amended. Commonwealth of Australia, Canberra

Commonwealth of Australia (2012) Basin plan. Commonwealth of Australia, Canberra

Connell D (2007) Water politics in the Murray-Darling Basin. The Federation Press, Leichardt

Connell D (2011) The role of the commonwealth environmental water holder. In: Connell D, Grafton RQ (eds) Basin futures: water reform in the Murray-Darling Basin. ANU E Press, Canberra, pp 327–338

CSIRO (2008) Water availability in the Murray-Darling Basin. A report from CSIRO to the Australian government. CSIRO, Canberra

Falkenmark M, Molden D (2008) Wake up to the realities of river basin closure. Water Resources Development 24(2):201–215

Foerster A (2008) Managing and protecting environmental water: lessons from the Gwydir for ecologically sustainable water management in the Murray Darling Basin. Environmental Planning and Law Journal 25(1):130–153

Garrick DE (2015) Water allocation in rivers under pressure. Edward Elgar, Cheltenham

Grafton RQ (2015) Water: risks, opportunities and public policy. In: Paper presented at the 18th annual conference on global economic analysis, Melbourne, 19 June 2015

Grafton RQ, Horne J (2014) Water markets in the Murray-Darling Basin. Agric Water Manag 145:61–71. doi:10.1016/j.agwat.2013.12.001

Grafton RQ, Pittock J, Davis R, Williams J, Fu G, Warburton M, Udall B, McKenzie R, Yu X, Che N, Connell D, Jiang Q, Kompas T, Lynch A, Norris R, Possingham H, Quiggin J (2012) Global insights into water resources, climate change and governance. Nature Clim Change 3 (4):315–321. doi:10.1038/nclimate1746

Grafton RQ, Pittock J, Williams J, Jiang Q, Possingham H, Quiggin J (2014) Water planning and hydro-climatic change in the Murray-Darling Basin. Australia. AMBIO:1–11. doi:10.1007/s13280-014-0495-x

Herron N, Davis R, Jones R (2002) The effects of large-scale afforestation and climate change on water allocation in the Macquarie River catchment, NSW. Australia. Journal of Environmental Management 65(4):369–381

Hollander R, Curran G (2001) The greening of the grey: national competition policy and the environment. Australian Journal of Public Administration 60(3):42–55. doi:10.1111/1467-8500.00223

Kingsford RT, Walker KF, Lester RE, Young WJ, Fairweather PG, Sammut J, Geddes MC (2011) A Ramsar wetland in crisis—the Coorong, Lower Lakes and Murray Mouth, Australia. Mar Freshw Res 62:255–265

Leblanc MJ, Tregoning P, Ramillien G, Tweed SO, Fakes A (2009) Basin-scale, integrated observations of the early 21st century multiyear drought in southeast Australia. Water Resour Res 45(4):W04408. doi:10.1029/2008wr007333

Leblanc M, Tweed S, Van Dijk A, Timbal B (2012) A review of historic and future hydrological changes in the Murray-Darling Basin. Global Planet Change 80–81:226–246. doi:10.1016/j.gloplacha.2011.10.012

Lukasiewicz A, Finlayson CM, Pittock J (2013) Identifying low risk climate change adaptation in catchment management while avoiding unintended consequences. National Climate Change Adaptation Research Facility, Gold Coast

Maheshwari BL, Walker KF, McMahon TA (1995) Effects of regulation on the flow regime of the river Murray, Australia. Regulated Rivers Res Manage 10(1):15–38

McHugh S (1996) Cottoning on: stories of Australian cotton-growing. Hale & Iremonger, Willoughby

MDBA (2010) Guide to the proposed Basin Plan: overview. Murray-Darling Basin Authority, Canberra

MDBA (2012a) Proposed Basin plan consultation report. MDBA publication no: 59/12. Murray-Darling Basin Authority, Canberra

MDBA (2012b) The proposed groundwater baseline and sustainable diversion limits: methods report. MDBA publication 16/12. Canberra, Murray-Darling Basin Authority

MDBA (2015) Water storage in the Basin. Murray Darling Basin Authority, Canberra

MDBC (2004) Native fish strategy for the Murray-Darling Basin 2003–2013. MDBC publication 25/04. Murray-Darling Basin Commission, Canberra

MDBMC (2001) Basin salinity management strategy 2001–2015. Murray-Darling Basin Ministerial Council, Canberra

MEA (Millennium Ecosystem Assessment) (2005) Ecosystems and human well-being: wetlands and water synthesis. World Resources Institute, Washington DC

Merz SK, CSIRO, Bureau of Rural Sciences (2010) Surface and/or groundwater interception activities: initial estimates. Waterlines report. National Water Commission, Canberra

Miller C (2005) The snowy river story: the grassroots campaign to save a national icon. ABC Books, Sydney

NCC (n.d.) National competition policy. Major areas for reform. National Competition Council, Canberra

NWC (2010) The coal seam gas and water challenge: national water commission position. National Water Commission, Canberra

NWC (2011) The national water initiative—securing Australia's water future: 2011 assessment. National Water Commission, Canberra

Pascoe B (2014) Dark emu: Black seeds agriculture or accident? Magabala Books, Broome

Pittock J (2013) Lessons from adaptation to sustain freshwater environments in the Murray-Darling Basin, Australia. Wiley Interdisc Rev Clim Change 4(6):429–438. doi:10.1002/wcc.230

Pittock J, Finlayson CM (2011) Australia's Murray-Darling Basin: freshwater ecosystem conservation options in an era of climate change. Mar Freshw Res 62:232–243

Pittock J, Hartmann J (2011) Taking a second look: climate change, periodic re-licensing and better management of old dams. Mar Freshw Res 62:312–320

Pittock J, Finlayson CM, Gardner A, McKay C (2010) Changing character: the Ramsar Convention on Wetlands and climate change in the Murray-Darling Basin, Australia. Environ Plann Law J 27(6):401–425

Pittock J, Finlayson CM, Howitt JA (2012) Beguiling and risky: "environmental works and measures" for wetlands conservation under a changing climate. Hydrobiologia 708(1): 111–131. doi:10.1007/s10750-012-1292-9

Pittock J, Hussey K, McGlennon S (2013) Australian climate, energy and water policies: conflicts and synergies. Aust Geogr 44(1):3–22. doi:10.1080/00049182.2013.765345

Pittock J, Grafton RQ, Williams J (2015a) The Murray-Darling Basin plan fails to deal adequately with climate change. Water J Aust Water Assoc 42(6):28–34

Pittock J, Hussey K, Dovers S (2015b) Ecologically sustainable development in broader retrospect and prospect: evaluating national framework policies against climate adaptation imperatives. Australas J Environ Manage 22(1):1–15. doi:10.1080/14486563.2014.999725

Productivity Commission (2010) Market mechanisms for recovering water in the Murray-Darling Basin, final report, March. Productivity Commission, Canberra

Qureshi M, Schwabe K, Connor J, Kirby M (2010) Environmental water incentive policy and return flows. Water Resour Res 46(4). doi:10.1029/2008WR007445

Russo TN (2000) US federal energy regulatory commission. Contributing paper prepared for Thematic Review IV.5: Operation, monitoring and decommissioning of dams. World Commission on Dams, Cape Town

SSCRRAT (2013) The management of the Murray-Darling Basin. Senate Standing Committee on Rural & Regional Affairs & Transport, Canberra

Steinfeld C, Kingsford RT (2013) Disconnecting the floodplain: earthworks and their ecological effect on a dryland floodplain in the Murray-Darling Basin, Australia. River Res Appl 29(2): 206–218

van Dijk A, Evans R, Hairsine P, Khan S, Nathan R, Paydar Z, Viney N, Zhang L (2006) Risks to the shared water resources of the Murray-Darling Basin. Murray-Darling Basin Commission, Canberra

Victorian Auditor General (2010) Irrigation water stores: Lake Mokoan and Tarago Reservoir. Victorian Government Printer, Melbourne

Water for Rivers (2009) Water for rivers. Water for Rivers, Albury

WCD (2000) Dams and development: a new framework for decision-making. The report of the World Commission on Dams. Earthscan, London

Weir JK (2009) Murray river country: an ecological dialogue with traditional owners. Aboriginal Studies Press, Acton

Weir JK (2011) Water planning and dispossession. In: Connell D, Grafton RQ (eds) Basin futures: water reform in the Murray-Darling basin. ANU E Press, Canberra, pp 179–191

WGCS (2012) Analysis of groundwater in the draft 2011 Murray-Darling Basin Plan. Wentworth Group of Concerned Scientists, Sydney

Chapter 4
Climate Change Adaptation, Water Infrastructure Development, and Responsive Governance in the Himalayas: The Case Study of Nepal's Koshi River Basin

Shahriar M. Wahid, Aditi Mukherji and Arun Shrestha

Abstract It is predicted that by the 2050s, the Koshi (Kosi) River basin, the largest Himalayan basin in Nepal, will be experiencing frequent and devastating flooding events and lower lean season flows due to climate change. This will threaten the livelihoods of millions of inhabitants. Development of water infrastructure has the potential to make water availability more consistent and secure. It could generate as much as 10,086 MW of economically feasible energy and irrigate approximately 500,000 ha of agricultural land. We argue that the challenges of water infrastructure development under climatic uncertainty can be overcome through systematic assessment of climatic and nonclimatic risks and responsive governance mechanisms that employ newer forms of stakeholder engagement and accountability, networks, partnerships and enhanced collaboration across sectors.

Keywords Himalayas · Koshi basin · River basin management · Infrastructure development · Responsive governance

4.1 Introduction

The Himalayas provide numerous ecosystem goods and services. As the "water towers" of Asia, the Himalayas sustain the region's ten major river basins—the Amu Darya, Brahmaputra, Ganges, Indus, Irrawaddy, Mekong, Salween, Tarim, Yangtze, and Yellow Rivers. These connect upstream and downstream areas in terms of culture, communication, trade, commerce, and resource management. However, the diverse topography, young geological formations, high degree of glaciation, strong monsoon influence, and critical resource governance deficit in the

S.M. Wahid (✉) · A. Mukherji · A. Shrestha
International Centre for Integrated Mountain Development (ICIMOD), Kathmandu, Nepal
e-mail: Shahriar.Wahid@icimod.org

C. Tortajada (ed.), *Increasing Resilience to Climate Variability and Change*, Water Resources Development and Management,
DOI 10.1007/978-981-10-1914-2_4

region mean that it is also a global climate change hotspot. Climate change is expected to have a direct impact on the water, food, and energy security of the countries in the Himalayan basins (Immerzeel et al. 2010; Molden et al. 2014).

A particular case in point is Nepal's Koshi (Kosi) River basin (Fig. 4.1). The basin is the largest Himalayan basin in Nepal and the most significant flow contributor to the Ganges River. The basin is one of the most vulnerable basins in Nepal in terms of likely impact of climate change on water availability, and the resulting effects on agriculture and livelihood options for over 30 million largely impoverished people (Bates et al. 2008; UNEP 2008; Gosain et al. 2011; Bharati et al. 2012). The large variation in seasonal water availability (Bharati et al. 2014) under climate change conditions renders the basin water insecure (Dixit et al. 2009). Bharati et al. (2012) suggest that the basin is likely to become more vulnerable to flooding in the future, with an increased frequency of high flow events. More than 50 % of the basin is projected to see frequent and devastating flooding events, and

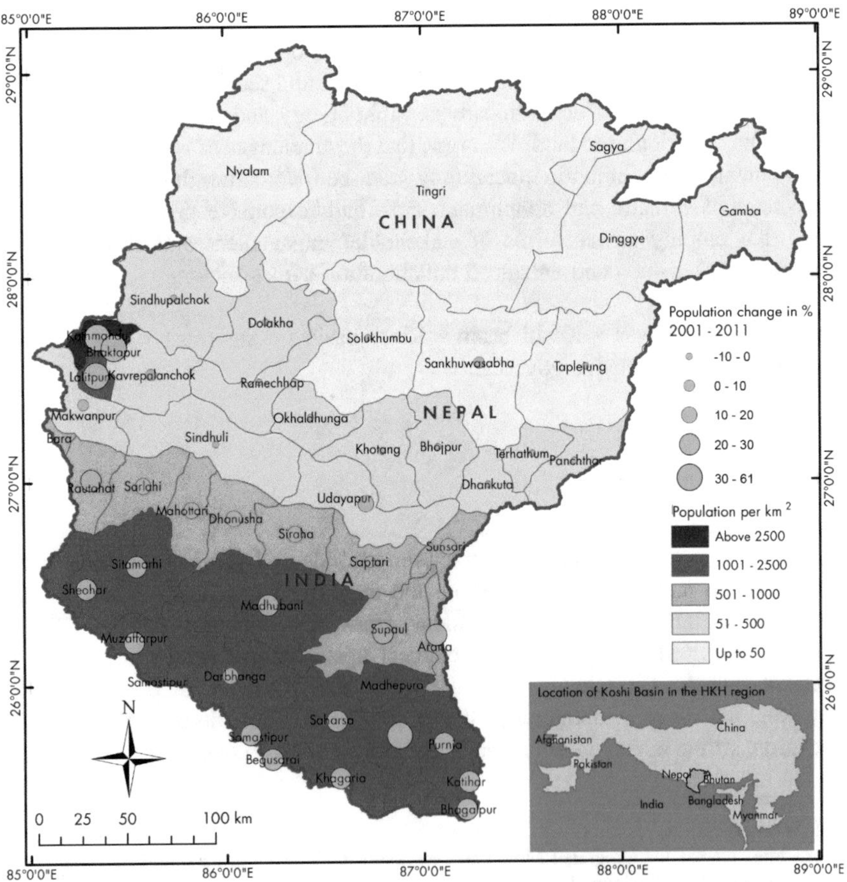

Fig. 4.1 Koshi River Basin

lower lean season flows by the 2050s. Variability in moisture conditions is also expected to increase, with 2–3 more drought weeks by the 2050s. Chen et al. (2013) reported that changes in precipitation patterns and extended durations of flooding, in addition to an increased frequency of landslide events, could make the basin highly erosion-prone and hinder infrastructure development.

Alternating periods of high and low water availability can threaten livelihoods, and investment in water infrastructure has the potential to make water availability more consistent and secure. The high flow gradient and deep valleys of the Koshi River basin provide great potential for water infrastructure development to generate hydropower to meet Nepal's growing energy demand, which is increasing at a rate of 9 % per year (NEA 2014), and expand irrigated areas. It is estimated that 48 billion m^3 of water is available annually in the Koshi basin; this can potentially generate about 10,086 MW of economically feasible power while also irrigating approximately 500,000 ha of agricultural land (GoN-WECS 1999).

The Government of Nepal clearly identifies the development of its rich hydropower potential and the integration into the South Asia regional power market as key strategies for ensuring universal access to a sustainable, reliable, and affordable power supply in Nepal, and a means to generate export revenues to sustain economic growth in the long term (GoN-WECS 2010). Despite protracted political transition, there is a broad consensus among political parties around hydropower development. Leaders of seven major political parties signed a joint statement of commitment on 9 April 2013 to improve hydropower-licensing procedures, share benefits with local communities, and facilitate development processes. The revisiting of projects such as Upper Tamakoshi and Arun III clearly suggests Nepal's recent focus on water infrastructure development for economic development in the Koshi basin (NEA 2014). However, traditional infrastructure development decision-making based on so-called stationarity fails to anticipate the future in an age of climatic uncertainty. The resulting dilemma for the proponents of infrastructure development, including government planners and the private sector, has hindered important investments in the water sector in Nepal (Bonzanigo et al. 2015).

Against this background, this chapter investigates how water infrastructure development could help Nepal to adapt to climate change in the Koshi basin and how responsive governance would greatly improve the sustainability of infrastructures, from transboundary issues through to individual projects. We argue that challenges of development and management of water infrastructures in view of climatic uncertainty should encourage us to understand how development trajectories might reallocate incomes, risks, and land and water resources; how they will have different consequences for different social groups; and how we will ensure that the benefits of development are shared equitably.

Methodologically, this chapter adopts a conceptual framework of sustainable infrastructure development and management consisting of a comprehensive analysis of climate change impacts, adaptation measures to account for vulnerabilities, and governance best practices developed from existing literature. Predicted climate change and water regime scenarios for the 2050s were obtained from statistical downscaling of two representative concentration pathways (RCP), RCP4.5 and

RCP8.5, conducted by Immerzeel et al. (2011). Hydrological simulations were carried out by the Koshi Basin Initiative at the International Centre for Integrated Mountain Development (ICIMOD 2015). The potential of infrastructure development for storage and power generation was investigated using a water evaluation and planning system adopted by the Koshi Basin Initiative (Chinnasamy et al. 2015).

A secondary literature review of the existing governance processes was made by covering strategic level documents (e.g. the National Water Resource Strategy [NWRS], National Water Plan [NWP], and Koshi Basin Strategic Management Plan [KBSMP]) that relate to infrastructure development in the Koshi basin. The gaps and inconsistencies found in the secondary procedural documents were explored to draw conclusions and propose recommendations to evolve flexible, robust, bottom-up, risk-based governance mechanisms and newer forms of institutional mechanisms. These would effectively move procedure from theoretical constructions of a participatory development approach to newer forms of stakeholder engagement and accountability, reliant on networks and partnerships, with enhanced collaboration across sectors, where government acts as enabler and coordinator rather than provider.

4.2 The Impact of Climate Change on Water in the Basin and Importance of Infrastructure as a Climate Change Adaptation Strategy

Originating in the high-altitude Tibetan Plateau and the Himalayas, the Koshi River and its tributaries flow through Nepal's mid-hills onto the Terai lowland region of Nepal. The climate in the basin varies from humid tropical in the south, through subtropical and temperate, to cold and arid in the north. The climate in the southern part of the basin is strongly influenced by the South Asian monsoon, whereas to the north, the Tibetan plateau lies in a rain-shadow area. In the Nepal portion, there are four distinct climatic seasons: pre-monsoon (March–May), monsoon (June–September), post-monsoon (October–November), and winter (December–February) (Rajbhandari et al. 2016). Analysis of precipitation data for 1951–2007 (Yatagai et al. 2012) shows high spatial and temporal variability. About 80 % of the precipitation occurred between June and September (the monsoon season). Analysis of extreme precipitation ($\geq$ 10.0 mm/day) for the two normal periods (1951–1980 and 1981–2007) indicates a slight increase in extreme precipitation, especially towards the western parts of the plain area between these two periods. The annual rates of evapotranspiration are generally less than 1000 mm (Kattelmann 1991). However, some parts of the basin, such as the Sun Koshi, have extremely high potential evapotranspiration and suffer from frequent droughts and soil erosion.

Bharati et al. (2014) reported a large temporal variation in the water balance components in the basin. The mean annual precipitation is highest (1920 mm) in the mountains and decreases in the hills and Indo-Gangetic Plains. Precipitation is the lowest (914 mm) in the trans-mountains. The annual mean actual evapotranspiration shows a different trend, with the minimum (231 mm) being in the trans-mountains, which increases up to the Indo-Gangetic Plains (660 mm). The annual mean water yield has a pattern similar to that of precipitation: minimum water yield occurs in the trans-mountains (569 mm), whereas the maximum occurs in the mountains (1442 mm). The range of the maximum to minimum values of precipitation, actual evapotranspiration, and net water yield is very large for the mountain region and lowest for the trans-mountains.

Water yields are high in the monsoon season and low in the dry period of the year. Large groundwater storage changes during the monsoon could be attributed to the high level of groundwater recharge, which is responsible for groundwater flow and ultimately base flow during the dry period of the year (Bharati et al. 2014). Also, a decreasing trend of groundwater storage from the monsoon to the dry period is very prominent and will impact agricultural productivity and rural livelihoods. July was found to be the wettest month, with a maximum precipitation of 393 mm; December was the driest with 9 mm.

A comprehensive water balance study conducted by the Koshi Basin Initiative (ICIMOD 2015) provided insight into the general hydrologic response of the Koshi River basin to the changes in climate. It showed that more than 50 % of the basin—especially in middle section where much of the infrastructures is developed—is projected to see an increase in precipitation and net water yield (and stream flow) during the monsoon and pre-monsoon seasons (and possibly the large spill energy in the monsoon season), which will lead to more frequent and devastating flooding events and lower precipitation during the winter by the 2050s under RCP 4.5 and RCP 8.5[1] climate change scenarios.

Evaluation of changes in the magnitude of flood peaks above the 99th percentile flow and flow duration curves indicates that peak stream flow might be higher in the RCP 4.5 climate change scenario than the RCP 8.5 by the 2050s (ICIMOD 2015). Variability in moisture conditions is expected to increase, and we can expect 2–3 more drought weeks by the 2050s. Temporal distribution of dependable lean season flows was assessed at six locations of the basin to guide adaptation planning and development of water storage facilities. Ninety percent exceedance flow is considered dependable assured water availability for storage development. Results indicate that dependable lean flow may reduce by the 2050s at many locations.

Large projected seasonal temporal and spatial changes in precipitation, water availability, and flow variability call for increasing water storage in the mountains through various means, as well as development and adaptive management of multipurpose water infrastructure to ensure future food and energy security (Chinnasamy et al. 2015). The potential risks of climate change on infrastructure development in mountainous regions like the Koshi basin does not imply that the investment itself is risky (World Bank 2015). In fact, the risks may be quite low and

[1]Greenhouse gas concentration trajectories adopted by the Intergovernmental Panel on Climate Change for its Fifth Assessment Report (Van Vuuren et al. 2010).

easily managed through identifying vulnerabilities, improving water-use efficiency, proper demand management, and responsive governance. The World Bank (2015) study indicated that there are risks to the economic viability of the proposed 335 MW Upper Arun Hydropower Project in the low-flow scenarios of the future. It suggested that it is enough to identify when and where the future variability is different from today to target water infrastructure planning and adaptation strategies.

Chinnasamy et al. (2015) evaluated 11 water infrastructures to meet domestic, agricultural, and industrial water and energy demands in the 2050s. Four of these 11 proposed infrastructures (Dudh Koshi, Sapta Koshi High Dam, Sun Koshi and Tamor) are considered storage-type hydropower developments, while the remaining (Arun III, Bhote Koshi, Lower Arun, Sundarijal, Sun Koshi HEP, Tama Koshi, and Upper Arun) are taken as run-of-the-river (ROR) types. Results revealed that 29,733 GWh energy could be generated annually through the implementation of these 11 water infrastructures. The generated energy could sustain Nepal's future power demand, as well as earning valuable foreign currency through the export of power to neighboring countries. In addition, these water infrastructures can store 8.4 billion m^3 of water, which would satisfy the projected water demands for the 2050s. Such infrastructures can also be used to regulate low-flow conditions during post-monsoon and winter months, and thus can promote positive upstream–downstream connectivity. Transboundary power trade and water-based connectivity is expected to boost regional economic cooperation.

4.3 Overview of Water Infrastructure Development in the Koshi Basin

Nepal has made remarkable progress in poverty reduction and human development in recent times. Its economic growth rate reached over 5.0 % in the fiscal year of 2014, slightly above the 4.7 % achieved on average over 2008–2012. To maintain the momentum, Nepal will need to exploit its demographic opportunities, increase public spending, and attract private investment, particularly for infrastructure, helping its reasonably educated youth to raise agricultural productivity and incomes, and strengthen non-farm livelihood opportunities, such as tourism service provision. The Government of Nepal is keen to develop water infrastructure for economic growth and human development because the country is endowed with rich water resources (GoN-MoE 2010). The National Electricity Crisis Resolution Action Plan and 10-year Hydropower Development Task Force of Nepal declared that it is essential to construct storage-type hydroelectric power plants that are able to supply electricity, but they also noted that the construction of storage-type hydroelectric power plants should be carried out systematically, taking into consideration the consistency of overall water development, hydrological and geological characteristics, environmental impact, etc. (WWF 2010).

The Japan International Cooperation Agency (JICA 1985) studied the infrastructure development potential of the Koshi basin, and proposed a "Master Plan Study on the Koshi River Water Resources Development." The plan outlined possible reservoirs and other ROR-type projects for hydropower as well as for irrigation in the downstream areas. The plan has been updated by JICA (2014) through "The Nationwide Master Plan Study on Storage Hydroelectric Power Development." At present, a number of hydropower plants, diversions, and other types of water infrastructure have been developed along the Koshi and its tributaries for river control and exploitation.

4.3.1 Hydropower Infrastructure

Currently, seven hydropower installations produce about 211.6 MW of power, and this accounts for nearly 37 % of the total power generated in the country (Table 4.1). Additionally, the basin has about 20 small and micro hydropower projects; while important locally, the total energy production from these projects is low.

There are a number of large projects that have either been surveyed or are already under construction. Fifty-two hydropower project sites have been identified within the Koshi basin alone, with a total production potential of 10,909 MW (World Bank 2015). Thirteen of these are priority projects with a total potential of 8473 MW. Plants scheduled to be commissioned include the Upper Tama Koshi (465 MW), and Kulekhani III (14 MW). In 2015, the Government of Nepal initiated the 335 MW Upper Arun Hydropower Project in the Koshi River basin. The project will need a 37 m dam construction in the Arun River. The development will be completely owned by the Government of Nepal. Interestingly, in order to address issues of greater social inclusion, a separate project called Ikhuwa Khola Hydropower Project (30 MW) has been identified and is proposed as an integral part of Upper Arun Hydropower in which local communities will be able to invest.

Feasibility studies have been completed for several projects, including Arun III (402 MW), Dudh Koshi (300 MW), Tamor-Mewa (101 MW), and Tama Koshi V (87 MW). The Dudh Koshi Project is a major storage-type infrastructure. The

Table 4.1 Existing major hydropower projects in the Koshi basin, Nepal (NEA 2014)

Name	Koshi sub-basin	Power generation (MW)
Khimti	Tama Koshi	60.0
Khulekhani I	Bagmati	60.0
Bhote Koshi	Bhote Koshi	36.0
Khulekhani II	Bagmati	32.0
Sun Koshi	Sun Koshi	10.5
Indrawati	Indrawati	7.5
Chatara	Sunsari Morang Irrigation project	3.2
Panauti	Bagmati	2.4

recommended scheme consists of a 160-m-high rock-filled dam on Dudh Koshi River, along with a powerhouse at Baikhu Khola, a tributary of Sun Koshi. The reservoir area is 10 km^2 at the full supply level (FSL) of 580 m a.s.l., with a gross storage volume of 687.4 million m^3 at FSL, a spillway of discharge capacity of 5500 m^3/s in the right abutment, and a headrace tunnel of 13,260 m length. The Tama Koshi V is a cascade development of Upper Tama Koshi project. It will feed on the discharge from the tailrace of the Upper Tama Koshi Project through an underground interconnecting arrangement and convey water to the headrace tunnel of the Tama Koshi V. Another promising major development is the Tamor Storage Hydropower Project. The 762 MW development, with a dam height of 550 m, is challenged by a high sediment load from the upstream catchment, but will create an effective storage of 1900 million m^3 for economic development. Another controversial project is the Sapta Koshi High Dam Multipurpose Project. The proposed 269-m-high concrete dam with gross reservoir storage capacity of 13.45 billion m^3 was investigated by the Government of India in 1981. The aim was to generate about 3000 MW of hydropower, developing irrigation, controlling or managing downstream flooding, and promoting water-based navigation.

4.3.2 Irrigation and Flood Control Infrastructure

Apart from the hydropower developments, one of most important forms of infrastructure in the basin is the Koshi Barrage and Embankment located at Bhim Nagar, Nepal, close to the India–Nepal border. The barrage is 1,149 m long and, along with its associated embankment system, was built for flood control and to provide water for irrigation. It also serves as a river gradient control measure through which sediments are deposited in the upstream reaches. The barrage was built between 1959 and 1962 following the signing of a treaty between the Governments of India and Nepal on 25 April 1954. The treaty, which was revised in 1966, entrusts India with the responsibility for the maintenance and operation of the Koshi Barrage (Dixit et al. 2009).

There are three irrigation systems linked to the barrage. The Sunsari Morang Irrigation System (SMIS)—one of the largest irrigation projects of Nepal—diverts water from the left bank of the Koshi River at Chatara, about 40 km upstream of the barrage, providing irrigation facilities for about 68,000 ha of land in Sunsari and Morang districts in Nepal (Dhungel 2009). Two more irrigation systems have been extended into India from the Koshi Barrage: the Eastern Koshi Main Canal, with a potential irrigation area of 612,500 ha in India, and the Western Koshi Main Canal, with a potential irrigation area of 356,600 ha in India and 24,480 ha in Nepal. Further irrigation projects include the Bagmati and Kamala projects, both of which are based on a barrage, and the Triyuga-Chandra Canal Irrigation Project. There is one small hydropower facility, the Chatara Hydropower Centre, on the irrigation canal of SMIS, which removes silt and sediment from the canal using dredgers.

The Koshi Barrage is closely related to the downstream Koshi Tappu Wildlife Reserve. The reserve is one of the most important wildlife reserves of Nepal. It was established in 1976 as a protected area of 175 km^2 and was designated the first Wetland of International Importance in Nepal by the Ramsar Convention in 1987, reflecting its high value in maintaining the genetic and ecological diversity on the region. The ecosystem services provided by the wetland area are estimated to have a total annual economic value of USD 16 million, equivalent to USD 916 per hectare or an average annual benefit of USD 959 per household (Sharma et al. 2015).

The Koshi embankment system linked to the barrage has a history of breaches, the most recent being the major breach in 2008. On 18 August 2008, there was a major breach of the eastern embankment of the Koshi River at Kusaha in Nepal, 12 km upstream of the Koshi Barrage. The river picked up an old channel it had abandoned over 80 years previously near the border with Nepal and India, which resulted in a major avulsion of the river. The new path through Sunsari District in Nepal was dominated by agricultural land, settlements, and infrastructure. The breach resulted in large-scale inundation of several districts in Nepal and India and devastation of the affected area. A population of 4.8 million was affected by the flood, 235 people died, as did 787 head of livestock, 322,169 houses were destroyed, more than 338,000 ha of agricultural land was damaged and displaced 45,000 people in Nepal (Mishra 2008).

In addition, groundwater is another source of irrigation in the Koshi basin floodplains (Terai) and, as in the rest of Nepal, Terai groundwater potential is greatly under-utilized (GWRDB 2016). According to Groundwater Resources Development Board of the Ministry of Irrigation, Nepal, Terai has a total dynamic groundwater reserve of 8800 million m^3, of which only approximately 1000 million m^3 is used for irrigation, drinking, and industrial purposes. Terai districts in Nepal have a little more than 120,000 shallow tube wells, of which half are in districts that lie in the Koshi basin of Nepal. There has been a rapid rise in groundwater-irrigated area in the past one decade (2005–06 to 2014–15) from 70,000 to 120,000 shallow tube wells. Yet, there is much larger scope to develop groundwater for irrigation through appropriate technologies like electric- and solar-powered pumps. Irrigation development in Koshi basin can therefore take advantage of large groundwater resources and use them in conjunction with surface water resources.

4.4 Challenges of Development and Management of Water Infrastructure Under Climatic Uncertainty

The development of water infrastructure related to agriculture and hydropower depends largely on a sustained quantity of water being provided by the Koshi River and its tributaries. However, developing and managing water infrastructures under climatic uncertainty can be challenging (UNECE–INBO 2015), partly because traditional decision-making, which is based on historical trends and notions of predictable variability in precipitation and river flows, lacks the understanding of long-term uncertainty in climate variability and change and partly because of concerns about the redistribution of the region's hydrology with concomitant issues of erosion and sedimentation, resettling displaced people, financial and institutional implementation modalities, and the overall contribution of infrastructure development to local and national benefit sharing. Uncertainties over climate change's effects on hydrology also complicate the assessment of how significant the environmental effects of infrastructure development will be.

4.4.1 Biophysical Challenges

Past experiences in the basin clearly illustrate some of the biophysical challenges related to climate change for water infrastructure development. For example, on 2 August 2014, a huge landslide dammed Sun Koshi River at Jure, Sindhupalchok district, and inundated the nearby Sanima hydropower installation. Investigation by the ICIMOD (2015) indicated that two unusually high intensity rainfalls (70 mm/day) during 30–31 July 2014 on the nearby highly weathered hill slopes, along with a high volume of seepage water from natural streams, springs, and a local irrigation canal, filled old, deep cracks, which increased pore water pressure triggering the massive landslide and damming the river. The 2014 Sun Koshi Landslide and its impact on water infrastructures in the sub-basin highlighted questions about biophysical risks of development in fragile ecosystems (Khanal et al. 2015), questions that are a long-standing source of dispute between development supporters and environmentalists.

The development of water infrastructure in the Koshi basin also faces challenges of land-use change as a result of human pressure, which in turn impacts seasonal flows, sediment yield, and biodiversity (Chettri et al. 2013; Uddin et al. 2016). A case study undertaken by the ICIMOD (2015) on the Kulekhani Hydropower Project in the basin looked at the potential impact of land-use change on hydrology and hydropower production. Results showed that excessively high flows and high sediment yield during the wet season can lead to flooding and excessive sediment deposition in the reservoir, whereas low flows in the dry season can increase water scarcity; both may reduce energy production. Results indicate that land-use change in terms of forest depletion and agricultural expansion, or forest land conversion to barren land, increases the average annual flow and average annual sediment yield. Seasonal flow is substantially increased during the wet season (June–September) and decreased during the dry season (October–May). The average annual power production decreases with decreased annual flow, but sediment deposition in the reservoir had less impact.

In the Koshi floodplains, the high silt deposits (Ries 1995; Andermann et al. 2012) and change in course of the Koshi River (Dixit et al. 2009) due to changing climate and water distribution has created many different types of aquatic habitats, such as small freshwater lakes, swamps, wetlands, lakes, oxbow lakes, and ponds, which are important habitats for a large fish population, and are also breeding and feeding areas for birds, carp, and other fish species. These aquatic habitats provide a critical resource for waterfowl, especially during winter (Baral 2013); this has led to increased hunting pressure by the local communities. Many aquatic habitats of smaller size, however, are less important for agriculture and fish farming (Ramsar 2016). The impact of water infrastructure development on freshwater ecosystems and environmental services has been a topic of intense debate over the years.

4.4.2 Management Challenges

Despite the various biophysical challenges related to climate change involved in developing and managing water infrastructure, recent advances in understanding predicted impact illustrate ample opportunity for investment to promote climate change adaptation through hydropower production, more timely irrigation water delivery and regulation of the extreme flows of the basin. Simultaneous investment is needed to develop and nurture governance and institutional systems for efficient, equitable, and sustainable governance and management of the water-related services provided by water infrastructures (Molden et al. 2014).

Contemporary Nepali policies, which often focus on a mix of indigenous micro- and medium-scale hydropower projects and externally funded, large storage projects (GoN-WECS 2002, 2005, 2010) without considering the whole range of different scenarios and possibilities, pose considerable challenges in the Koshi basin. The approach is towards “no bad infrastructure” as opposed to “no infrastructure.” Because of the large power shortages that already exist in Nepal—power cuts can be as much as 16 h a day—there is often a “power imperative” that favors major hydropower projects. Approaching large hydropower infrastructure development in this way may have negative socioecological outcomes in the long term. All but the smallest hydropower projects are likely have marked environmental and socioeconomic impacts for a considerable distance upstream and downstream, and multiple projects are likely to have an even more intense cumulative effect (Wang 2012).

Significantly, however, the current method of assessing the costs and benefits of hydropower development analyzes the impacts of each new infrastructure project on a project-by-project basis and thus does not account for the cumulative effect on hydrology, sediment movement and erosion, aquatic and terrestrial biodiversity, and other environmental aspects. This also impacts the potential for equitable distribution of benefits, as local and vulnerable people tend to receive the fewest benefits, unless governments and companies make provision to avoid this and to develop on a smaller scale (SWECO 2011). The resulting ad hoc or ambiguous management arrangements for benefit sharing, which lack adequate processes for defining how compensation will be determined and benefits shared, often fuel politicization of infrastructure development schemes by local, regional, and national interest groups, particularly for hydropower development.

There is also controversy over how or if large water infrastructures should be managed for flood control in the basin. Large upstream water storage facilities have long been considered the best way to control extensive riverine flooding that devastates vast floodplains of the basin. Though the World Bank-supported Ganges Strategic Basin Assessment (2011) concluded that at the basin-wide level, the storage potential of the system is simply too small to meaningfully regulate the full river system, several Nepali scholars have voiced opposing views (JVS 2012), and argued that all three ecosystem services (hydropower, low flow augmentation and flood control) need to be valued in order to negate social impacts of large storage infrastructure in the basin. Therefore, a comprehensive study needs to be undertaken in order to evaluate the role of storage infrastructure to regulate floodplain flooding and augment lean season water availability for hydropower and agricultural development. There is also a need to mainstream technical expertise and alternate views for river basin planning and the cumulative impact evaluation of infrastructure projects.

Interestingly, climate change has emerged as a policy and governance issue in the past decade or so, most notably through the National Action Plan for Adaptation (MoE 2010) and Local Adaptation Plan for Action; but it has gained increasing priority recently given the considerable impacts that Nepal is projected to experience. Improved monitoring and modelling of water flows is generally prioritized but detailed assessments of infrastructure-specific climate impacts at even the individual project level are scarce because of a lack of capacity. Similarly, climate impacts are inadequately addressed in any existing policies, plans, and programs (PPPs) (e.g. the Water Resources Act 1992, Water Resources Regulation 1993, Water Resources Strategy 2002, National Water Plan 2005, Hydropower Development Policy 2001, Electricity Act 1992, Ten Years Hydropower Development Plan 2009, Rural Energy Policy 2006). Synergistic impacts on flow rate changes are even further removed from current management considerations. The overall PPP framework in Nepal shows considerable promise, but it still could benefit from the introduction of strong guidelines in the area of environmental baselines and thresholds (Wang 2012), knowledge-sharing efforts between research and management communities, the establishment of comprehensive information systems for investment decision support, and disaster preparedness at the basin level.

4.5 Responsive Water Infrastructure Governance Under Climatic Uncertainty

The challenges of development and management of water infrastructure in Nepal's Koshi basin under climatic uncertainty should not fuel an a priori anti-infrastructure stance, but it should instead encourage us to understand how development trajectories might reallocate the land and water resources, incomes, and risks, and therefore have different consequences for different social groups (Gyawali 2008). The concomitant governance complexity, where community and jurisdictional boundaries often become porous and unstable, calls for effective cooperation (GoN-WECS 2002, 2005, 2010) and integration across boundaries (public–private) and communities for the creation of a climate-aware responsive governance mechanisms. Decision-makers and practitioners need to put in place a governance structure for developing and managing infrastructure in the basin that: (1) embraces innovative forms of stakeholder engagement, reliant on networks and partnerships and enhanced collaboration across sectors, where government acts as enabler and coordinator rather than provider (Tropp 2005); (2) is flexible and polycentric in organizational and accountable (political, administrative, professional, and output) forms to reduce risks from climate change uncertainty, face water-related hazards and provide beneficial outcomes at basin scale (UN 2015); and (3) nurtures distributed leadership and is able to acknowledge and understand the special nature of mountain resources governance.

The transformation from traditional public service governance to responsive governance will entail organization of different stakeholders and resource users to initiate and sustain coordinated and collaborative actions and more diverse forms of accountability to manage the risks, costs, and benefits of development (Tropp 2007). More attention is required to attend to the structures and interrelationships of organizations; sharing strategies, sophisticated monitoring and communication mechanisms; and coordinated response structure at multiple layers from river basin to smaller-scale water storage, catchment, or irrigation-system levels.

A river basin management approach provides a framework for an integrated assessment of cumulative development impacts and trade-offs in the basin (Speed et al. 2013). Three of Nepal's policy documents already contain elements of river basin management: the NWRS, the NWP, and the KBSMP. The KBSMP was prepared by the Government of Nepal to operationalize the National Water Plan 2005. The strategy aimed to establish a River Basin Office/Koshi at the basin level, sub-basin offices in seven sub-basins (Indrawati, Sun Koshi, Tama Koshi, Likhu, Dudh Koshi, Arun, and Tamor) and Integrated Resource Management Committees at catchment level to coordinate programmes and activities at basin, sub-basin, and key catchments, respectively. However, apart from the establishment of the Koshi Basin Office of the Department of Hydrology and Meteorology in Dharan and Biratnagar, no other institutions have been established to address issues of climate change and water management at basin and sub-basin level. Thus, while several policies introduced some promising aspects of river basin management, in practice, these aspects have not been implemented under adequate legal or regulatory regimes.

More recent strategies on water, climate change, and biodiversity legislation overtly aim to integrate the objectives of preceding PPPs. Actual acts and laws still focus on specific sectors, in particular the energy sector. The NWRS and NWP aimed to assert crossover of sectoral interests and skills, but little material reform has been enacted. Fortunately, there has recently been a general promotion of the unbundling of these issues at basin scale. Knowledge support from national and regional centers (e.g., ICIMOD) and funding support from international financing institutions have paved the way to undertaking the Participatory Watershed Management and Local Governance Project in Nepal (JICA 2014) and initiating the Nepal Power Sector Reform and Sustainable Hydropower Development Project in close collaboration the Water and Energy Commission Secretariat and Nepal Electricity Authority (World Bank 2015).

More attention is required to develop catchment-level or irrigation-system-level governance structures and institutions for promoting and managing smaller-scale water storage under climate change uncertainty. A pilot exercise by the Koshi Basin Initiative at ICIMOD (2014–2015) to revive traditional springs and ponds—critical to sustaining domestic needs and the rural economy in Nepali mid-hills—identified the problem of finding a "voice of authority" in a highly fractured village system with a lack of capacity at the lowest units of governance such as wards within village development committees. Over the last 4 years, some progress has been seen in ward-level planning, budgeting, and implementation following a directive by the Ministry of Federal Affairs and Local Development to establish ward-level citizens' forums ('*nirdeshika*' or '*woda nagrik manch*'). The presence of legitimate, elected local officials is crucial; ad hoc committees, while useful for a while, are not sustainable in the long run. Although there are some good examples at the community level (e.g., the development displayed by farmer-managed irrigation systems), more and different institutional arrangements across scales are needed. This is particularly important given the increasing spatial and temporal variability of available water due to climate change, and the need for basin-level upstream and downstream user interaction and institutional arrangements for risk and benefit sharing of infrastructure investment. For instance, although research has recognized the importance of local institutions in infrastructure projects, developers are only required to engage in informing and consulting activities. Because consultation is only required at the interim stages of planning when many key decisions have already been made, meaningful local-level participation in the development process is usually lacking. While the need for such participatory approaches is well recognized (GoN-WECS 2002), moving from theoretical constructs of participatory approaches to their application in cost, benefit and risk sharing needs to be ensured.

Although many of the large infrastructure development options in the Koshi basin could be pursued by Nepal, some require effective transboundary cooperation. In the past, such transboundary cooperation has faced operational challenges (Dhungel 2009; Birnie et al. 2009; Rahaman 2012), leading to something as serious as the Koshi embankment breach and devastating floods in 2008. Shrestha et al. (2010) listed the various shortcomings of operational aspects of the Koshi Treaty. For example, while a large portion of the embankments are on the Nepali side, maintenance responsibility lies with Government of India, which is done through the Government of Bihar (Shrestha et al. 2010). So, when the embankment was breached, it was challenging for India to repair the embankment within Nepali territory. Similar gaps were seen in existing institutional mechanisms, for example, the Joint Committee on Water Resources between Nepal and India which oversees several projects and a number of rivers, including Koshi, is supposed to meet every year, but had not met for 4 years prior to the devastating floods of 2008; hence, there was inadequate knowledge sharing about the status of the embankment. Local governments are also bypassed, so even though the local people and the government in Nepal had information about the impending breach, the bureaucratic and centralized mechanisms for that information to flow into local-level decision-making processes proved to be less productive. The Asia Foundation (Surie and Prasai 2015) reported that even as governments in the South Asia region have made progress towards greater cooperation and benefit sharing in trade, energy and food security, data, and information sharing on water remains highly nationalistic, technocratic, and zealously securitized.

Many such significant opportunities for cooperative development (e.g., the proposed 3000-MW Sapta Koshi High Dam) have been discussed for decades with little regional consensus or action on the ground (IPPAN/CII 2006; Shrestha et al. 2010). Nepal and India have contrasting views of the value of the freshwater ecosystem services that would be created by the construction of storage infrastructures. However, India's recent efforts to improve bilateral relations with neighboring countries, including transboundary cooperation, have opened new windows of opportunity and generated initial optimism in Nepal. While the existing political leadership in both riparian countries seems to be committed, as reflected by the Indian Prime Minister's first visit to Nepal in 17 years in 2014, institutional constraints will have to be overcome to create a policy and institutional environment that will lead to benefit sharing through cooperative approaches. Important steps have been made in this direction, such as the Koshi River Basin Strategic Management Plan (Nepal) and the Ganga River Basin Management Plan (India). Similarly, in the climate change adaptation arena, countries of the Koshi basin have developed national adaptation strategies that offer much scope for integration and the associated benefits that this could yield. These efforts must be extended across the transboundary basin to reap the potential benefits of infrastructure development.

4.6 Conclusions and Recommendations

Nepal's Koshi basin has experienced significant climate change. Most recently, large temporal and spatial changes in water availability and flow variability have been predicted. Developing water infrastructure is considered crucial for sustainable growth under climatic uncertainty but is constrained by the effectiveness of water governance and management mechanisms. In this context, this chapter adopted a robust sustainable infrastructure development and management framework to focus on potential climate change impacts on water resource availability for water infrastructure development, as well as the challenges and issues of addressing climatic uncertainty and responsive governance.

There are deep uncertainties surrounding climate change, and historical experience of developing water infrastructure in the basin can only provide a partial analogue for dealing with future climatic uncertainty. For decision-makers, investors, and communities at large, who do not know what to expect in the upcoming rainy season, managing the impacts of long-term climate change impacts will require robust decision analysis, identification of system vulnerabilities, and assessment of alternatives for ameliorating those vulnerabilities. Drawing on these findings, important recommendations can be made to develop and manage water infrastructure and adapt to climate change in the Koshi basin.

Given the substantial uncertainty regarding the impact of climate change on infrastructure, a flexible, robust, bottom-up, risk-based approach needs to be adopted. This should be used to set criteria and screen infrastructure development options in policy and practice to assess future risks of climate change on water resource availability and the potential for multiple uses of water infrastructure at a river basin scale. The approach should identify (through modelling and "what-if" scenarios) the response of important performance indicators to parametrically varied changes in climate (e.g., runoff elasticity of hydropower generation and the economic internal rate of return). It can then use available climate change projections to assess which of these responses are more or less likely, then evaluate the way people report and deal with uncertainty.

The success of the responsive governance approach will depend on the Nepali government ensuring transparency, self-regulation, and stakeholder communication. This can be achieved by developing and operationalizing a basin-wide comprehensive water information system to create new opportunities for administrative integration, coordination, and cross-sectoral cooperation, and improve service delivery responsiveness and effectiveness. A further requirement is the identification and capacity development of a network of knowledge-based competence centers that involves scientists and practitioners, building upon local government, and local-level community-led groups, which are able to support Nepali government in taking sound climate-aware decisions.

River basin planning and management involving the private sector, civil society, and public–private partnerships can provide alternative forms of river basin management to transform state-centric ways of making decisions on a complex system more inclusive and pluralistic. Such arrangements can meditate the conflicting interests of different actors, mobilize communities to manage resources at local levels, form acceptable partnerships, and institutionalize the processes of stakeholder dialogues and partnerships, with inclusive decision-making and participation. At the same time, they can select and apply appropriate technology to ensure the sustainability of the ecosystem and equitable and efficient use of water with benefit sharing at local, basin, and regional levels (Wang 2012). For example, the national governments are in the best position to take the lead in managing political accountability, as well as risks due to stream flow variability, sediment-load changes, and potential glacier lake outburst floods or landslide events (Molden et al. 2014). The community should be encouraged to adopt a more holistic "watershed approach," which takes into consideration upstream and downstream interactions, direct and indirect impacts, and recognizes the need for institutional mechanisms for cost, benefit, and risk sharing (Dixit and Basnet 2005).

Investment in climate-sensitive and sustainable infrastructure development needs to be smart. It should include investment in institutions and human capital (the enabling environment), such as through better water management, operation of existing assets, or use of green or multipurpose infrastructure. It is essential to make the best use of existing sources of finance by proper project design, planning, and sequencing. To do this, new sources of finance such as the private sector, Green Climate Fund, long-term investors, or philanthropies should be harnessed.

Consideration of the regional context of infrastructure development is crucial to ensuring investment and sharing of benefits. The Nepali government can utilize regional umbrella mechanisms (e.g., ICIMOD) to address regional knowledge issues (Crow and Singh 2009). Recent regional development knowledge-generation programmes like the Koshi Basin Initiative and the Himalayan Climate Adaptation Programme initiated at ICIMOD provide new avenues to bring together diverse stakeholders on a common platform for basin-wide learning, provoking social action and advising policymakers on how to arrive at more socioecologically robust and egalitarian governance transformations. These initiatives can be further strengthened through the establishment of an adequately mandated regional group/body—hosted by a relevant organization—to independently facilitate and coordinate regional dialogue and strategic processes of climate change adaptation and infrastructure development in the Himalayas.

Acknowledgments This chapter is partially based on the results of case study completed under the Koshi Basin Initiative at the International Centre for Integrated Mountain Development (ICIMOD) supported by the Department of Foreign Affairs and Trade (DFAT), Government of Australia through Sustainable development investment portfolio for South Asia (SDIP) and core funds of ICIMOD contributed by the governments of Afghanistan, Australia, Austria, Bangladesh, Bhutan, China, India, Myanmar, Nepal, Norway, Pakistan, Switzerland, and the United Kingdom.

All views and interpretations expressed in this paper are those of the author(s). They are not attributable to ICIMOD or the Australian Government and do not imply the expression of any opinion concerning the legal status of any country, territory, city, or area of its authorities, or concerning the delimitation of its boundaries.

References

Andermann C, Crave A, Gloaguen R, Davy P, Bonnet S (2012) Connecting source and transport: suspended sediments in the Nepal Himalayas. Earth Planet Sci Lett 351–352:158–170

Baral HS (2013) An assessment of the impact of Koshi floods to birds and mammals. Nepalese J Biosci 2. doi:10.3126/njbs.v2i0.7482

Bates BC, Kundzewicz ZW, Wu S, Palutikof JP (eds) (2008) Climate change and water. Technical paper. IPCC Secretariat, Geneva

Bharati L, Gurung P, Jayakody P (2012) Hydrologic characterization of the Koshi Basin and the impact of climate change. Hydro Nepal: J Water Energy Environ 11(1):18–22

Bharati L, Gurung P, Jayakody P, Smakhtin V, Bhattarai U (2014) The impact of climate change on water resources availability and development in the Koshi Basin, Nepal. Mt Res Dev 34:118–130

Birnie P, Boyle A, Redgwell C (2009) International law and the environment, 3rd edn. Oxford University Press, Oxford

Bonzanigo L, Brown C, Harou J, Hurford A, Karki P, Newmann J, Ray P (2015) South Asia investment decision making in hydropower: decision tree case study of the Upper Arun Hydropower Project and Koshi Basin Hydropower Development in Nepal. Report No.: AUS 11077. GEEDR South Asia, The World Bank, New York

Crow B, Singh N (2009) The management of international rivers as demands grow and supplies tighten: India, China, Nepal, Pakistan, Bangladesh. India Rev 8(3):306–339. doi:10.1080/14736480903116826

Chen NSh, Hu GSh, Deng W, Khanal N, Zhu YH, Han D (2013) On the water hazards in the trans-boundary Koshi basin. Nat Hazards Earth Syst Sci 13:795–808

Chettri N, Uddin K, Chaudhary S, Sharma E (2013) Linking spatio-temporal land cover change to biodiversity conservation in the Koshi Tappu Wildlife Reserve, Nepal. Diversity 5:335–351

Chinnasamy P, Bharati L, Bhattarai U, Khadka A, Dahal V, Wahid S (2015) Impact of planned water resource development on current and future water demand of the Koshi river basin, Nepal. Water Int 40(7):1004–1020

Dhungel D (2009) Historical eye view. In: Dhungel D, Pun S (eds) Nepal-India water resources relationship: challenges. Springer, Dordrecht, pp 11–68

Dixit A, Basnet S (2005) Recognising entitlements and sharing benefits: emerging trends in Nepal's hydropower terrain. IUCN, Gland, Switzerland

Dixit A, Upadhya M, Dixit K, Pokhrel A, Rai DR (2009) Living with water stress in the hills of the Koshi Basin, Nepal. International Centre for Integrated Mountain Development, Kathmandu, Nepal

Government of Nepal–Ministry of Environment (GoN-MoE) (2010) National Adaptation Programme of Action (NAPA) to climate change, Nepal. GoN–MoE, Kathmandu

Government of Nepal, Water and Energy Commission Secretariat (GoN-WECS) (1999) Basin wise water resources and water utilization study of the Koshi River basin. GoN–WECS, Kathmandu

Government of Nepal, Water and Energy Commission Secretariat (GoN-WECS) (2002) Water resource strategy, Nepal. GoN–WECS, Kathmandu

Government of Nepal, Water and Energy Commission Secretariat (GoN-WECS) (2005) National water plan, Nepal. GoN–WECS, Kathmandu

Government of Nepal, Water and Energy Commission Secretariat (GoN-WECS) (2010) Koshi River basin management strategic plan (2011–2021). GoN–WECS, Kathmandu

Gosain AK, Rao S, Mani A (2011) Hydrological modelling: a case study of the Koshi Himalayan basin using SWAT. Soil hydrology, land use and agriculture. CAB International, Wallingford, UK

Groundwater Resources Development Board (GWRDB) (2016) Hydrogeological studies. http://www.gwrdb.gov.np/hydrogeological_studies.php. Accessed 19 Feb 2016

Gyawali D (2008) Epilogue (Re-imagining Nepal's water: institutional blind spots, developmental blind alleys and the lessons of the century past). In: Dhungel D, Pun S (eds) Nepal-India water resources relationship: challenges. Springer, Dordrecht

International Centre for Integrated Mountain Development (ICIMOD) (2015) Water availability and agricultural adaptation options of the Koshi Basin under global environmental change. ICIMOD, Kathmandu

Immerzeel WW, van Beek LPH, Bierkens MFP (2010) Climate change will affect the Asian water towers. Science 328:1382–1385

Immerzeel WW, van Beek LPH, Konz M, Shrestha AB, Bierkens MFP (2011) Hydrological response to climate change in a glacierized catchment in the Himalayas. Clim Change 110:721–736

Independent Power Producers' Association Nepal/Confederation of Indian Industry (IPPAN/CII) (2006) Research on Nepal–India Cooperation on Hydropower (NICOH). IPPAN/CII, New Delhi

Japan International Cooperation Agency (JICA) (1985) Master plan study on the Kosi River water resources development: final report. Japan International Cooperation Agency, Tokyo

Japan International Cooperation Agency (JICA) (2014) Nationwide master plan study on storage-type hydroelectric power development in Nepal: final report. Japan International Cooperation Agency, Tokyo

Jalsrot Vikas Sanstha (JVS) (2012) Consultative workshop report: Ganges Strategic basin assessment, Nepal. JVS, Kathmandu

Kattelmann R (1991) Hydrologic regime of the Sapt Kosi basin, Nepal. In: Hydrology for the water management of large river basins. Proceedings of the Vienna symposium, Vienna, August 1991

Khanal NR, Hu J-M, Mool P (2015) Glacial lake outburst flood risk in the Poiqu/Bhote Koshi/Sun Koshi River basin in the Central Himalayas. Mt Res Dev 35(4):351–364

Mishra DK (2008) Bihar floods: the inevitable has happened. Econ Polit Wkly, 6 Sept 2008

Ministry of Environment (MoE) (2010) National adaptation programme of action (NAPA) thematic working group summary report. Ministry of Environment, Kathmandu, Nepal

Molden D, Vaidya R, Shrestha A, Rasul G, Shrestha M (2014) Water infrastructure for the Hindu Kush Himalayas. Int J Water Resour Dev 30:60–77

National Electricity Authority (NEA) (2014) A year in review-fiscal year-2013/2014. NEA, Kathmandu

Rahaman MM (2012) Hydropower ambitions of South Asian nations and China: Ganges and Brahmaputra Rivers basins. Int J Sust Soc 4(1/2):131–157

Rajbhandari R, Shrestha AB, Nepal S, Wahid S (2016) Projection of future climate over the Koshi River basin based on CMIP5 GCMs. Atmos Clim Sci 6(02):190

Ramsar (2016) Directory of Asian Wetlands. www.ramsar.wetlands.org. Accessed 10 May 2016

Ries JB (1995) Does soil erosion in the high mountain region of the eastern Nepalese Himalayas affect the plains? Phys Chem Earth 20:51–269

Sharma B, Rasul G, Chettri N (2015) The economic value of wetland ecosystem services: evidence from the Koshi Tappu Wildlife Reserve, Nepal. Ecosyst Serv 12:84–93

Shrestha RM, Ahlers R, Bakker M, Gupta J (2010) Institutional dysfunction and challenges in flood control: a case study of the Kosi flood 2008. Econ Polit Wkly XLV(2):45–53

Speed R, Li Y, Le Quesne T, Pegram G, Zhiwei Z (2013) Basin water allocation planning. Principles, procedures and approaches for basin allocation planning. UNESCO, Paris

Surie M, Prasai S (2015) Strengthening transparency and access to information on transboundary rivers in South Asia. The Asia Foundation, New Delhi

SWECO (2011) Draft case study reports on benefit sharing and hydropower. Reports prepared for the World Bank, SWECO, Stockholm

Tropp H (2005) Developing water governance capacities. Feature article. UNDP Water Governance Facility/SIWI, Stockholm

Tropp H (2007) Water governance: trends and needs for new capacity development. Water Policy 9(2):19–30

Uddin K, Murthy MSR, Wahid SM, Matin MA (2016) Estimation of soil erosion dynamics in the Koshi basin using GIS and remote sensing to assess priority areas for conservation. PLoS ONE 11(3)

United Nations (UN) (2015) Responsive and accountable public governance. https://publicadministration.un.org/en/Research/World-Public-Sector-Reports. Accessed 10 May 2016

United Nations Economic Commission for Europe-International Network of Basin Organizations (UNECE–INBO) (2015) Water and climate change adaptation in transboundary basins: lessons learned and good practices. United Nations, Geneva and INBO, Paris

United Nations Environment Programme (UNEP) (2008) Freshwater under threat South Asia: vulnerability assessment of freshwater resources to environmental change. http://www.unep.org/pdf/southasia_report.pdf. Accessed 25 Jan 2016

Van Vuuren DP, Edmonds J, Thomson A, Riahi K, Kainuma M, Matsui T, Hurtt GC, Lamarque J-F, Meinshausen M, Smith S, Granier C, Rose SK, Hibbard KA (2010) The representative concentration pathways: an overview. Clim Change 109(1–2):5–31

Wang C (2012) A guide for local benefit sharing in hydropower projects. Social development working papers no. 128. The World Bank Social Development, Washington, DC

World Bank (2015) Project appraisal document on a proposed credit to Nepal, Power sector reform and sustainable hydropower development project, September 2015. World Bank, Washington, DC

World Wildlife Fund (WWF) (2010) From policy to practice, Koshi River Basin Management. WWK Nepal, Kathmandu

Yatagai A, Kamiguchi K, Arakawa O, Hamada A, Yasutomi N, Kitoh A (2012) Constructing a long-term daily gridded precipitation dataset for Asia based on a dense network of rain gauge. Bull Am Meteorol Soc, Sept 2012

Chapter 5
Building Pakistan's Resilience to Flood Disasters in the Indus River Basin

Guillermo Mendoza and Zarif Khero

Abstract Pakistan's losses from the 2010 flood were the worst in the country's history, despite the magnitude being less than the major flood event of 1976. The reasons for the losses were an inadequate early warning system, the intensity of the rainfall, and the capacity of available waterways. The flood, which affected all provinces in Pakistan, killed 1600 people, caused damage totaling more than USD 10 billion, and inundated an area of approximately 38,600 km^2. Building resilience to future flood events will require investment in an integrated water resources management and infrastructure strategy. Water supply management affects water storage options and sediment management, which in turn affects the options for flood management. This chapter details a resilience strategy that could help Pakistan to prepare, absorb, adapt, and recover from extreme flood events. Resilience to flood disasters will require infrastructure (absorb) that is properly maintained with coordinated management (prepare), an ability to forecast crises and coordinate actions (prepare), and a pragmatic approach to future uncertainties, such as climate change (adapt). Finally, improved governance will be required to implement fundamental change and aid recovery after disasters.

Keywords Pakistan · Floods · IWRM · Integrated water resources management · Climate change · Decision scaling · Resilience · Infrastructure · Governance

---

G. Mendoza (✉)
US Army Corps of Engineers Institute for Water Resources, International Centre for Integrated Water Resources Management (ICIWaRM), Alexandria, VA, USA
e-mail: Guillermo.f.mendoza@usace.army.mil

Z. Khero
Irrigation Department, Government of Sindh, Karachi, Pakistan
e-mail: zarifkhero@gmail.com

C. Tortajada (ed.), *Increasing Resilience to Climate Variability and Change*, Water Resources Development and Management,
DOI 10.1007/978-981-10-1914-2_5

5.1 Introduction

Monsoonal rains are the most significant flood-causing factor in the Indus basin. Methods for understanding and predicting the monsoon are poor, but research is ongoing to explain its origin, processes, strength, variability, distribution, and predictability (Awan and Maqbool 2010). Catchment characteristics and land use channel water from the monsoon and Himalayan snowmelt along the Indus River and its tributaries the Jhelum, Chenab, Ravi, Beas, and Sutlej Rivers (Fig. 5.1). The hydrology of the Indus River basin in Pakistan has been heavily altered by one of the most ambitious engineering and political feats in history (Briscoe and Qamar 2006) following the Indus River Basin Treaty of 1960. Briscoe and Qamar (2006: extracted from pp. 23–29) wrote:

> *Pakistani engineers, together with their Indian counterparts and The World Bank, negotiated the Indus Waters Treaty, giving Pakistan rights in perpetuity to the waters of the Indus, Jhelum, and Chenab rivers, which comprise 75 percent of the flow of the whole Indus system.* [This treaty had a fundamental engineering challenge] *that there was now a mismatch between the location of Pakistan's water (in the western rivers) and the major*

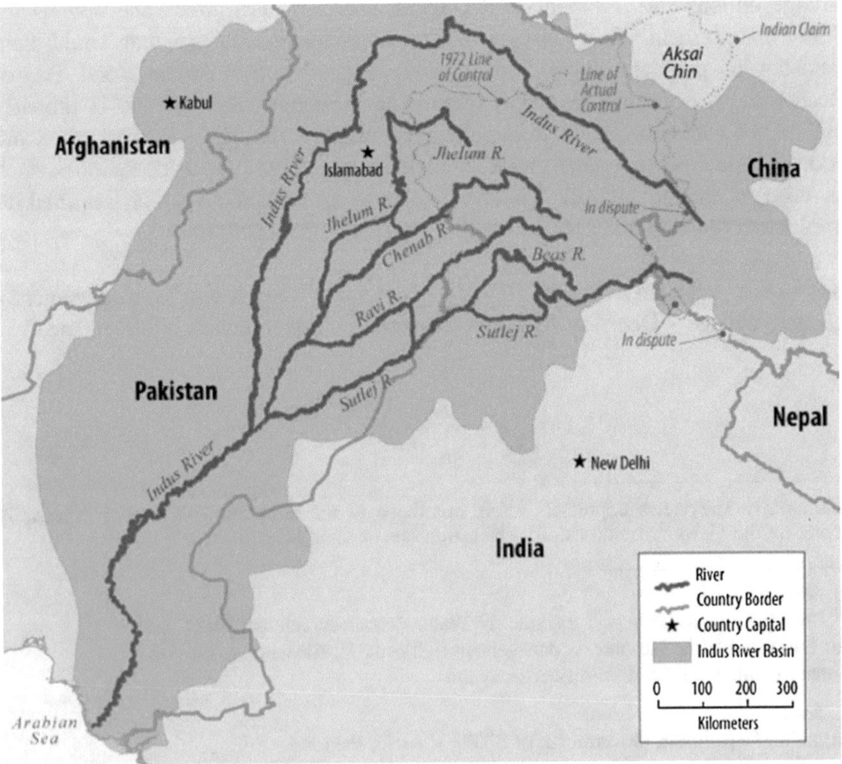

Fig. 5.1 Map of the Indus River (U.S. Congress Report 2011)

> *irrigated area in the east. Again Pakistan's water engineers were up to the task, building the world's largest earthfill dam, the Tarbela on the Indus, and link canals, which ran for hundreds of miles and carried flows ten times the flow of the river Thames.*

The hydraulic system, which included 3 major reservoirs (Chashma, Mangla and Tarbela), 19 barrages (Ferozepur, Sulemanki, Islam, Balloki, Marala, Trimmu, Panjnad, Kalabagh, Sukkur, Kotri, Taunsa, Guddu, Chashma, Mailsi, Balloki, Sidhnai, Rasul, Qadirabad and Marala), and a number of bridges on the rivers of Indus basin, was developed for water security and the economic development of Pakistan. These infrastructure investments allocate treaty waters from the Western Indus basin to Eastern Indus valley farmland and support the largest contiguous irrigation system in the world (Fig. 5.2). The Indus River basin now irrigates more than 14.9 million ha in Pakistan through 2 headworks, 12 interlinking canals, 45 irrigation canal systems, and 140,000 watercourses, in addition to the millions of farm channels and field ditches. The total length of main canal system is estimated to be about 585,000 km and that of tertiary irrigation ditches and field channels exceeds 1.5 million km (Khalil et al. 2011).

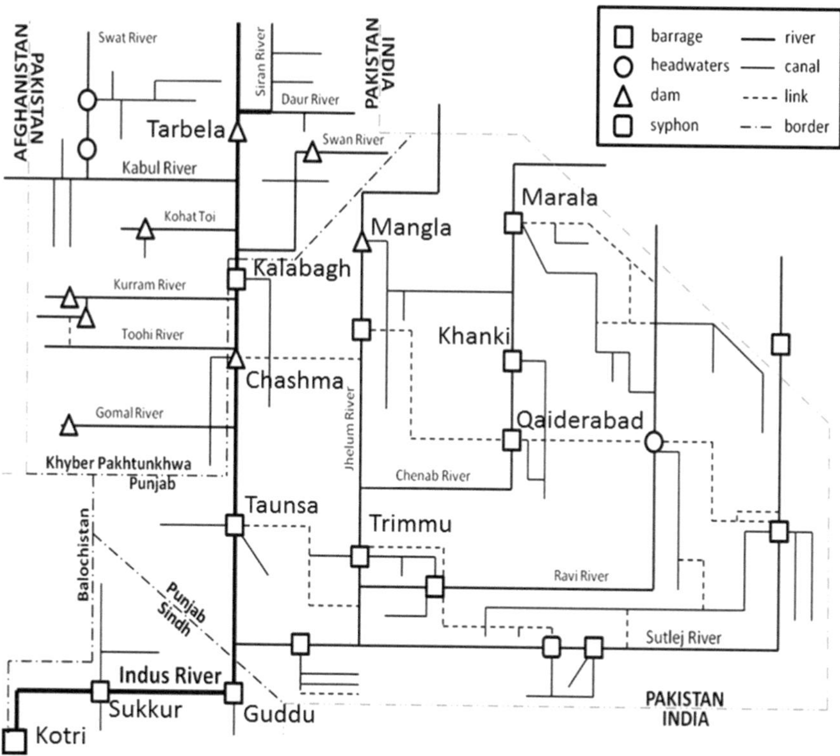

Fig. 5.2 A schematic of the Indus River irrigation network. Note the diversion links from barrages on the Indus, Jhelum, and Chenab (Pakistan Indus Treaty rivers) on the west toward the eastern Punjab (modified from PMPIU 2011)

This water security infrastructure has a direct impact on flood-risk management. During a flood event, Pakistan's response includes regional options. In the Punjab province, the reaches above the confluence of the Sutlej and Chenab with the main stem of the Indus, the intentional breaching of levee sections at barrages and bridges is practiced to protect critical irrigation infrastructure. This involves a trade-off between losses from an intentional breach and losses of critical infrastructure meant to feed the nation. In the Lower Indus basin, the region below the confluence of tributaries, flood risk is managed by a continuous system of embankments up to the Arabian Sea. A flood bypass study has been completed, but the options are very limited because the flat topography and river channel aggradation pose a risk of unmanaged breaching that would affect highly populated areas before flowing into the Arabian Sea.

Since the Indus Valley Civilization, cities and towns have developed in one of the world's most arid regions, with an average rainfall of 240 mm per year. Pakistan's population and economy are heavily dependent on an annual influx that starts at the headwaters of the Indus. Livelihoods and culture are coupled with the river. Unfortunately, this relationship is periodically disrupted by annual flooding of Indus and its tributaries, resulting in economic damage that is a major burden on Pakistan. Flooding causes massive damage to infrastructure facilities and agricultural lands, diverting limited national resources to flood relief operations. Moreover, the flat topography and slow drainage in Lower Indus basin results in long periods of inundation of valuable land and other properties. Economic resilience from flooding is low.

In 2010, a 5-year drought came to an abrupt end with the largest natural disaster in Pakistan's history. In August and September, approximately one-fifth of Pakistan was flooded. Thousands of villages were completely destroyed and millions of people were displaced. In addition to completely destroying millions of acres of cropland, large amounts of silt were deposited on the land and in the irrigation canals. Hundreds of thousands of livestock, which are needed to plant the crops, were killed. The 2010 flood, which affected all of the provinces of Pakistan, killed 1600 people, caused damage totaling over USD 10 billion (almost 6 % of the gross domestic product [GDP] in 2010) and inundated an area of about 38,600 km^2. This chapter is about the actions and challenges of building resilience to flood disasters in Pakistan.

5.1.1 The Floods of 2010 and 2014

Before 28 July 2010, the Indus basin was undergoing a 5-year drought and water shortages were being experienced below Taunsa Barrage (Fig. 5.2). Water could not meet the crop water requirements in Sindh. Tensions were raised between the governments of Sindh and Punjab when the Sindh representation in the Indus River System Authority resigned in protest over the Punjab opening a Chashma-Jhelum diversion further reducing flows into Sindh. During the third week of July, the nature of the disaster changed. Forecasters began predicting substantial rainfall (Table 5.1).

Table 5.1 Pakistan weather forecast in July 2010

Forecast	Nature of forecast
22 July 2010, issued by Pakistan Meteorology Department	Rainfall is expected within 24 h in different parts of country
27 July 2010, Bulletin No. A-043 issued by Pakistan Meteorology Department	Widespread thunderstorms and rain with isolated heavy falls is expected over Punjab, Khyber-Pakhtoonkhwa and Upper Sindh
3 August 2010, S. No. 21 issued by Chief Meteorologist Flood forecasting Division Lahore	Significant flood forecast for River Indus at Gudu and Sukkur. Gudu: River Indus at Gudu is likely to attain exceptionally high flood level ranging b/w 9–10 lac cusecs on 5 and 6 August. Sukkur: according to latest hydrological condition River Indus at Sukkur is likely to attain exceptionally high flood level ranging b/w 9–10 lac cusecs on 6 and 7 August 2010

Flood Forecasting Division (FFD) Lahore, an affiliate of the Pakistan Meteorological Department; http://www.pmd.gov.pk/FFD/cp/floodpage.htm

By the beginning of August, continued localized rainfall throughout the country and high antecedent moisture led to high inflows and flash floods in the Kabul River and below Taunsa Barrage around the Suleiman Range (Fig. 5.2). Estimates of flows were unpredictable and the rise was abrupt. Conditions changed quickly from drought to super flood, which was exacerbated due to a build-up of large sediment deposits against the embankments resulting in unprecedented river channel aggradation. Given the increase in water in the Indus headwaters, high flows persisted. The persistent flows quickly filled the floodplain areas between levees (~12 km). In addition, the continual heavy rainfall hindered flood-fighting actions, and accessibility to sites and evacuation became difficult. At Tarbela, these flows increased from 190,000 cfs[1] on 28 July to a maximum of 603,000 cfs by the morning of 30 July before subsiding.

The second component of the flood disaster was peak run-off events from the several watersheds contributing to the Kabul River downstream of Tarbela. On 28 July, an 8.62 ft (2.63 m) flood stage was recorded by gauge of Kabul River producing 57,500 cfs of discharge. The combined flows from continuous watershed peak run-off events and high sustained antecedent flows resulted in maximum peaks of around 1 million cfs at the Kalabagh, Chashma, and Taunsa Barrages at 20.00 h on 30 July, 16.00 h on 1 August, and 22.00 h on 2 August, respectively (Table 5.2).

On 2 August 2010 at 22.00 h, a discharge of 959,991 cfs breached the levee on the upstream left side of Taunsa Barrage (Table 5.2). This breach created a difficulty in the estimates of flood flow toward Sindh, which is mainly dependent upon the flows passing downstream of Taunsa Barrage, and the flows from eastern rivers passing downstream of Panjnad Barrage. Forecasting flow synchronizations has a

[1] 1 cfs = 0.02832 m^3/s.

Table 5.2 Flows measured at select locations 28 July to 3 August 2010 along the Upper Indus River

Date	Time (h)	Tarbela Dam outflow (m^3/s)	Kabul River		Kalabagh Barrage upstream (m^3/s)	Chashma/Barrage upstream (m^3/s)	Taunsa Barrage upstream (m^3/s)
			Gauge (m)	Discharge (m^3/s)			
28-Jul	06.00	5380.15	2.63	1628.20	7996.66	8035.34	6854.59
29-Jul	06.00	10,145.83	5.43	5663.32	11,678.21	9781.03	7657.96
30-Jul	06.00	**17,074.90**	Gauge overflowed due to excessive rainfall on the mountains		19,438.74	21,067.25	8480.87
	20.00	13,756.19			**26,545.46**	25,879.43	9761.80
31-Jul	06.00	14,438.62			23,339.83	27,586.41	12,017.30
1-Aug	06.00	11,892.96			22,017.78	28,470.76	17,488.38
	16.00	10,216.62			20,739.15	**29,417.33**	22,613.65
2-Aug	06.00	10,148.66	Discharge from Kabul river may be the difference of Tabella Outflow and the Kalabagh upstream		20,880.73	27,042.64	22,613.65
	22.00	9647.46			17,054.11	22,340.37	**27,183.66**
3-Aug	06.00	10,205.86			14,659.01	21,991.56	21,728.76

Bold numerals represent maximum flows (Government of Sindh Irrigation Department)

pivotal role in understanding the duration and intensity of floods in Sindh province (Table 5.3).

After the breach upstream of Taunsa Barrage, the discharge at Gudu Barrage reached and remained around 1,000,000 cfs after a few days. On 4 August, the FFD Lahore was able to start providing information from the only watershed stream gauge (CHACHARAN), which was essentially the only gauging station for the cumulative flows into Sindh through Gudu Barrage from the confluence of the upstream Indus, Chenab, Jhelum, Ravi, and Sutlej, and peak runoff from an ungauged catchment of the Suleiman/Mazari Hills on the right bank of the Indus River downstream of Taunsa Barrage.

The flow of eastern rivers, after passing down Panjnad Barrage, merged with the Indus River flow. Being unmetered, the torrential flows emerging out of Mazari Hills (Fig. 5.3) were not measured at Gudu Barrage. This further intensified the flood situation in Sindh at Gudu Barrage and further downstream. In addition, a breach at Tori—some 50 km upstream of Indus Delta in District Thatta—also occurred. Therefore, during 2010, the flood management and forecasting for Sindh province is imprecise. The lack of information on water flows, exacerbated by limited hydraulic information due to river channel aggradation into Sindh, made it very challenging to manage the floodwaters.

On 8 August 2010, a second breach occurred on the downstream right side of Gudu Barrage at the Tori levee (Fig. 5.4). In addition to the Tori breach, five breaches occurred in the left marginal bund of Gudu Barrage and two breaches in the districts of Thatta and Sujawal near the Arabian Sea.

Table 5.3 Flows measured (cfs) at select locations 3 August to 21 August 2010 at the Confluence of Western and Eastern Rivers and along the lower Indus River (Government of Sindh Irrigation Department)

Date	Time (a.m.)	Taunsa	Panjnad	Chachran		Gudu	Sukkur	
		D/S	D/S	Gauge	Discharge	U/S	U/S	D/S
3-Aug	6:00	767,351	38,315			354,773	215,660	183,810
4-Aug	6:00	790,021	64,951	13.4		462,922	252,232	224,132
5-Aug	6:00	743,466	78,083	14.0		700,800	311,600	283,550
6-Aug	6:00	692,981	138,442	**14.5**	**1,165,000**	962,678	600,710	577,910
7-Aug	6:00	614,418	158,629	14.5		962,678	871,682	850,682
8-Aug	6:00		175,479	14.5		1,128,854	1,049,285	1,040,070
9-Aug	6:00	566,351	203,721	14.3		**1,148,738**	1,130,220	**1,124,720**
10-Aug	6:00	614,015	258,732	13.8		1,087,870	**1,130,995**	1,108,795
11-Aug	6:00	741,614	261,780	13.8		1,036,752	1,130,995	1,108,795
12-Aug	6:00	761,774	283,586	13.9		996,978	1,113,210	1,089,510
13-Aug	6:00	**773,612**	298,588	14.1		976,870	1,083,660	1,053,820
14-Aug	6:00	769,110	**310,117**	14.2		996,873	1,010,327	975,087
15-Aug	6:00	769,450		14.4		1,026,708	1,008,377	975,087
16-Aug	6:00	644,338	273,105	14.4		1,056,862	1,021,220	987,890
17-Aug	6:00	597,848	244,274	14.3		1,076,728	1,025,630	993,880
18-Aug	6:00	569,026	229,825	14.1		1,037,610	1,025,455	993,880
19-Aug	6:00	527,753	192,803	13.5		993,447	1,001,710	965,100
20-Aug	6:00	475,309	189,053	13.1		946,718	990,535	954,550
21-Aug	6:00	437,368	158,629	12.8		875,907	974,453	945,758

1 cfs = 0.02832 m^3/s

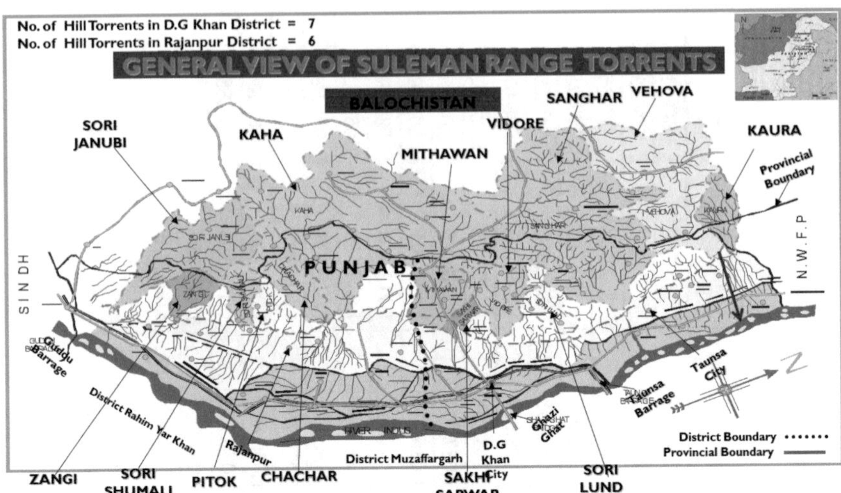

Fig. 5.3 Suleiman Range Hill Torrents. Sindh province flooding damage was largely a result of unquantified peak runoff from 13 watersheds on the right bank of the Indus River downstream of Taunsa (*Source* Irrigation Department of Sindh Province)

Fig. 5.4 **a** The *left panel* shows floodwaters leaving the Indus River and flowing across the land into Manchar Lake. **b** The levee breach at Tori

Four years later, another major flood hit Pakistan. In 2014, a flood propagated (Fig. 5.5) from intense rainfall in the Chenab river catchment above Marala Barrage in Indian territory and in the Jhelum catchment above Mangla Dam (Fig. 5.6). However, in this case, it was largely the uncoordinated reservoir operation and the limited flood storage capacity of Mangla Dam that caused the loss of life and property around Jhelum River. Anticipated inflows and actual reservoir operation during the 2014 flood illustrate the need for efficient reservoir operation during the floods.

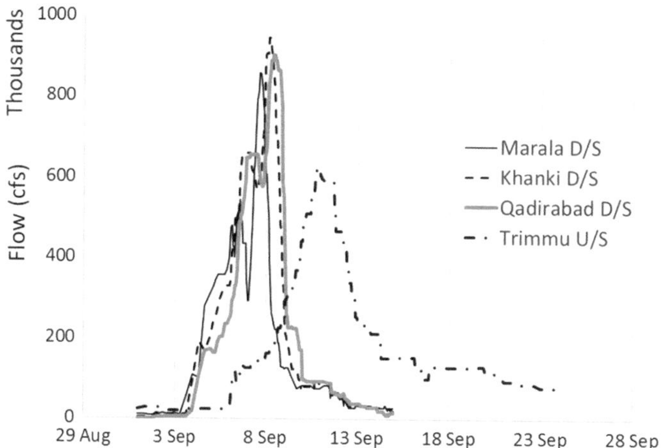

Fig. 5.5 Flood propagation and attenuation from 29 August 2014 to 28 September 2014 (*Data source* Government of Sindh Irrigation Department) (1 cfs = 0.02832 m^3/s)

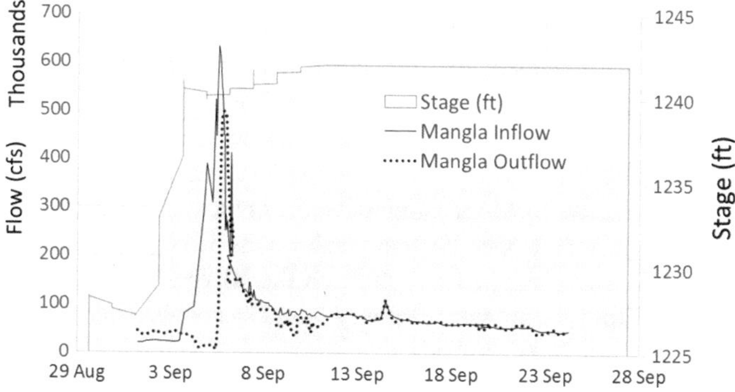

Fig. 5.6 Reservoir operation at Mangla during the flood of 2014. The figure shows that the flood peaks were not mitigated (1 cfs = 0.02832 m^3/s)

5.2 Building Resilience

This chapter presents four key actions for resilience to flood disasters. These actions aim to enhance the ability to prepare, absorb, recover, and adapt to future adverse events. The majority of the discussion is around the ability to absorb adverse impacts through infrastructure and management. This is a specific challenge in the Indus system due to the many dependencies and diverging interests. The abilities to recover, prepare, and adapt have strong governance and institutional challenges.

5.2.1 Prepare Through Flood Forecasting and Early Warning

Following the devastating floods of 1992, a national flood early warning system was reinforced through the creation of the Lahore-based Floods Forecasting Division (FFD), affiliated with the Pakistan Meteorological Department (PMD) as a national focal point for monsoon early warnings and disseminating data on flood levels in various rivers. The PMD maintains the Tropical Cyclone Warning Centre in Karachi and is linked to the World Meteorological Organization's Regional Warning Centre in New Delhi, India. Similarly, PMD benefits from tsunami early warnings from the Japan Meteorological Agency and US Tsunami Early Warning Centre in Hawaii. PMD also undertakes drought and seismological hazards monitoring.

Vulnerable regions with early warning coverage are the Upper Khyber Pakhtunkhwa, along the Kabul and Swat Rivers, Gligit Baltistan, Upper Azad Jammu, and Kashmir and where the Chenab enters Pakistan in Eastern Punjab (Fig. 5.7). The Kashmir catchment mainly feeds the four eastern rivers: Jhelum, Chenab, Ravi, and Sutlej (Fig. 5.1). Early warning for major floods in these areas is usually on the order of 6–18 h.

However, the flood early warning coverage of the Kabul and Indus catchments in Upper Khyber Pakhtunkhwa is poor. This is where most people died during the floods of 2010. There is no effective early warning system in place for flash floods, which result from intense localized rainfall events upon saturated soil. They tend to occur in Gligit Baltistan, Upper Khyber Pakhtunkhwa, Upper Azad Jammu, and Kashmir, along the Indus Right Bank through the Khyber Pakhtunkhwa, Punjab, and Baluchistan districts. Only one telemetry-based system is deployed for the early warning of flash floods in Islamabad.

Pakistan's National Disaster Management Authority recognizes that the inefficient early warning system and poor communication are the leading causes of community vulnerability to floods (NDMA 2013). Current forecasting and early warning systems lack the capacity to monitor multi-hazards and disseminate actionable information to the vulnerable population and local institutions in a timely manner. There has been much progress in investing in technologies for flood early

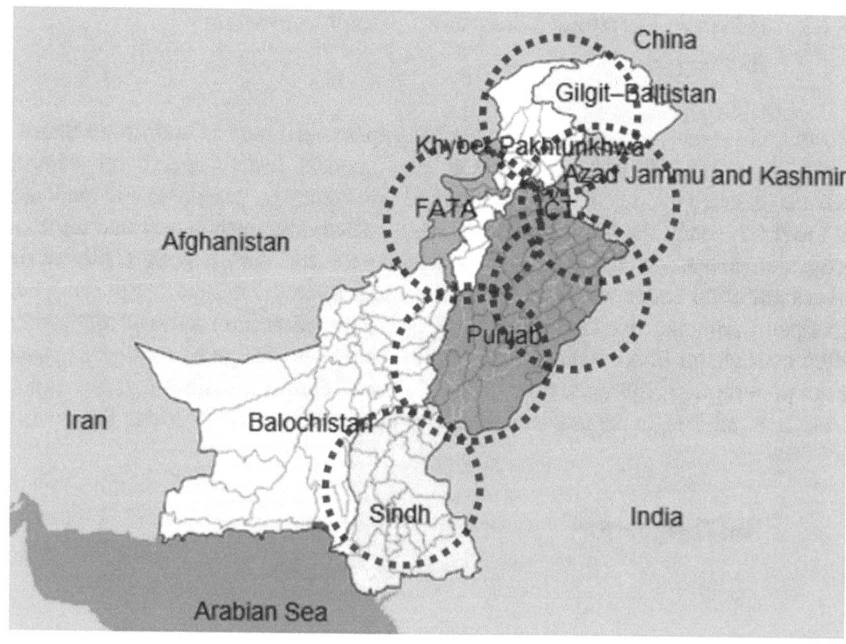

Fig. 5.7 Pakistan provincial governments and approximate ranges of existing PMD weather radar systems (modified from PMD with map from Wikipedia)

warning systems, which now cover more than 39 districts of the 88 that are vulnerable to floods (Ahmad 2015). The challenge is that in addition to monitoring and warning services, flood early warning systems must systematically collect data and make risk assessments, communicate risk information, and build national and community response capabilities (ISDR 2006).

The World Bank/Asian Development Bank 2010 Flood Preliminary Damage and Needs Assessment highlighted the need to strengthen and build capacity in national- and community-response capabilities (World Bank 2010). Moreover, at the community level, there is generally low disaster risk awareness in Pakistan, especially among women and children. Disaster management agencies have reported that people tend to ignore flood warnings until they are hit by the hazard (Ahmad 2015). This might be expected given that the characteristics of large floods in Pakistan have historically had unique spatial distributions (Fig. 5.13). Building resilience through flood forecasting will require an integrated approach that considers not only the technical requirements but also communication, science, capacity building, and outreach.

5.2.2 *Absorb Through Integrated Water Resources Management*

Historically prone to disasters, Pakistan has experienced major floods more than 20 times since 1947 (Fig. 5.8). The state continuously takes incremental steps to enhance disaster management structures and emergency preparedness and response. In the Indus basin, the traditional flood management approach is centered on flood protection levees. In the upper-reach catchments, the design peak flows of the levees are often underestimated due to the sharp hydrographic peaks; in the lower floodplains, drainage backwaters and channel aggradation from sediment deposition often overwhelm flood management performance. The design of the levee system lacks pre-emptive solutions, operating only when danger becomes real and imminent. It is ad hoc in nature and does not comprehensively consider the basin's

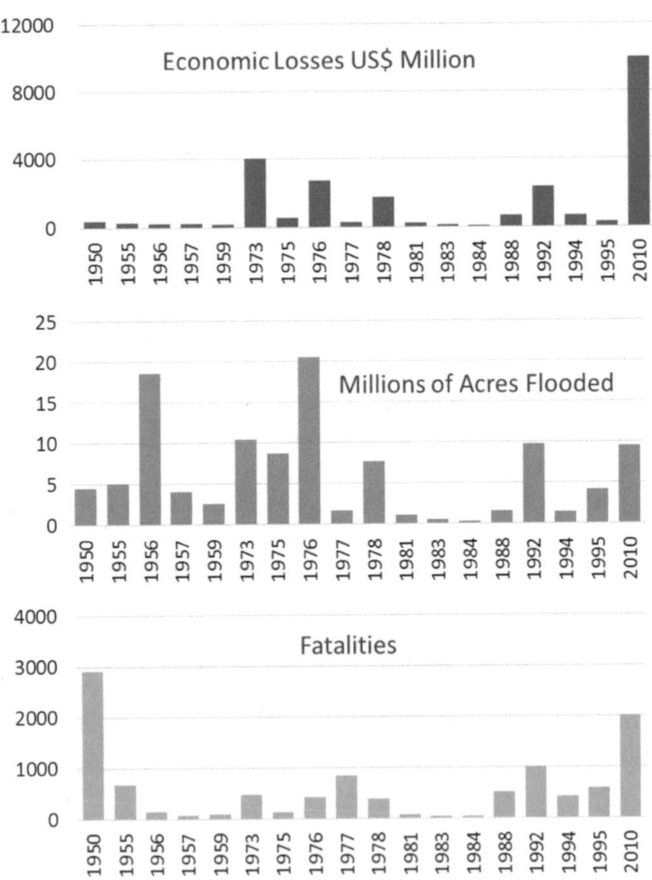

Fig. 5.8 Flood damage in Pakistan since 1950 (modified from Kazi 2005)

hydro-climate realities, physical setting, and development needs. Moreover, it lacks effective policies, planning, and institutional backing (Ali 2013).

To deal holistically with floods, a comprehensive and practical national flood-risk mitigation strategy along with an operational plan incorporating measurable targets and clear timelines are required. Equivocally, for the Upper (western and eastern) and Lower Indus basins in Pakistan, flood management requires a different mindset due to their geography and morphology.

Along the Indus and its tributaries, the entire system is interdependent. Human infrastructure for irrigation is tightly coupled with floodwater hydraulics. Levee design factors, such as discharge, water surface slope, water velocity, and depth and width of the channel impact are affected by the barrages, storage reservoirs, and diversions for irrigation. For example, the reduction of flows into the lower reaches in Sindh has resulted in greater sediment deposition, thus raising the river bottom within the leveed system and increasing flood risk. In other words, different approaches to flood management are required along different reaches of the river.

5.2.2.1 Storage: Conflict of Flow Extremes

Ironically, one of the flood-risk reduction challenges of the Indus is the increasingly low-energy flow during non-flood periods. The Indus River has one of the largest sediment loads in the world, which under natural flow conditions has formed an extensive delta on the high-energy coast of the Arabian Sea. Water and sediment discharge have been drastically altered in the Indus since the early 1960s, when several barrages were built along the river to supply the world's largest irrigation system (Giosan et al. 2006). A digital terrain model based on detailed nineteenth-century surveys assessed the morphology of the Indus shelf. Comparison of the digital terrain model to a 1950s Pakistani bathymetric survey allowed an estimation of the natural sedimentation regime before the extensive human-induced changes. Over the last few years, water flow below Kotri (Fig. 5.2) has been effectively cut down to approximately 2 months per year of active flow during August and September (Asianics 2000; Inam et al. 2004). Digital analysis of the Indus Delta coastline based on satellite imagery was used to explore the effects of the drastic decrease in sediment delivery following extensive dam building. The damming lead to dramatic reductions of water and sediment discharge by more than 70 and 80 %, respectively, after the 1950s (Milliman et al. 1984). These reductions in flows have led to increases in sediment deposition at the lower reaches of the Indus before Guddu and Sukhur Barrages (Fig. 5.2), creating channel aggradations and thus increasing riverine flood risk, as well as a decrease of sediment deposition at the delta that increases coastal erosion and coastal flood risk.

In 2005, the Government of Pakistan commissioned an International Panel of Experts (IPOE) to review studies on water flow requirements past Kotri Barrage as part of Pakistan's Water Apportionment Accord of 1991 to address water management coordination throughout the provinces. The IPOE recommended that flows of 5000 cfs (141.58 m^3/s) at Kotri Barrage were required throughout the year to

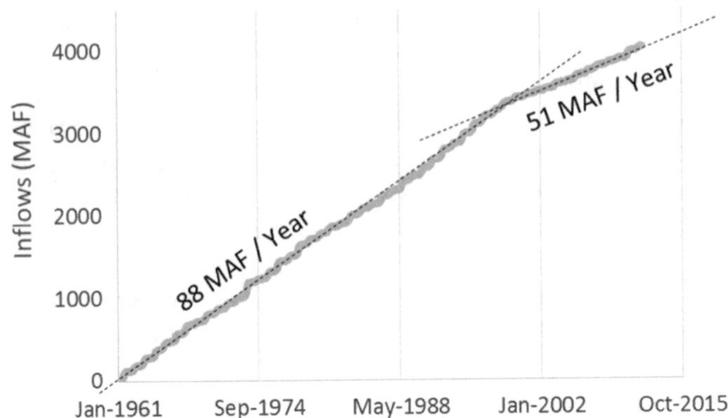

Fig. 5.9 Cumulative flows through the Sukhur Barrage (Government of Sindh Irrigation Department) (1 million acre ft (MAF) = 1233.5 million m^3 (MCM))

check seawater intrusion, accommodate the needs of fisheries and environmental sustainability, and maintain the river channel (González et al. 2005). The IPOE report underscored the problem of sea intrusion/coastal erosion occurring in the Indus Delta area as a national problem. In effect, there has been a substantial reduction in sediment delivery through Kotri over time, due to the diversion of most of the water for irrigation. Because most of the sediment is supplied during peak flows, the IPOE recommended managed peak flow discharges through Kotri at a total volume of 25 million acre ft (MAF)[2] over a 5-year period (about 5 MAF at a minimum per year).

An analysis of the 50-year data was made to analyze the condition of inflows approaching the head of Lower Indus basin at Sukkur Barrage (Fig. 5.2). In 1960–1999, annual flow discharges were 88 MAF. After 1999–2012, they declined to 51 MAF (Fig. 5.9). This declining trend is consistent across the Indus River basin.

One problem is that many provincial governments in Pakistan need to share this resource. In the apportionment of the waters of the Indus River system between the provinces of Pakistan, the annual shares between the provinces of Sindh and Baluchistan (Fig. 5.7) are 48.76 and 3.87 MAF, respectively. This totals 52.63 MAF, which is almost the current total annual flow and which continues to decline toward Sukkur Barrage in the Lower Indus basin. Because Baluchistan has no direct access to the Indus River, it acquires the allocated flows from the barrages of Sindh province (i.e., Gudu and Sukkur Barrages). This paints a grim picture. Under an increasingly closed basin, the allocation of flows for nonproductive use by the government of Sindh would comprise at a minimum about 10 % of the total apportionment.

To address increasing water scarcity (Fig. 5.9) and increasing demands for irrigation and hydropower energy, Pakistan has proposed several new dam projects

[2]1 million acre ft (MAF) = 1233.5 million m^3 (MCM).

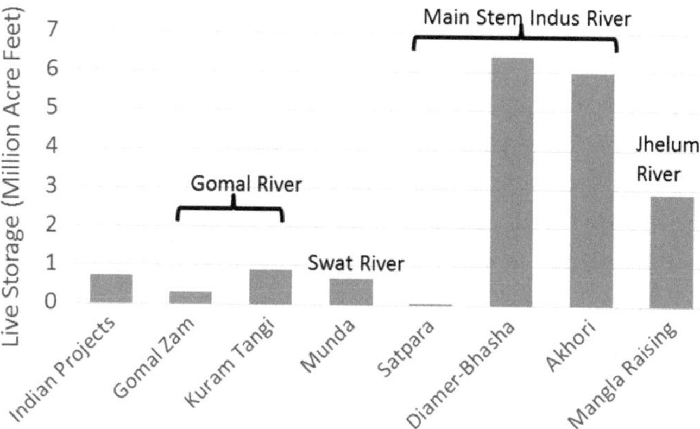

Fig. 5.10 Under construction and proposed storage volume in the Indus Upper Basin (MAF). Note that new storage projects are located in the western rivers of the Indus for support of the irrigation system (Government of Sindh Irrigation Department) (1 million acre ft (MAF) = 1233.5 million m^3 (MCM))

(Fig. 5.10). These new investments are geared toward building resilience to droughts and supporting economic growth. However, they could potentially further disrupt the sediment balance, which would accelerate sea intrusion and coastal erosion along the Indus Delta.

5.2.2.2 Levees and Embankments

Levees have played an important role in managing flood damages throughout Pakistan's history. Some of the earliest levees were constructed by the Indus Valley Civilization (around 2600 BC). Today, there are approximately 6000 km of flood protection embankments in Pakistan, including approximately 5300 km in the provinces of Punjab or Sindh (Kazi 2005). However, all components of the levee system are integral to flood-risk reduction. On the one hand, floods in Pakistan have historically shown unique patterns in the Indus basin (Fig. 5.11), which makes all components important. Figure 5.12 illustrates that the site of 2010 flood event was the main Indus and its western tributaries, whereas the 2014 floods (Fig. 5.13) originated in the eastern rivers. On the other hand, floodwaters contained by levees are essentially transferring risk downstream.

Therefore, any strategy to contain floodwaters within the floodplain must include all of the Indus River tributaries and be part of an integrated operation strategy. Sediment deposition will continue to cause channel aggradation, reducing the capacity of the levees below Taunsa Barrage and increasing risk for the coastal communities of the delta.

A resilience strategy to enhance design capacities and operations of levees also has regional implications. The geography of Punjab province means that the river

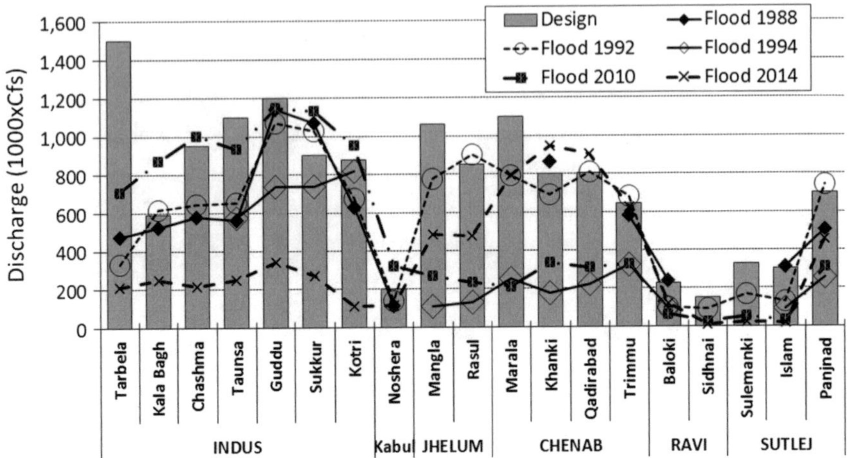

Fig. 5.11 Indus basin structure design capacities versus several maximum peak flood discharges (Government of Sindh Irrigation Department) (1 cfs = 0.02832 m^3/s)

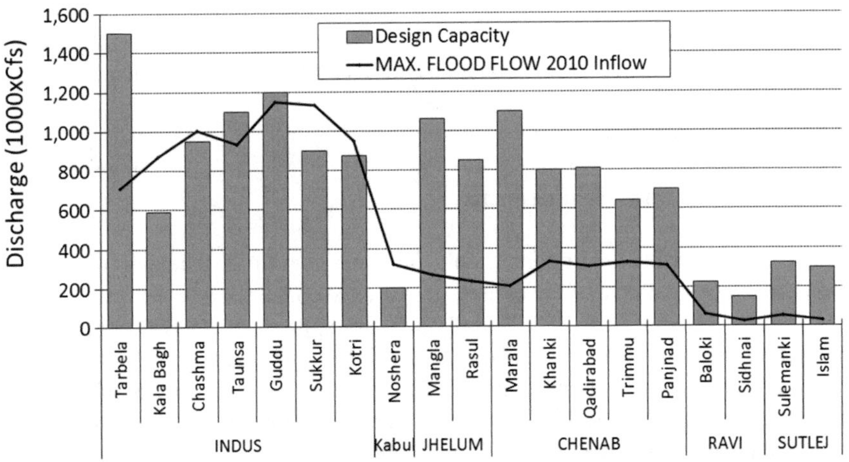

Fig. 5.12 Indus basin structure design capacities versus 2010 maximum peak flood discharges (Government of Sindh Irrigation Department) (1 cfs = 0.02832 m^3/s)

flow is lower than the surrounding land. In exceptionally high floods, discharges are mitigated by managed breaching of levees at predetermined sites to protect the structural integrity and safety of barrages and bridges. There are a total of 17 intentional breaching sites: 12 are operated by Punjab Irrigation Department, 4 by the Railway Department, and 1 by the Highway Department. This multiagency levee breaching operation protocol is somewhat indicative of uncoordinated flood-risk reduction planning.

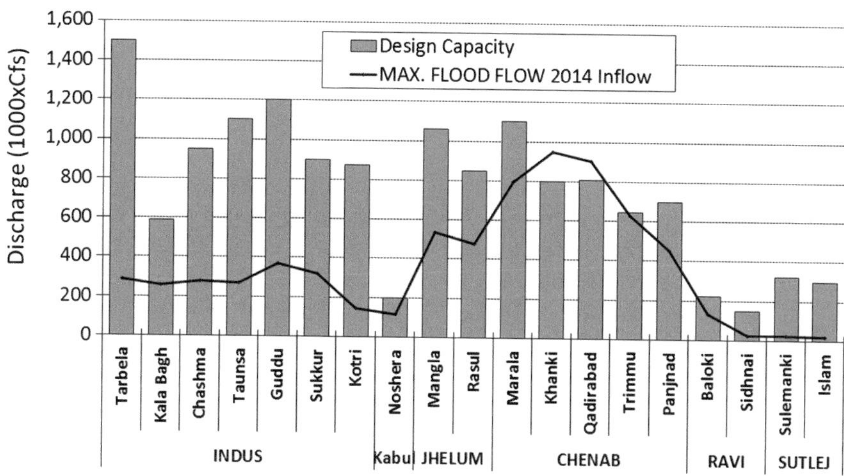

Fig. 5.13 Indus basin structure design capacities versus 2014 maximum peak flood discharges (Government of Sindh Irrigation Department) (1 cfs = 0.02832 m^3/s)

Downstream from Punjab in Sindh province, levee requirements differ because of the unique topographic conditions that require provision of continuous lines of embankments along both sides the Lower Indus River. The low relief topography means that breaches in flood embankments cause the inundation of large areas with limited potential for the return of escaped floodwater to the river. The use of emergency breaching sections for bypassing flood peaks around undersized structures or to relieve pressure against downstream flood embankments is not feasible. Extensive agricultural and infrastructure developments exist on the land side of bunds along both river banks. Intentional breaching of bunds would primarily threaten human life, cause inundation of valuable agricultural lands, and damage irrigation facilities and other valuable infrastructure involving huge investments of private and public sector funds. The co-existence of similar requirements for intentional breaching on both sides at multiple places complicates the decision-making for its timely operation.

Moreover, river channel aggradation in Sindh province due to sediment deposition before reaching the delta is slowly reducing the effectiveness of the Sindh levees. As such, maintenance of the Sindh levees must include infrastructure investments to help maintain sediment-carrying flows during normal years, managed flow releases or higher allocations of water to Sindh, and periodic dredging. The best options to provide the needed sediment deposition in the delta are the first two.

After the 2010 flood, the Flood Emergency and Reconstruction Project (FERP) was implemented by the Irrigation Department Government of Sindh through Asian Development Bank loan #44372-013. The FERP raised levees in Sindh by 1.8 m above the 2010 flood mark. It is anticipated that the rehabilitated embankments will sustain floods of the magnitude of 2010. This is an immediate and required investment that can be maintained through integrated management of water and sediment flows in the Indus River system.

5.2.2.3 Flood Bypass

The conveying capacity of the Indus River is limited by the conveying capacities of the existing barrages and bridges. At Sukkur, the design is constrained to a 10-year return period (Fig. 5.14). The flood of 2010 had the flow of an approximately 30-year return period event at Sukkur Barrage (Fig. 5.14). Figure 5.15 illustrates that the 2010 flood flow event through Sukkur Barrage was not unprecedented. This suggests that more extreme events are likely.

The Lower Indus River has flowed along many courses over time. An old river course historically discharged in Manchar Lake, which was the 2010 flood breach route (Fig. 5.4). A flood bypass to Manchar Lake will require an expensive construction project with a spillway just above Guddu Barrage, but it would be a one-off expenditure that would cost less than repairing the damage of the 2010 flood. This alternative flood bypass would likely have a similar path, albeit controlled, as that of the Tori breach in 2010 (Fig. 5.4).

The identification of a flood retention area, such as Manchar Lake, in the Thar Desert of Sindh could mitigate floods by retention and managed releases. This would cause an effective reduction in peak flow discharge. The safety of life and other facilities in diversion and storage is a major concern. The Irrigation Department in the Government of Sindh published a feasibility study entitled ‘The Flood Escape Routes to Divert Excessive Floodwater along River Indus’ in 2014; the report identified probable escape routes, which require stakeholder engagement at the regional and national levels. The identified routes are quite sensitive to the

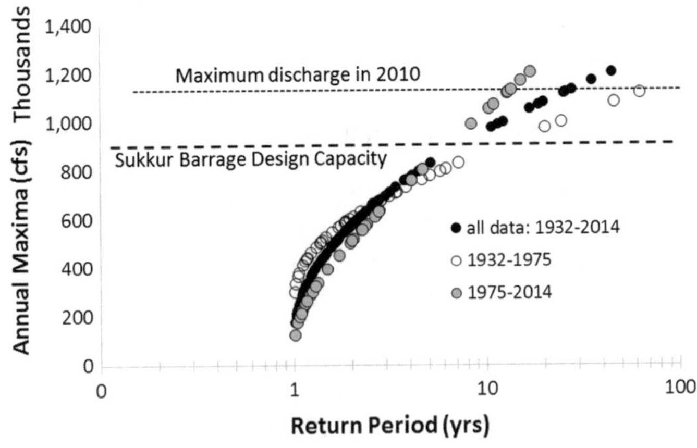

Fig. 5.14 Peak discharge flow return periods at Sukkur Barrage using the log Pearson III probability distribution method (1 cfs = 0.02832 m^3/s)

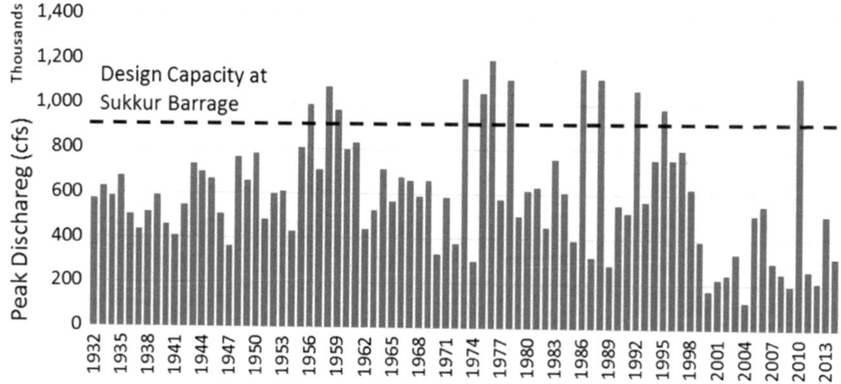

Fig. 5.15 Maximum annual flow events (1932–2014) at Sukkur Barrage (Government of Sindh Irrigation Department) (1 cfs = 0.02832 m^3/s)

duration of flood peak discharges. The 2010 flood peaks prevailed for about 16 days. Under this situation, stakeholders would be anxious about receiving outside floodwaters.

5.2.2.4 Reservoir Operation and Storage

Water reservoirs on the Indus primarily support irrigation and the generation of hydropower energy. There are large reservoirs at the Tarbela and Mangla Dams and Chashma. The Tarbela Dam and the Chasma are located on the main stem of the Indus River; the Mangla Dam is located on the Jhelum River, which drains into the Chenab tributary of the Indus. Although the potential storage of these three principal reservoirs amounts to 18.94 MAF, effective live storage is estimated to be 14.8 MAF as a result of sedimentation. This represents a loss of almost 22 % due to the accumulation of silt. To build resilience against droughts and to secure economic growth, there are several dam proposals and ongoing construction that would increase live storage capacity by approximately 24.3 MAF (Kazi 2005).[3]

Table 5.4 lists proposed dam projects where construction has started. All proposed new dams have an agricultural component (AG) and the majority have hydropower purposes (HP). Only a few of the projects have flood-risk reduction (FC) components. Although these dam projects build resilience to water scarcity, an integrated approach could also contribute to flood resilience.

First, additional flood storage volumes can help to mitigate peak flows along the main stem and tributaries of the Indus. Figure 5.16 compares the potential reduction of total flood volume by proposed new dams that would be observed at the Taunsa and Guddu Barrages due to cumulative upstream flood storage (Table 5.4). The

[3] 1 million acre ft (MAF) = 1233.5 million m^3 (MCM).

Table 5.4 Dam proposals and ongoing dam construction projects on the Pakistani Indus River Basin

Dam	Height (m)	Live storage (MAF)	2025/Ongoing	Energy (MW)	Irrigation area (acres)	Purpose
Akhori	122	6.97	2025			AG
Diamer-Bhasha	272	6.40	2025	4500		AG, HP
Kalabagh	79	6.10	2025	3600		HP
Munda	213	1.30	2025	740		AG, HP, FC
Tank Zam	88	0.10	2025			Unknown
Kurram Tangi	89	0.31	2025	83.5	1716	FC, AG, HP
Gomal Zam	133	1.14	Ongoing	17.4		AG, HP, FC
Satpara	39	0.06	Ongoing	17.16	15,500	AG, HP, M
Sabakzai	35	0.01	Ongoing			AG
Bara	92	0.03*	Ongoing	5.8	83,450	AG, M, HP, FC
Garuk	56	0.02*	Ongoing	0.3	26,000	AG, HP
Hingol	54	0.48*	Ongoing	3.5	80,000	HP, AG
Naulong	51	0.08*	Ongoing	4.4	47,000	FC, HP, AG
Pelar	40	0.02*	Ongoing	0.3	16,000	FC, AG, HP
Darawat	36	0.04*	Ongoing	0.45	50,000	AG, HP, fisheries
Daraban	47	0.02*	Ongoing	0.75	30,690	AG, FC, HP, M
Ghabir	42	0.02*	Ongoing	0.15	30,000	FC, HP, AG
Nai Gaj	59	0.10*	Ongoing	4.2	80,000	AG, HP, FC
Papin	32	0.02*	Ongoing	0.3	40,000	AG, FC, HP, fisheries
Winder	31	0.01*	Ongoing	0.3	20,000	AG, HP, M, GW recharge

Source Pakistan agency websites about dam projects
*These represent gross storage capacity
1 million acre foot (MAF) = 1233.5 million m^3 (MCM)

analysis is unrealistic because it assumes a mostly flood-control purpose of all the new dams and does not take into account the persistent high flows that occurred days before the maximum peaks. However, it provides insight into maximum possible reductions at Taunsa, if the cumulative flood control storage of new dams was 100, 50, and 25 %. The flood volumes (volumes in excess of banks) at Guddu would not decrease significantly because most projects are planned on the western rivers to benefit from waters guaranteed in the Indus River Treaty. Effective flood-risk reduction operations would require improvements in real-time operations and monitoring, improvements in forecasting, and coordination between the different dams such that pre-flood releases are made on a timely basis and operators offset flood peak synchronization.

Second, water stored in reservoirs provides the best method of controlling water flows throughout the year. In effect, the ability to store water can provide some

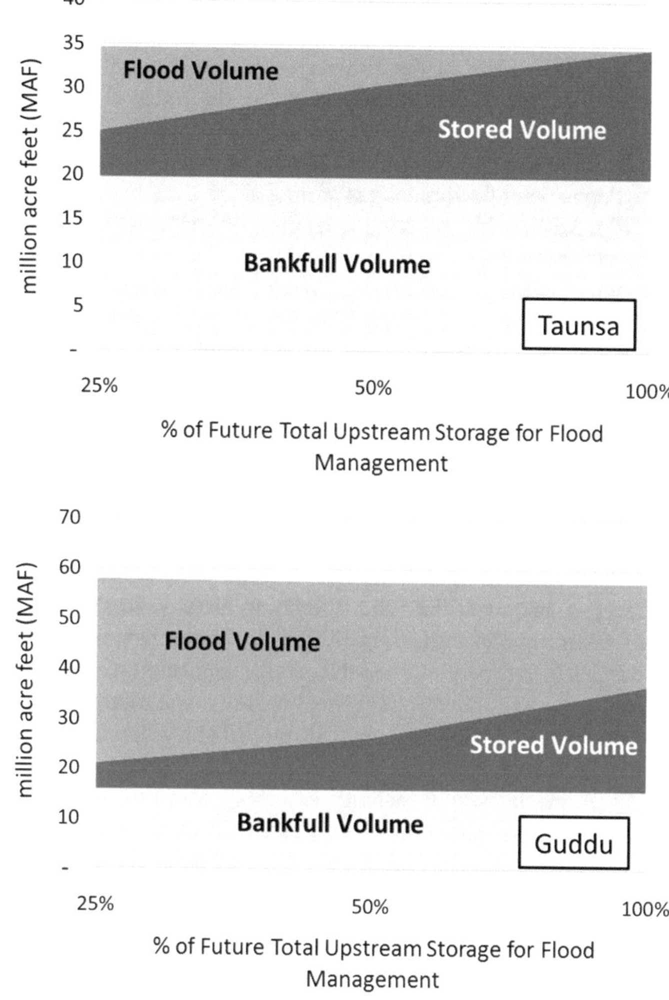

Fig. 5.16 Flood management potential of proposed large reservoir allocation of the total flow volume for 5 August to 13 September 2010 with respect to impacts at Taunsa and Guddu

flexibility in the release of water to manage the sediment balance, reduce aggradation of the river channel, reduce flood risk, and reduce the erosion of the coastal delta in Sindh province. Moreover, the large sediment loads of the Indus will continue to be trapped by existing and proposed dams. Pakistan's reservoirs can be managed as a renewable resource or as an exhaustible resource, depending on whether sediments can be managed effectively (Annandale 2013). Trapped sediment can be removed by flushing, sluicing, or bypasses. Storing water for these purposes builds sustainability and resilience.

5.2.2.5 Building Room for the River

The Indus is the oldest river in the Himalayan region. It used to carry the fifth largest suspended sediment load in the world (on the order of 480 million tons annually) and deposit it in the Arabian Sea (Holeman 1968). Pithawalla (1959) predicted a rise in bed level below the Taunsa Barrage of 0.3 m per century. Moreover, the deposition of sediment has promoted the dynamic meandering of the Indus River (Fig. 5.17), with the wide spreading of sediment material that comprises the alluvial plain.

The development of bridges and other infrastructure crossing the river constrains the river and, during low flows, funnels sediment into narrow channels. In the Lower Indus basin, the floodplain has an average width of 12 km, which is constricted to about 1 km in regions of bridges, barrages, and other development. Sediment carried during low flows eventually loses energy and is deposited within these narrow areas, resulting in very elevated river channel aggradation. During floods, this critical infrastructure and development is at higher risk due to the elevated river.

The river's tendency to meander has been controlled by the erection of embankments. The construction of barrages and introduction of canals have now virtually resulted in an end to the natural flooding activity. Irrigation canals have replaced flood channels and agriculture has replaced the natural floodplain. The taming of the Indus River has restricted the natural deposition of alluvium over the entire floodplain and now the present channel is rising—an increase of 1.5 m over the last 50 years. Human settlements have developed along the natural courses and have obstructed the natural drainage. The result is a detrimental rise in waterlogging to about 60 % of the land, with salinity assuming alarming proportions in the region. The protection of the floodplain effectively builds room for the river, reduces risk by keeping people and property out of harm's way, provides flood storage, dissipates the energy of floodwaters, improves the dynamics of sediment deposition, and protects the fertility of agricultural land.

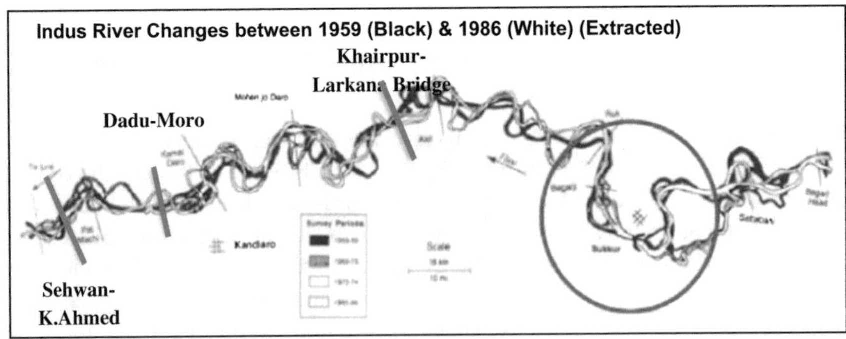

Fig. 5.17 The Indus River's change in course between 1959 and 1986 (*Source* Jorgensen et al. 1993)

5.2.3 *Adapt for Future Uncertainties and Climate Change*

A significant quantity of water in the Indus River basin is stored naturally in the glaciers of the Western Himalayas. Although there is high uncertainty in climate projections, they consistently show increases in temperature for Pakistan across all emission scenarios and time horizons (Fig. 5.18). In the short term, the most optimistic model predicts temperature increases of up to 2 °C that could reach up to 3 °C without changes in global emissions. In the long term, the range of model predictions is even greater: from no change to over 8 °C. In effect, this means that monthly average temperatures could increase anywhere from a minimal change to 8 °C in the next 100 years and up to 3 °C within the next 10 years. The projected changes for precipitation are far more uncertain, ranging from wetter to drier futures in similar magnitudes.

Furthermore, the impact on basin yields will likely be dynamic, resulting in short-term increased flows (exacerbating flooding) as the Indus glaciers melt, and in long-term reductions to base flow yields as glacier volume is reduced (Fig. 5.19).

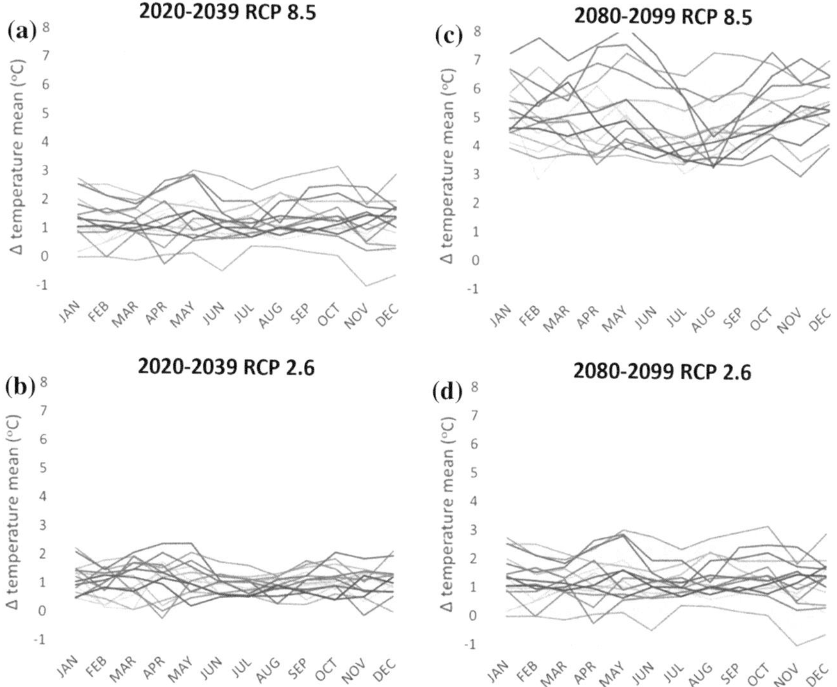

Fig. 5.18 Global climate model projections of changes in monthly temperatures (°C) from observed averages. All projections are included for 2020–2039 (**a**, **b**), 2080–2099 (**c**, **d**), optimistic emission scenario RCP 2.6 (**b**, **d**), and business-as-usual emissions through the twenty-first century (**a**, **c**). There are many intermediate ensembles (World Bank Climate Change Knowledge Portal)

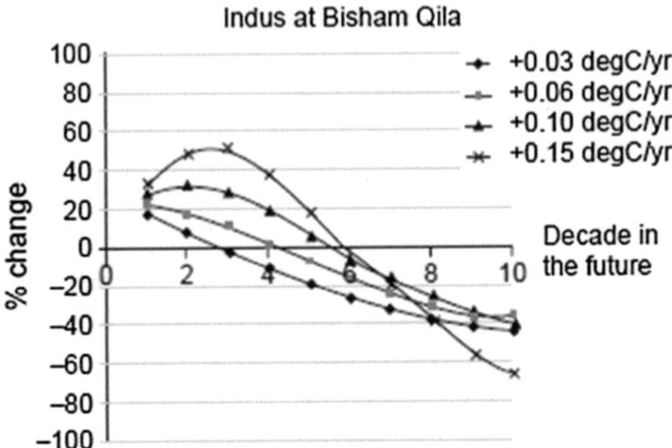

Fig. 5.19 Projected changes in Indus River flows at Tarbela due to changes in glacier melt (Rees and Collins 2005). Although this modelled forecast is outdated, it is consistent with the expected dynamic forecast. As temperatures warm, glaciers melt. This results in increased spring and summer runoff until the glacial water reservoir is diminished, resulting in reduced overall yields

This dynamism and uncertainty pose significant challenges for the planning and design of sustainable water resources. Namely, what are future irrigation demands and Indus basin yield under future precipitation and temperature change uncertainty? How do managers and decision-makers plan and design a cost-effective and robust water resources system? What are the planning horizons? What future scenarios and model should be selected? A planning construct that is driven by raw and downscaled climate projections will lead to cascading uncertainties and decision bottlenecks (Brown and Wilby 2012).

The World Bank has recognized this planning and decision-making challenge. The organization has endorsed a bottom-up decision-scaling approach for planning and design under future uncertainties (Ray and Brown 2015). This construct begins by defining critical thresholds or breaking points in system performance through engagement with sector interest groups (Fig. 5.20). Once these are defined, a system model is used to identify external drivers, such as precipitation variability or temperatures, which "break" the system's performance threshold. These external drivers are developed through weather generators that allow an exploration of the entire vulnerability domain, unconstrained by climate projection capabilities or ranges. It is only after a "bottom-up" vulnerability domain is identified that all sources of evidence—such as trends, science, and climate projections—are used to assess the plausibility of futures that will "break" the system. Planning and adaptation strategies are evaluated by cost-effective measures of robustness to maintain performance across a range of futures or by improvements in system performance resilience.

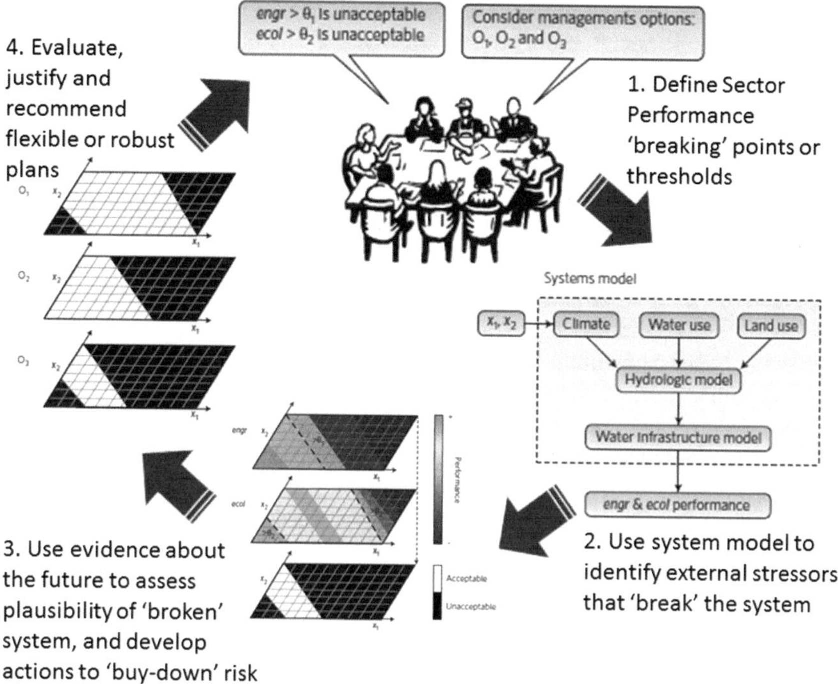

Fig. 5.20 Decision scaling/bottom-up vulnerability assessment framework (modified from Poff et al. 2015)

5.2.4 *Prepare and Recover: Reversing Extemporaneous Governance*

The previous sections outlined a series of challenges that are further compounded by the extemporaneous governance of Indus water resources. Responses to flood disaster events are poorly prepared and lessons were not learned from previous experiences. Exhaustive response efforts are required every time, starting from scratch, even though these events recur periodically. As shown in Fig. 5.15, the flood of 2010 was not unprecedented. A huge amount of experience is available, yet there is no proactive planning.

The lack of information has long been an outstanding issue. On 10 December 1920, Sir Thomas Ward, Inspector General of Irrigation in India, stated that "almost all the controversies which have up to date taken place in India in question of water rights have been directly attributable to the fact that adequate figures of flows were not forthcoming due to unavailability of gauging" (Dharmadhikary et al. 2005).

However, even when information is available, it is often not shared in a timely manner. A fundamental problem for information sharing across administrative boundaries is a lack of trust within Pakistan and between Pakistan, India, and Afghanistan, which are at the headwaters of the eastern and western Indus reaches, respectively. Flood events that originate from the eastern rivers in India or the western rivers in Afghanistan have never been reported in a timely manner to assist Pakistan in preparedness. In the case of Afghanistan, resources and capacity are lacking. This lack of transboundary coordination requires international assistance for arbitration. However, a Pakistani authority that might be able to coordinate, plan, and respond internally is still required. For example, the Suleiman hill catchments at the head Sindh province were ungauged giving very limited response time to the breaches in Sindh during the 2010 floods. They are still ungauged today. This is another example of lessons not being learned.

One of the challenges for internal coordination within Pakistan's provinces and sectors is conflict. The carrying capacity of Indus River basin is limited (Fig. 5.9) and priorities need to be assigned to different uses. Considerations of a hierarchy of needs suggest that the basic human requirements will be ensured. The demand for water from the Indus basin rapidly increased after the partition of India and Pakistan in twentieth century. Water resources are overexploited, but the demands are still increasing. The consequent reduction in the capacity to meet different demands has resulted in conflicts between different water uses and between upstream and downstream uses. The Indus River system is already considered to be in *water stress* status and is expected to achieve *water scarcity* by 2035 (Briscoe and Qamar 2006). Not surprisingly, irrigation authorities in the upstream provinces have little incentive to discharge water for flood storage or for adequate sediment transport to reduce aggradation in Sindh. Additional reservoir infrastructure is sorely needed to reduce conflict.

Pakistan needs an approved water policy and a comprehensive flood management plan or river plain regulatory framework to promote governance in the basin. The Indus River System Authority, which is part of the Water Apportionment Accord 1991, defines the institutional setup for the distribution of surface waters among the provinces of Pakistan. This could be enhanced to integrate these additional requirements. The Sindh and Punjab Irrigation and Drainage Authorities Acts 1997 allows the participation of water users from selected parts of the provinces in pilot schemes for integrated management. However, these have not proved as successful and some have caused damage to irrigation systems due to lack of education and maturity of the communities coordinating the farmer organizations.

More than 12 organizations at the national and provincial levels are responsible for flood preparedness and response. The irrigation authorities operate and maintain the major flood infrastructure. Flood management is a provincial subject with inadequate funding (Kazi 2005). Moreover, flood forecasting and early warning is the responsibility of the PMD through FFD Lahore. Rescue and relief is coordinated at the district level. In addition, procurement to deal with emergencies is managed by the Public Procurement Authorities, which have no role in flood mitigation and management, making it difficult to fund emergency preparedness activities. The coordination of all of these pillars is in itself a bigger task than the floods. Water resource investments in Pakistan need to transition from single-purpose infrastructure investments to integrated investments in multipurpose infrastructure and management and institution building (Briscoe and Qamar 2006).

The need for governance in the basin arises because the noncoordinated use of resources is inefficient as well as damaging. The time has come to create an Indus River basin organization. This organization will need to have a clear mandate, be well funded and accountable, have highly trained staff, and have authority across provincial and multisectoral interests. In addition, this basin organization must be able to implement an adaptive management framework to tackle future uncertainties, such as climate change.

5.3 Conclusion

The security and economic future of Pakistan is largely tied to how Pakistan can build resilience to natural disasters. Building flood resilience in Pakistan's Indus River requires an integrated effort that must be coordinated with the national effort to build resilience to water scarcity. Efforts must involve significant coordinated infrastructure investments in levees, flood-plain management, water storage and allocations, monitoring, and early warning. At a minimum, reservoirs should control water to provide a steady supply and to help manage flood risk due to sediment deposition and coastal erosion. Ideally, they would also allow for storage to control floods. More or bigger dams may be required, perhaps on the eastern rivers for flood control to reduce flood risk. Resilience to flooding in Pakistan should have many potential management and infrastructure redundancies to act as fail-safes, rather than relying on a few strategies. History has shown that Pakistan floods all have different characteristics. An integration of strategies and diverse infrastructure investments will reduce Pakistan's flood risk.

Effective governance is also needed to achieve cooperation and political commitment within the country and among the basin nations of India, Pakistan, and Afghanistan. Looking to the problems of sea intrusion at the Indus Delta, the pressures on the environment have surpassed the levels of sustainability. As a consequence, vulnerability to extreme natural events has increased over the last 50 years. Conflicts between different water uses and between upstream and downstream users may cause irreparable loss to the nations in the Indus basin.

The capacity of the Indus basins to meet growing basic human needs, such as drinking water, is decreasing rapidly. For example, the province of Sindh—already struggling to meet its irrigation needs—is now facing water scarcity and uncertainty due to climate change in the basin, coupled with a burgeoning population of Karachi through migration from Afghanistan and northern Pakistan.

An integrated approach covering all waters of the basin (surface and subsurface) is needed to solve these conflicts. It should consider temporal and spatial distribution of the water quantity and quality, as well as the interaction of water with land use and land cover. Sediment transport processes ensuring the adequate supply of the sediment and water to the delta are of pivotal importance.

The basin's ability to receive return flow is limited due to the excessive pumping involved in groundwater mining. Therefore, including regenerated flows in the total available resource would be an error. In the Indus basin, people rely on the ecosystem. There is also an intrinsic social value through aesthetic, cultural, and heritage links. The aim should be to maintain the capacity of the ecosystem to deliver goods and services over a long timeframe.

Pakistan is highly disaster prone, with two major disasters in the past 5 years alone, yet major losses are not inevitable. Excellent management policies exist to minimize the impact of disasters, but they are not being implemented on the ground. During the floods, huge gaps became apparent. This is not good enough. It undermines efforts to help the economy to grow, minimize food insecurity, and improve social and political stability (Azad 2011). The priority in Pakistan's water sector has been to address water security. The consideration of flood risk must now be considered in parallel with investments in water security.

Natural disasters are inevitable in Pakistan, and they will worsen with climate change, mass migration of Afghan refugees into Sindh province, rapid population growth, and poor urban management. Moreover, the floods of 2010 and 2014 were not particularly rare events (note in Fig. 5.15 that maximum flows through Sukkur Barrage in 2010 were similar to several events within the last 60 years). A national policy for flood-risk reduction must be developed as part of national efforts that are meant to build economic prosperity and security. Pakistan is still recovering from the 2010 to 2014 floods; in addition to preparing for the next disaster, it must enhance investments that mitigate flood impact. In Pakistan, the Indus River is central to economic growth and development but also to natural disasters that have extremely heavy tolls in terms of lives and GDP. It is essential that lessons be learned from the 2010 floods and resilience strategies be implemented for water resources, which integrate development objectives with diverse strategies to reduce risk. In Pakistan, water is critical to growth and prosperity. However, without an integrated strategy directed by a strong river basin organization, water will continue to destroy the benefits obtained from harnessing it.

References

Ahmad N (2015) Early warning systems and disaster risk information. Pakistan National Briefing. Lead House, Islamabad, Pakistan

Ali A (2013) Indus Basin floods mechanisms, impacts, and management. Asian Development Bank, Mandaluyong City, Philippines

Annandale G (2013) Quenching the thirst: sustainable water supply and climate change. CreateSpace, Charleston, SC

Asianics (2000) Tarbela Dam and related aspects of the Indus River Basin, Pakistan. A world commission on dams case study, Agro-Dev. International (Pvt.) Ltd., Cape Town

Awan JA, Maqbool O (2010) Application of artificial neural networks for monsoon rainfall prediction. In: 6th international conference on emerging technologies (ICET), Islamabad, 18–19 Oct 2010

Azad A (2011) Ready or not: Pakistan's resilience to disasters one year on from the floods. Oxfam Briefing Paper #150, 26 July. Oxfam International, Oxford, UK

Briscoe J, Qamar U (2006) Pakistan's water economy: running dry. The World Bank, Washington, DC

Brown C, Wilby RL (2012) An alternate approach to assessing climate risks. Eos Trans Am Geophys Union 93(41):401–402

Dharmadhikary S, Sheshadri S, Rehmat (2005) Report of a study of the Bhakra Nangal Project. Manthan Adhyayan Kendra, Madhya Pradesh. Available via www.narmada.org/misc/Bhakra11.pdf. Accessed 5 May 2015

Giosan L, Constantinescu S, Clift PD, Tabrez AR, Danesh M, Inam A (2006) Recent morphodynamics of the Indus shore and shelf. Cont Shelf Res 26:1668–1684

González FJ, Basson T, Schultz B (2005) Final report of IPOE for review of studies on water escapages below Kotri Barrage. http://s3.amazonaws.com/zanran_storage/www.pakistan.gov.pk/ContentPages/19050336.pdf. Accessed 8 May 2016

Holeman JN (1968) The sediment yield of the major rivers of the world. Water Resour Res 4 (4):737–747

Inam A, Khan ATM, Amjad S, Danish M, Tabrez AR (2004) Natural and man made stresses on the stability of Indus deltaic ecoregion. In: Extended abstract, the fifth international conference on Asian marine geology, Bangkok, Thailand, 13–18 Jan 2004 (IGCP475/APN)

Jorgensen DW, Harvey MD, Schumm SA, Flam L (1993) Morphology and dynamics of the Indus River: implication for the Mohenjo Daro site. In: Schroder JF Jr (ed) Himalaya to the sea. Routledge, London, pp 288–326

Kazi AH (2005) Flood control and management. World Bank Country Water Resources Assistance Strategy Background Paper #14, section 3.3. World Bank, New York

Khalil H, Sarwar MW, Raza HA, Akhtar P, Hassan Z (2011) Irrigation system of Pakistan. Project and report, Department of Agriculture Engineering and Technology, University of Agriculture, Faisalabad

Mendoza G, Matthews J, Jeuken Ad (eds) (2016) Water resources planning and design for uncertain futures: climate risk informed decision analysis. ICIWaRM Press, Lima, Peru

Milliman JD, Quraishee GS, Beg MAA (1984) Sediment discharge from the Indus River to the ocean: past, present and future. In: Haq BU, Milliman JD (eds) Marine geology and oceanography of Arabian sea and coastal Pakistan. Van Nostrand Reinhold, New York, pp 65–70

National Disaster Management Authority (NDMA) (2013) National Disaster Management Plan (NDMP) 2012 and DRR Policy 2013. NDMA, Lahore

Project Management and Policy Implementation Unit (PMPIU) (2011) website. http://wspakistan.org/index.php?Type=SDIBIS&Page=Schematic_Diagram_of_Indus_Basin_Irrigation_System. Accessed 8 May 2016

Pithawalla MB (1959) A physical and economic geography of Sind, the Lower Indus Basin. Sindhi Adabi Board, Karachi (print)

Poff L, Brown CB, Grantham TE, Matthews JH, Palmer MA, Spence CM, Wilby RL, Haasnoot M, Mendoza GF, Dominique KC, Baeza A (2015) Sustainable water management under future uncertainty with eco-engineering decision scaling. Nature Clim Change. doi:10.1038/nclimate2765

Ray P, Brown C (2015) Including climate uncertainty in water resources planning and project design: decision tree methodology. World Bank Group, Washington, DC

Rees HG, Collins DN (2005) Regional differences in response of flow in glacier-fed Himalayan rivers to climatic warming. Hydrol Process 20(10):2157–2169

UN Inter-Agency Secretariat for International Strategy for Disaster Reduction (ISDR) (2006) Developing early warning systems: a checklist. In: EWC III third international conference on early warning from concept to action, Bonn, Germany, 27–29 March 2006

US Congress Report (2011) Avoiding water wars: water scarcity and Central Asia's growing importance for stability in Afghanistan and Pakistan. A majority staff report, S. Prt 112-10, Congress twelfth session, US Government Printing Office, Washington, DC

World Bank (2010) Pakistan floods 2010 preliminary damage and needs assessment. Joint report, World Bank and the Asian Development Bank, New York

Chapter 6
Yellow River: Re-operation of the Infrastructure System to Increase Resilience to Climate Variability and Changes

Yangbo Sun and Xinfeng Fu

Abstract The Yellow River has generated huge social and economic benefits. Large-scale infrastructure has been developed along the river since the 1950s. However, the construction and operation of infrastructure projects also led to significant changes in flow regime and water allocation, and the river often ran dry before it reached the sea. Reoperation of the infrastructure system is now crucial in optimizing flow and sediment transportation. With focuses on water scarcity, flood discharge, water quality, environment flow, and ecological restoration, the new strategy substantially improved the river's adaptive capability to climate variability and other changes.

Keywords Reservoir operation · Infrastructure · Climate change · Flow and sediment regulation · Water allocation

6.1 Introduction

Water infrastructure has supported the evolution of all human societies and will remain an integral part of socioeconomic development and modernization, especially in rapidly developing countries (Muller et al. 2015). Built infrastructure is positive for increasing resilience to climate variability and change, and it is an essential element for sustainable development and an ecologically balanced civilization (Tortajada 2014; Muller et al. 2015).

Dam construction provides water for irrigation, municipal, and industrial supply; it also can be used for electricity generation as well as flood protection (Hering et al. 2015). In addition, adaptive operation of infrastructure can be defined as "a continuous systematic process of learning and optimizing the planning and strategies" (Xia et al. 2014). Adaptive operation aims to improve water management policies

Y. Sun (✉) · X. Fu
Yellow River Conservancy Commission, Zhengzhou, China
e-mail: sunyangbo@yrcc.gov.cn

C. Tortajada (ed.), *Increasing Resilience to Climate Variability and Change*,
Water Resources Development and Management,
DOI 10.1007/978-981-10-1914-2_6

and practices, enhance adaptability and multiple functions, and achieve sustainable development.

The Yellow River basin is the cradle of the Chinese civilization, which has a long history of living with the river and coping with frequent devastating floods. The construction of dikes and irrigation facilities along the river commenced two thousand years ago. By the end of 2015, a total of 30 large dams and hydropower projects had been constructed on the main stream of the Yellow River, with a total storage capacity of 69.7 billion m^3, exceeding the river's annual discharge by 20 %. These infrastructure projects have enhanced the capability for development and utilization of the Yellow River water resources. They have played a significant role in flood and drought management and water supply for industrial, agricultural, urban and rural domestic uses, as well as power generation, thus generating huge social and economic benefits.

As the economy and society developed, conflicts among different users of the Yellow River's water resources have increased, and multiple stresses have threatened the river's ecosystem. Additionally, the construction and operation of infrastructure projects has also led to changes in flow regime and sediment runoff, with significant impacts on the lower reaches. Concurrently with the basin-wide infrastructure construction and operation, the total water abstraction capacity and total water demand also increased, causing serious water depletion in the lower Yellow River. In 1972, the Yellow River failed to reach the sea; in 1997, the river mouth dried up for 226 days.

In response to concerns about water scarcity and the declining quality of ecosystem services that the river provides to society, Li (2004) formulated a policy framework for keeping the Yellow River healthy. The vision encompasses both socioeconomic and ecological health, which can be expressed in terms of the total amount of water resources, flood discharge capacity, sediment carrying capacity, water quality, and the capacity to maintain ecosystems.

In 1999, the Yellow River Conservancy Commission (YRCC) commenced an integrated approach to regulate the flow and sediments through the reoperation of the infrastructure system on the Yellow River, together with administrative, legislative, and economic measures. The major change is that the frequent occurrences of the river drying up in the second half of the twentieth century have ended and water scarcity in the lower reaches has been successfully alleviated, with environmental flow and ecological restoration also addressed.

Enormous infrastructure systems were built on the Yellow River in the past 50 years, with great achievements in flood safety and water supply. However, from the perspective of sustainable river basin management, there are still various challenges due to climate, environment, human society, or even the infrastructure itself. In response to these concerns, water governance and infrastructure operation rules for the Yellow River have gradually changed in the past two decades. This case study reviews the process of the reservoir's reoperation and analyzes the reasons, situations, countermeasures, and impacts to demonstrate how the Yellow River can respond to climate variability and change through integrated river basin management.

6.2 The Yellow River and Its Flow and Sediment Regulation System

The Yellow River is the second longest river in China. It flows 5464 km from the Tibetan plateau in Qinghai to the Bohai Sea in Shandong Province, through a catchment of about 795,000 km^2 (Fig. 6.1). It carries just 2.6 % of the natural runoff of the entire country (annual mean of 58 billion m^3), yet nourishes 12 % of the country's population, irrigates 15 % of the country's arable land, and supplies water to more than 400 cities and towns along its banks.

The Yellow River is a high-sediment laden river with a mean annual sediment load of 1.6 billion tons, which ranks it first among the world's rivers in terms of sediment load. The average suspended sediment concentration is within the range of 24–35 kg/m^3; however, hyperconcentrated floods with sediment concentration over 100 kg/m^3 are not unusual. The lower Yellow River is prone to flooding due to the elevated nature of the riverbed, running between dikes above the broad plains surrounding it, which is now above the surrounding ground because of silt accumulated during 2000 years of levee construction.

The basin lies in two different climatic zones: arid and semi-arid continental monsoon in the northwest and semi-humid in the southeast. More than 60 % of annual precipitation falls during June to September, and precipitation varies greatly within and between years; thus, the Yellow River has severe flood and drought problems. Average rainfall during 1954–2000 was 454 mm over the entire basin, with 370 mm in the upper, 530 mm in the middle, and 670 mm in the lower basin. During the 1990s, because of prevailing drought conditions, average precipitation dropped 7.5 % below the long-term average. The average total surface runoff of the Yellow River is estimated at 58 billion m^3. Water demand in the basin sharply

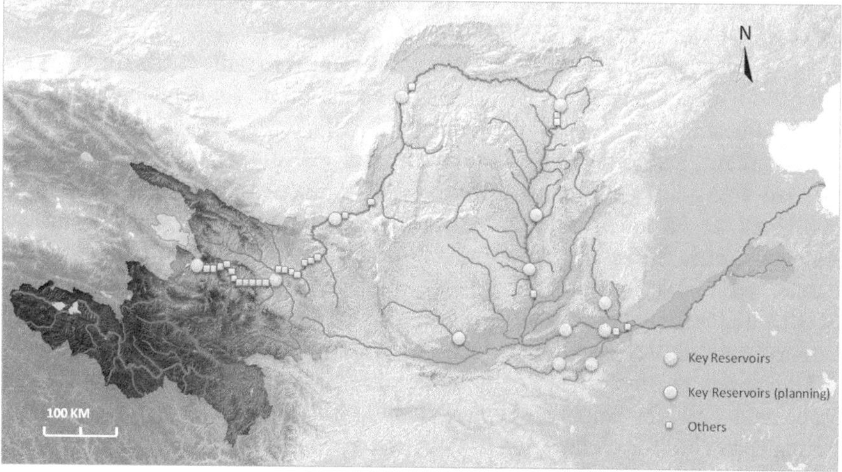

Fig. 6.1 Layout of the Yellow River flow and sediment regulation system

increased from 10 billion m^3 in 1949 to 37.5 billion m^3 today. In addition, a booming economy coupled with rapid industrialization and population growth has had a direct impact on water quality. Therefore, it is a major challenge for river basin management to control and improve the situation of water and sediment regulation.

6.2.1 Yellow River Flow and Sediment Regulation System

According to the revised *Master Plan of Yellow River Harnessing and Development* (YRCC 1997) (hereafter the "Master Plan"), a total of 36 cascade dam projects were planned on the main stream (Table 6.1). In 1957, the construction of Sanmenxia dam announced the beginning of cascade hydropower development on the main stream of the Yellow River. By 2015, a total of 30 dam projects and hydropower stations were built on the main stream, among which 24 dams were located in the upper reaches of the river, including Longyangxia, Liujiaxia, and Laxiwa; and six in the middle reaches, including Wanjiazhai, Sanmenxia, and Xiaolangdi. These six named reservoirs have a total storage capacity of about 62.5 billion m^3 on calibration water level, a total storage capacity of about 59 billion m^3 on normal storage level, and an effective storage capacity of about 36.3 billion m^3, accounting for 58 % (59 of 101 billion m^3) and 72 % (36.3 of 50.3 billion m^3) of all 36 reservoirs. In the Yellow River upper reaches, more than 10 hydropower stations (including Longyangxia, Lijiaxia, Liujiaxia, Yanguoxia, Bapanxia, and Qingtongxia) form the largest group of cascade hydropower stations in China, which also makes the Yellow River one of the most developed rivers in China. In addition, more than 2600 large, medium, and smaller-sized reservoirs have been constructed on the tributaries, with a total storage capacity of over 13 billion m^3.

According to the Master Plan, six more water control projects and hydropower stations are to be constructed, with Shanping and Daliushu in the upper stream and Qikou, Guxian, Ganzepo, and Taohuayu in the middle stream. These six reservoirs will have a total storage capacity of about 42.2 billion m^3 on normal storage level and an effective storage capacity of about 14.1 billion m^3, accounting for 42 % (of 101 billion m^3) and 28 % (of 50.3 billion m^3), respectively, of all 36 reservoirs. The installed electricity generation capacity of these six reservoirs will account for 26.5 % of the 36 reservoirs' total installed capacity. When all 36 reservoirs are completed, their total storage capacity will be 1.8 times the Yellow River's annual runoff. These projects are currently being operated and used as designed and have brought tremendous comprehensive utilization benefits in terms of flood control, ice-flood control, water supply, irrigation, and power generation.

The seven reservoirs of Longyangxia, Liujiaxia, Daliushu, Qikou, Guxian, Sanmenxia, and Xiaolangdi were planned as key controlling projects, constituting the main part of the Yellow River flow and sediment regulation system. These seven projects have a total storage capacity of 92 billion m^3 and an effective storage capacity of 47.05 billion m^3, accounting for 91.3 and 93.2 % of the 36 cascade

Table 6.1 Yellow River master plan of Yellow River harnessing and development—dam projects (YRCC 1997)

No.	Dams/reservoirs	Catchment area (km^2)	Normal water level (M.S.L)	Total storage capacity (km^3)	Effective storage capacity (km^3)	Total installed capacity (MW)
1	Longyangxia	131,000	2600	247.0	193.5	1280
2	Laxiwa	132,000	2452	10.0	1.5	4200
3	Nina	132,000	2235	0.3	0.1	160
4	Shanping	132,000	2219	1.2	0.1	160
5	Lijiaxia	137,000	2180	16.5	0.6	2000
6	Zhiganglaka	137,000	2050	0.2	/	192
7	Kangyang	137,000	2033	0.2	0.1	283
8	GongboXia	144,000	2005	6.2	0.8	1500
9	Suzhi	144,000	1900	0.3	0.1	225
10	Huangfeng	144,000	1880	0.7	0.1	225
11	Jishixia	147,000	1856	4.2	0.4	1020
12	Dahejia	147,000	1783	0.1	/	220
13	Sigouxia	147,000	1748	1.0	0.1	240
14	Liujiaxia	182,000	1735	57.0	35.0	1690
15	Yangguoxia	183,000	1619	2.2	0.1	472
16	Bapanxia	216,000	1578	0.5	0.1	252
17	Hekou	216,000	1558	0.1	/	74
18	Cajiaxia	221,000	1550	0.2	/	96
19	Xiaxia	225,000	1499	0.4	0.1	230
20	Daxia	228,000	1480	0.9	0.6	324
21	Wujinxia	229,000	1436	0.2	0.1	140
22	Daliushu	252,000	1380	107.4	57.1	2000
23	Shapotou	254,000	1240	0.3	0.1	120
24	Qingtongxia	275,000	1156	5.7	0.1	324
25	Haibowan	311,000	1076	4.1	1.5	60
26	Sanshengong	314,000	1055	0.8	0.2	
27	Wanjiazhai	395,000	977	9.0	4.5	1080
28	Longkou	397,000	898	1.8	0.7	420
29	Tianqiao	404,000	834	0.7	/	128
30	Qikou	431,000	785	125.7	27.9	1800
31	Guxian	490,000	645	165.7	47.7	2100
32	Ganzepo	497,000	425	4.4	2.4	440
33	Sanmenxia	688,000	335	96.4	/	410
34	Xiaolangdi	694,000	275	126.5	51	1800
35	Xixiayuan	695,000	134	1.5	0.3	140
36	Taohuayu	715,000	110	17.3	11.9	/

projects in the Master Plan, respectively. Among these seven key reservoirs, Longyangxia, Liujiaxia, Sanmenxia, and Xiaolangdi are operated and used as designed and have played a vital role in flood control (including ice-flood control), sedimentation reduction, industrial and agricultural water supply, and power generation, providing strong support for the economic and social development of the Yellow River Basin and the Huang-Huai-Hai Plain.

The Xiaolangdi dam, one of the key structures, has played a significant role in the flow and sediment control of the lower river since 1999. The Xiaolangdi reservoir has a total storage capacity of 12.65 billion m^3, in which the sediment deposit storage is 7.55 billion m^3 and the long-term effective storage capacity is 5.1 billion m^3. Mean annual flow into the reservoir is 27.9 billion m^3 and average annual sediment deposit is 1.3 billion tons. As a multipurpose project, Xiaolangdi was mainly designed for flood control (including ice-jamming prevention) and siltation reduction, water supply, irrigation, and power generation, rather than for water resources management and environmental benefits. However, during construction in 1992–2001 and trial operation in 2003, the operation of Xiaolangdi was focused on water scarcity and drought management because of climate change and rapidly increasing water demand due to socioeconomic growth, as well as increasing public awareness of environmental protection.

6.2.2 *Nonstructural Measures*

Nonstructural measures have been improved, mainly by developing and applying a hydrological monitoring system, a forecasting and warning system, a remote sensing and remote control system, and by implementing laws, regulations, policies, and economic approaches. These measures include managing river channels, allocating integrated water resources, regulating flow and sediment, and establishing flood management schemes for extremely large floods.

Based on the Integrated Water Allocation Scheme approved by the State Council in 1987, provinces in the upper reaches are allocated 39 % of the available flow, provinces in the middle reaches receive 22 %, and provinces in lower reaches get 39 %. The allocation is revised annually to reflect seasonal variations in water resource availability. Since 2000, according to the most recent approach adopted by the Ministry of Water Resources, water management and related development activities in the Yellow River basin have aimed to integrate the interests of all regions and sectors and to balance available water supply and demand of various sectors. The YRCC is responsible for developing the annual water allocation scheme based on hydrological conditions, reservoir storage, and supply and demand patterns. Water abstraction permits are issued to users to ensure adequate supply for priority areas, especially in the case of drought (State Council of China 2006). Furthermore, the YRCC has established regulations encouraging household users to install water-saving devices, farmers to adopt water-efficient practices, and industry to promote techniques that minimize water use and waste discharge.

The YRCC also prepares and implements the basin water development plan, decides the allocation of water resources at provincial level, and is in charge of constructing and maintaining structures for water resources development and flood prevention. The YRCC manages the Yellow River in the following main areas: constructing and maintaining engineering structures along the main course and large tributaries, including dikes, reservoirs, canals and water diversion projects; preventing and controlling floods in the basin; and proposing and operating water allocation plans at provincial level through diversions, lifting stations along the main course, and large tributaries for agricultural and domestic uses.

6.3 Stages of Yellow River Reoperation

By the end of 2000, a total of 22 large dams and reservoirs had been built on the main stream of the Yellow River. Operation of the reservoirs changed the natural flow regime, dramatically reducing both water and sediment flows over the past 50 years; seasonal peak flows almost disappeared and the river ran dry before reaching the sea for much of the year in the 1990s (Liu et al. 2006; Jiang 2009). The YRCC is responsible for restoring the flow regime of the river, achieved through improved governance and reoperation of the infrastructure system, so as increase the river's adaptability to large challenges (Table 6.2).

Table 6.2 Yellow River reoperation schemes

Items	Phase 1 Before 1999	Phase 2 1999–2006	Phase 3 2006–2015
1. Flow and sediment regulation system			
1.1 Structures			
Reservoirs (nos.)	22	28	30
Key reservoirs (nos.)	3	4	4
1.2 Nonstructures			
Hydrology monitoring network	✓	✓	✓
Forecasting and warning system		✓	✓
Reservoir operation decision support system			✓
2. Institution and governance			
2.1 Setup of water resources department		✓	✓
2.2 Water governance			
Initial water rights	✓	✓	✓
Regulations and policy		✓	✓
Water market			✓
2.3 Stakeholder participation		✓	✓
3. Reservoir reoperation			

(continued)

Table 6.2 (continued)

Items	Phase 1 Before 1999	Phase 2 1999–2006	Phase 3 2006–2015
Joint reservoir operation	Independent	Lower YR	Basin wide
Flood safety	✓	✓	✓
Irrigation	✓	✓	✓
Hydropower	✓	✓	✓
Water supply	✓	✓	✓
Flow continuity		✓	✓
Water quality improvement		✓	✓
Ecological water demand		✓	✓
Delta wetlands restoration		✓	✓
Sediment flushing (artificial floods)			✓
Estuary aquatics			✓

6.3.1 *Phase 1: Before 1999*

Since the 1950s, the basin witnessed infrastructure construction for flood control, hydropower, and irrigation as the government invested in water development projects. A soil conservation campaign was initiated in new terrace croplands on the Loess Plateau along the middle reaches, irrigation diversions were substantially expanded in the lower-reach provinces, and, notably, large dams were constructed in the upper basin.

Much of the infrastructure on the Yellow River is financed and owned by local authorities or power companies and operated independently. The YRCC only controls key reservoirs such as Longyangxia, LiujiaXia, Sanmenxia, Luhun, and Guxian for flood operation, while the others are usually independently operated by the dam authorities (Li et al. 2012).

Although the reservoir system works effectively for Yellow River flood protection and power generation, recurrences of the river drying up in the lower Yellow River began in 1972. In the 27 years from 1972 to 1998, the Lijin Hydrological Station in the lower Yellow River suffered from cutoffs in as many as 21 years; during the 88 cutoffs, the cumulative cutoff was a total of 1050 days, or an average of 50 days for each cutoff year. During the 1990s, cutoffs occurred every year; the most serious was in 1997, during which Lijin Station had an overlong cutoff of 226 days (Liu et al. 2006).

6.3.2 *Phase 2: 1999–2006*

The highest priority for Yellow River reoperation in 1999–2006 was restoration of the river flow. The serious water scarcity and drying up of the river in the 1990s were very much a governance problem rather than results of climate variations. More than 90 % of water had been diverted for irrigation in the upper and middle reaches with low efficiency and low productivity, which endangered water security in the lower reaches. Therefore, governance and operation of the Yellow River infrastructure system required improvement, with the following actions taken for this purpose.

A new pattern was set up for integrated river basin management. By the end of the 1990s, dams on the Yellow River were still basically operated independently, except in cases of flooding emergency. Differing from the previous distributed management, a new integrated Yellow River management pattern was established, under which the central government was responsible for consultation and decision making, the river basin authority was responsible for organization and implementation, and provinces were responsible for water use and water transportation. Total runoff and cross-section discharge are now monitored and controlled, while key infrastructures and reservoirs are regulated in an integrated way (Fig. 6.2).

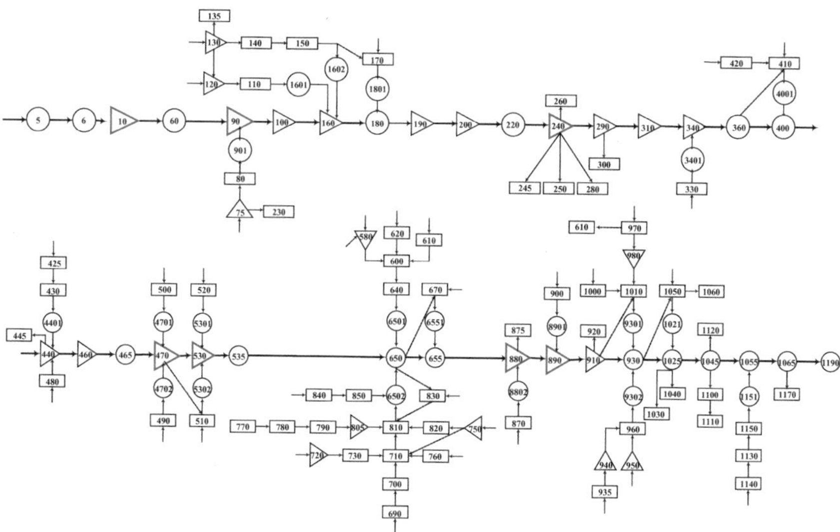

Key reservoirs	Water users	Tributaries
10 Longyangxia	5 Sichuan	1601 Huangshui
90 Longyangxia	6 Gansu	1602 Datong
240 Daliushu	180 Lanzhou	1801 Zhuanglang
470 Qikou	1025 Henan	4001 Dahei
530 Guxian	1045 Shangdong	6501 Fenhe
880 Sanmenxia		6502 Weihe
890 Xiaolangdi		6551 Sushui
		9301 Qinhe

Fig. 6.2 Schematic model of the Yellow River flow and sediment regulation system (*source* Jiang 2009)

Legislative support for Yellow River water allocation is also important. In addition to temporary reservoir operation rules and regulations issued by the YRCC, a state-level ordinance was enacted in 2006 to regulate and control water resources in the Yellow River. This ordinance put Yellow River water resources under state control, with aims to satisfy water demand and improve environmental conditions by ensuring sufficient river flow, especially in the lower reaches. This ordinance identified the accountability of flow volume control at specific river sections; established the legal status of a yearly water allocation plan, a monthly scheme, and a 10-day scheme; and defined the low-flow warning standards at cross-boundary and key river sections and minimum flow standard for tributaries, which improve the effectiveness and accuracy of Yellow River flow regulation.

To promote socioeconomic development in the basin, the ordinance foresees an integrated water allocation scheme. It vests in the YRCC the responsibility of drafting the annual water-use plan in consultation with 11 provinces and autonomous regions. The plan sets the quota for each province according to river flow forecasts. These quotas are then updated on a monthly basis, taking into consideration actual water availability in the river. The provincial governments are responsible for allocation of water resources in their jurisdiction, within the limits of their respective quotas.

Xiaolangdi dam was constructed and operated as a controlling structure for the lower reaches. Xiaolangdi dam controls 90 % of the total catchment area and has a total storage capacity of 12.65 billion m^3, with sediment deposit storage of 7.55 billion m^3 and a long-term effective storage capacity of 5.1 billion m^3. This project was mainly designed for flood and sediment management rather than water resources management and environmental benefits. However, its operation has focused on water scarcity and drought management due to climate changes and increasing water demand from socioeconomic growth, as well as the increasing public awareness of environmental protection.

6.3.3 Phase 3: 2006–2015

In addition to the problem of the Yellow River drying up, declining river health also became a major issue. Li (2004) formulated a policy framework for sustainable management of the Yellow River, listing the major issues to be resolved, including geomorphology and sedimentation, declining water availability, degradation of riverine wetlands, and water quality.

Flushing of Sediment through Dams Because of the extremely high sediment loads of the Yellow River, reservoir operation has to include sediment management, thus prolonging the operating life of reservoirs and also benefiting downstream areas by mitigating sediment deposition and restoring the flood-carrying capacity. A wide range of sediment management techniques are used to preserve reservoir capacity. Two important issues in reservoir operation are flushing sediment from dams and forming density flow (NHI 2012). In the process of sediment flushing,

Fig. 6.3 Sediment flushing (by density currents) in Xiaolangdi dam, Yellow River (*source* YRCC)

simultaneous flushing is achieved by releasing a flushing pulse from the upstream reservoir first and then opening lower-level gates of the dam immediately before the pulse reaches the downstream reservoir, thus releasing the sediment. After sediment flushing is finished, the reservoir is refilled and clear water is released from upper-level gates to rinse deposited sediment in the downstream channel.

Density currents are important for the transport and deposition of sediment in reservoirs. When inflowing water with high sediment concentration forms a distinct current with higher density, that current will flow along the bottom of the reservoir towards the dam without mixing with the overlying, lower-density waters, thereby forming density currents. Both Sanmenxia and Xiaolangdi reservoirs on the Yellow River vent such density currents, along with flushing to discharge sediments (Fig. 6.3).

Environmental Flow The objective of reservoir reoperation is to restore a more natural and seasonally variable flow pattern. This is to reconnect the river to its floodplain within the existing levee system; increase the areal extent, duration, and frequency of seasonal inundation of delta wetlands; identify the magnitude, frequency, duration, and timing of both high and low flows that would most benefit native fish species, including pelagic, estuarine, anadromous, and catadromous species; and protect surface water quality (Gippel et al. 2011a, b).

Climate Variation The reoperation scheme also aims to make the Yellow River more resilient to climate change. An expected result of climate change could be more severe hydrologic extremes—namely, larger and more frequent floods and more severe and longer droughts. With the incorporation of engineering solutions to create more storage to capture the floods and buffer droughts, the reoperation scheme could also realize greater flood storage capacity and additional water supply.

6.4 Impact Assessment

6.4.1 Social Benefits

The direct incentive for Yellow River reservoir reoperation is to keep the river flowing in its 5000-km length. The Yellow River started to suffer from drying up in 1972. This occurred for 226 days in 1997, denying water to 7.4 million acres of farmland and producing a dry riverbed for more than 700 km. The joint reservoir reoperation scheme that started in 1999 has restored river flow and kept the river from drying up in the twenty-first century. At the beginning, minimal flow standards were used to operate reservoirs and maintain river flow, followed by the promulgation of the Regulations of Yellow River Water Regulating in 2006 (State Council of China 2006) that improved the enforcement and accuracy of flow regulation. The flow volumes of river sections were thereby guaranteed and the river now runs to the sea without drying up, even during periods as dry as the 1990s (Table 6.3).

The joint operation of the multi-reservoir system also changed the water consumption pattern of major water users. During the dispatch period, the upstream Inner Mongolia reduced water usage by 7.1 % from 128 % prior to the dispatch to 120 % of its water quota; the downstream Shandong Province reduced its water usage by 8.5 % from 86 % prior to the dispatch to 77.9 % of its water quota. Through optimizing water allocation, improving domestic water supply to riparian provinces and river sections, and alleviating water use conflicts between the upper and lower reaches, right and left banks, and different regions, water utilization disputes have been greatly reduced and water use efficiency and sustainable water management have been effectively promoted (Fu 2010). River depletion especially endangers the sanitary and safe drinking water supply to local communities, who would have to use high-fluoride or brackish water as drinking water. Thus, the regulation ensures the safety of drinking water.

Table 6.3 Flow continuity at the Yellow River mouth during 1995–2003 (Jiang 2009)

Phase II			Phase I			
Year	Minimum flow at river mouth (Lijin) (m^3/s)	Number of zero-flow days	Year of equivalent annual runoff	Minimum flow at river mouth (Lijin) (m^3/s)	Number of zero-flow days	Dry riverbed (km)
1999	3.11	0	1995	0	118	579
2000	8.55	0	1997	0	129	704
2001	6.92	0	1997	0	129	704
2002	25.50	0	1991	0	56	303
2003	51.20	0	1997	0	129	704

Geomorphology and sediment management are key issues for flood safety of the Yellow River. The severe sedimentation in the lower river channels and shrinkage of the riverbed changed the river morphology and introduced the risk of large floods. Xiaolangdi dam is used to simulate a favorable flow and sediment regime to create artificial floods, create a harmonious flow and sediment relationship to reduce river channel siltation, and flush sediments in the lower river channel. Since 2003, water and sediment regulation has been carried out every year. As a result, the riverbed in the lower reaches has been lowered by 1.5 m on average, and bank-full discharge has been improved from 1800 m^3/s in 2002 to 4000 m^3/s.

6.4.2 *Environmental Benefits*

Water pollution was a severe problem in the basin. In 1997, only 17 % of the course of the Yellow River was fit for drinking water. This had a direct negative impact on human health and the basin's ecosystems. Water quality in the main stream of the Yellow River has improved considerably through reservoir reoperation.

The reservoir operation adopted an integrated water quality and water quantity management approach by setting a maximum limit for pollutant discharge, strengthening water quality monitoring on provincial boundaries, and preparing for emergency water release in case of pollution accidents.

According to standard GB 3838-2002 Environmental Quality Standards for Surface Water (MEP 2002), the official target water quality grade for the lower Yellow River is Grade III. The concentration value of each parameter is tested monthly through sampling and compared with the limits associated with each grade in GB 3838-2002. The management target Grade III was achieved 30–80 % of the time prior to 1999, and the year 2000 was the worst year for water quality. From 2003 onwards, water quality has had an improving trend. According to the most recent data, water quality in the Yellow River shows a high degree of compliance with the Grade III standard, with the target achieved 90 % of the time at all stations (Fig. 6.4).

The regulated river flow has also significantly improved the ecological status of the Yellow River delta, which is an important indicator of the river's health. Groundwater level in the delta has continued to rise, the continuous fresh water supply has prevented the delta ecological environment from further deterioration, and the river channel wetlands (which were destroyed due to the river drying up) have regained their previous healthy status.

The ecological value of the internationally important wetlands in the lower Yellow River is heavily dependent on the health of vegetation communities. Vegetation is a source of food as well as habitat; it can improve water quality and influence the hydrology, hydraulics, and movement of sediment in the river. A joint China–Australia research team evaluated the health status of the lower Yellow River in the past decade; they developed an index for wetland vegetation in the delta through analysis of data from Landsat satellite images (Fig. 6.5).

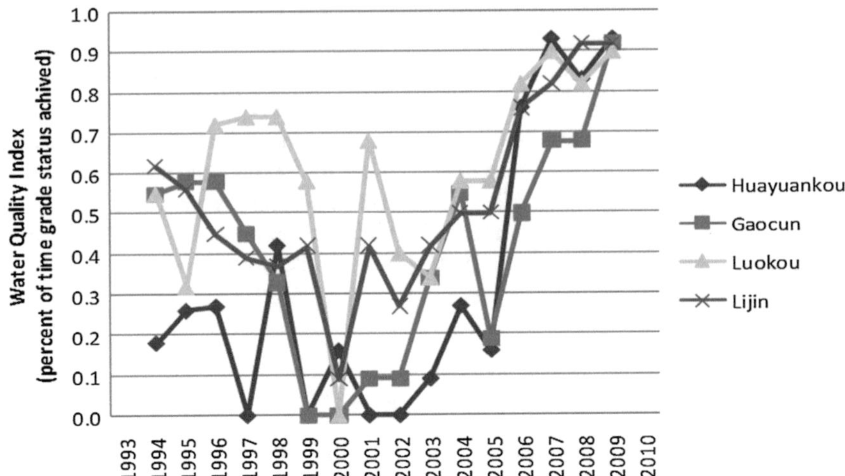

Fig. 6.4 Time series of percent of time that water quality Grade III was achieved in the lower Yellow River during 1994–2009 (*source* Gippel et al. 2011a)

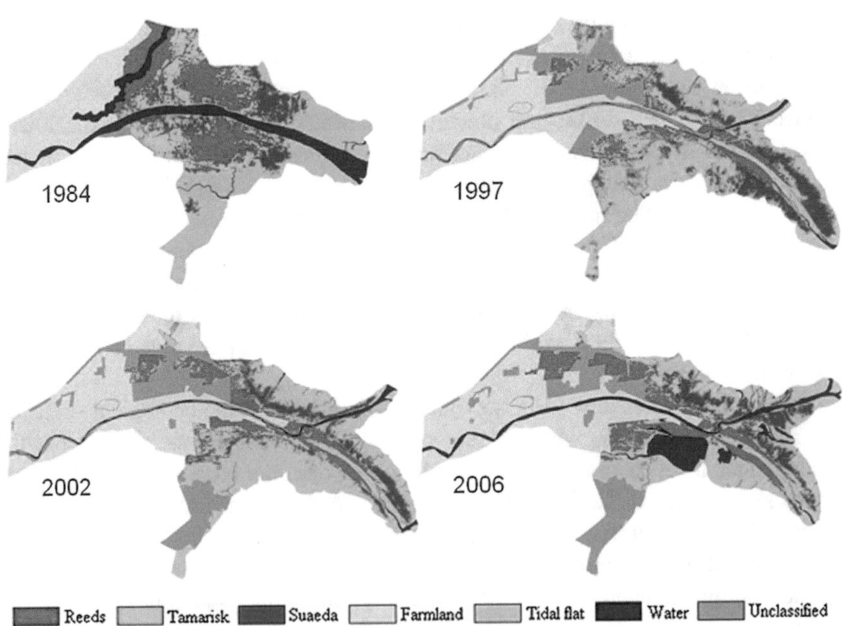

Fig. 6.5 Distribution of main land types on the Yellow River delta over time (*source* Gippel et al. 2011a)

Focusing on three main wetland vegetation types in the delta, the loss of wetland area to agriculture was assessed relative to the conditions in 1984. The results for 1997–2006 suggested a trend of declining and then improving health of the vegetation in the delta, with the improvement coming from an increase in the area of reeds (Gippel et al. 2011a).

During the past decade, the reed area in the Yellow River delta has increased to 52,000 ha, 4200 ha wetlands have regained natural status, 459 wild bird species had nested there by 2011, and the number of species of state-protected birds increased from 187 to 283. Since the spring of 2004, first-grade state-protected endangered species of birds, including the red-crowned crane, white crane, white stork, and black stork, have gradually reappeared in the wetland rehabilitation area within the estuary.

The migration of fish is another key indicator of river health. The maintenance of the environmental flow in the lower Yellow River and the large increase of the delta wetland area ensured a favorable ecological environment for breeding of Yellow River fishes and prompted the resurrection of fish resources. In recent years, the fish population has gradually increased. The northern bronze gudgeon saury that disappeared 20 years ago has reappeared in the Yellow River delta, swimming in healthy shoals.

Flow variations influence the nutrients entering the sea, which are very unevenly distributed throughout the year; they mostly enter the sea in the flood season, with only a small proportion in the nonflood season. Flow regulation changes the annual distribution of nutrients entering the sea, especially increasing the amount of inflow nutrients in nonflood season, thereby ensuring the nutrient supply for growth of aquatic plants and spawning of fish.

6.5 Conclusion

For a long time, the development of the Yellow River had focused primarily on flood control and water supply for agriculture and industry and ignored ecological issues. However, the reservoir reoperation scheme, supported by the administrative, legislative, and technical tools, brought drastic changes, as follows.

The drying up of the river, which was frequent in the second half of the twentieth century, has ended. The Yellow River flow regime has been restored since 1999 and water supply to the riparian cities has been guaranteed for domestic water use, agricultural irrigation, and industrial demand. Water scarcity in the lower reaches was successfully alleviated, while river health and environmental flow have been significantly improved. The marginal economic gain of the increased industrial and domestic water supply has increased productivity by an estimated 2.5–3.5 %. The industrial water supply benefit in the Yellow River basin is estimated at 4.44–9.06 Yuan/m^3, with an average of 6.78 Yuan/m^3. Each year, the increased industrial water supply of 476 million m^3 could contribute an economic benefit of 3.7 billion Yuan to the basin.

Irrigation represents 90 % of the total water usage in the Yellow River basin. Effective water allocation promotes the transfer of water resources towards users with greater efficiency and higher productivity, and optimization of water-saving irrigation and farming patterns has been adopted to sustain agricultural production with reduced water consumption and higher water use efficiency. Since 1999, an estimated annual average of 1.16 billion m^3 of water has been saved from irrigation.

The restoration of flow continuity of the Yellow River generated significant social, political, and economic benefits, as well as enormous indirect benefits. The environmental water use of the lower Yellow River is guaranteed through reservoir operation, including 1.5 billion m^3 of water for sediment transportation in flood season, 0.5 billion m^3 of base flow in nonflood season, and 0.1 billion m^3 of water loss.

Experience with the Yellow River also indicates that the integrated river basin management is crucial to increasing resilience to climate variability and change, and guaranteeing water resources sustainability. River basin authorities play a key role not only for planning, constructing, and maintaining infrastructure but also for operating the river system, including the infrastructure. In a centralized water management system, the YRCC is the governing agency appointed by the Chinese government to manage the river, which makes it possible to develop a basin-wide strategy for sustainable development and climate change adaptation, as well as to build and operate dikes, reservoirs, canals, and water diversions in collaboration with regional governments and various sector stakeholders.

References

Fu X (2010) Social indicators system—river health assessment of the Yellow River section from Xiaolangdi reservoir to delta. Yellow River Press, Zhengzhou

Gippel CJ (2010) River health monitoring, assessment and applications. Technical report 1. ACEDP—river health and environmental flow in China. ACEDP Activity No: P0018. International Water Centre Pty Ltd, Brisbane

Gippel CJ, Jiang X, Cooling M, Kerr G, Close P, Jin S, Li L, Wang L, Sun Y, Pang H, Song R, Sun F, Shang H (2011a) Environmental flows assessment for the Lower Yellow River: site, assets, issues and objectives. ACEDP technical report 6. Australia–China environment development partnership, river health and environmental flow in China. Yellow River Conservancy Commission, Ministry of Water Resources and the International Water Centre, Brisbane

Gippel CJ, Jiang X, Zhang D, Cooling M, Kerr G, Close P, Jin S, Li L, Wang Z, Ma Z, Wang L, Sun Y, Pang H, Song R, Sun F, Shang H (2011b) Environmental flows assessment for the lower Yellow River: ecohydraulic modelling and flow recommendations. ACEDP technical report 7. Australia–China environment development partnership, River Health and environmental flow in China. Yellow River Conservancy Commission, Ministry of Water Resources and the International Water Centre, Brisbane

Hering JG, Sedlak DL, Tortajada C, Biswas AK, Niwagaba C, Breu T (2015) Local perspectives on water. Science 349(6247):479–480

Jiang XH (2009) Yellow River controlling project's impact assessment to the downstream ecosystem and regulation recommendations. Yellow River Press, Zhengzhou

Li G (2004) Keep healthy life of rivers—a case study of the Yellow River. In: Proceedings of the ninth international symposium on river sedimentation. Yichang, China, 18–21 Oct 2004

Li XD, Lou SJ, Li LX (2012) Analysis of the Yellow River's flood management and reservoir operation. China Flood Drought Manage 22(10)

Liu C, Zheng H (2002) Hydrological cycle changes in China's large river basin: the Yellow River drained dry. In: Beniston M (ed.) Climatic change: implications for the hydrological cycle

Liu X, Zhang Y, Zhang J (2006) Healthy Yellow River's essence and indicators. J Geog Sci 16 (3):259–270

MEP (Ministry of Environment Protection) (2002) The national standards of the People's Republic of China. Environmental quality standard for surface water. GB 3838-2002. MEP, Beijing

Muller M, Biswas A, Martin-Hurtado R, Tortajada C (2015) Built infrastructure is essential. Science 349(6248):585–586

NHI (Natural Heritage Institute) (2012) Best management practices for managing sediment flows through reservoirs. Distillation of presentations and consensus of discussion at the reservoir sediment management workshop in Zhengzhou, China, on "harvesting the global state of knowledge on techniques for managing sediment flows through reservoirs", Sept 2012

State Council of China (2006) Regulations of Yellow River regulating. Beijing, 1 Aug 2006

Tortajada C (2014) Water infrastructure as an essential element for human development. Int J Water Resour Dev 30(1):8–19

Xia J, Peng SM, Wang C (2014) Climate change's impact to the Yellow River water resources and adaptive management. Yellow River 36(10)

Xu ZX, Tacheuchi K, Ishidaira H, Zhang XW (2002) Sustainability analysis for Yellow River water resources using the system dynamics approach. Water Resour Manage 16:239–261

YRCC (1997) Master plan of Yellow River harnessing and development (revised version). Yellow River Press, Zhengzhou

Chapter 7
The Durance-Verdon River Basin in France: The Role of Infrastructures and Governance for Adaptation to Climate Change

Emmanuel Branche

Abstract The Durance and Verdon Rivers form a major river basin located in the south of France (Mediterranean climate). The hydroelectric infrastructures in the Durance and Verdon valleys contribute to the supply of water and renewable energy across the region and foster territorial development. Water resources are under a high level of pressure due to significant abstractions for various water uses (irrigation, hydropower, drinking water, industrial use, and recreational activities) and the need to maintain ecological services. The hydropower reservoir is managed to reconcile the needs resulting from all forms of water use and the safety of individuals and property. Electricité de France actively invests in coordinated efforts with all water stakeholders to determine shared action programs for the benefit of all water users, the river, and its ecosystems. The effects of climate stress and the evolution of demographic and socioeconomic territories on water resources and water demand requires the development of adaptation strategies. Among the strategies for climate change adaptation used in the Durance-Verdon river basin are the R2D2 2050 project, the value creation methodology, and water-saving agreements. The Durance-Verdon river basin is an excellent example of how a collaborative approach involving all stakeholders can lead to sustainable water management in the face of climate change as a result of robust water infrastructures.

Keywords Durance-Verdon · Climate change impact study · Water resource · Water uses · Governance · Socioeconomic impacts · Resilience · Value creation · Sharing benefits · Economical approach · Multipurpose infrastructure

---

E. Branche (✉)
Electricité de France (EDF), Sustainable Development Department, Savoie Technolac, 73373 Le Bourget du Lac Cedex, France
e-mail: emmanuel.branche@edf.fr

C. Tortajada (ed.), *Increasing Resilience to Climate Variability and Change*, Water Resources Development and Management,
DOI 10.1007/978-981-10-1914-2_7

7.1 Description of the Durance-Verdon River Infrastructure

This section presents a brief description of the Durance-Verdon valleys in terms of different water users and other stakeholder relations (Branche 2014, 2015). The Durance and Verdon Rivers form a major river basin located in the south of France. The hydroelectric infrastructures in the Durance and Verdon valleys—and the Serre-Ponçon and Sainte-Croix water reservoirs, in particular—contribute to the supply of water and renewable energy across the region and foster sustainable territorial development.

7.1.1 The Development of the Durance-Verdon Infrastructure: Adaptation as a Principle from the Beginning

7.1.1.1 Introduction

The Hydropower Management Unit of the Durance and Verdon valleys runs 30 plants and 17 dams based in the Hautes-Alpes, Alpes de Haute-Provence, Var, Vaucluse, and Bouches-du-Rhône Departments (southeast France). It produces an average of 6.5 billion kWh per year using clean and renewable energy, equal to the annual residential consumption of a city of more than 2.5 million inhabitants. It accounts for 40 % of the electricity produced in the Provence-Alpes-Côte d'Azur (PACA) region. In particular, the Serre-Ponçon and Sainte-Croix water reservoirs—and, more generally, the hydroelectrical infrastructures in the Durance and Verdon valleys—help to meet the various demands for water and renewable energy across the region and in turn foster territorial development. The Unit guarantees fully synchronized operations capable of generating up to 2000 MW power in under 10 min, and thus supplies customers in a competitive and responsive manner.

Construction was started in 1955 by Electricité de France (EDF), who have operated the Unit ever since. The Serre-Ponçon dam and all of the hydroelectric production structures in the Durance and Verdon regions form a source of solidarity-based water management in the Provence–Alps–Côte d'Azur Region. From its very inception, the Serre-Ponçon site—an immense water reservoir holding 1.2 billion m^3 of water—was designed to capture and store water resources and redistribute them to respond to various demands. These include renewable energy, drinking water supply, agricultural and industrial water supplies, and tourist activities developed around the reservoirs that contribute to the region's business activities and attractiveness. A structure unparalleled anywhere in France in size, the Serre-Ponçon dam secures regional water management by averting the consequences of droughts for the Provence region and regulating the formidable flooding to which the Durance has been subject in the past.

7.1.1.2 Initial Studies: A Long-Term Process

The Durance is a large river with a mean annual flow of 180 m^3/s at Mirabeau's Bridge, and 6000 m^3/s when it flooded in 1856 and in 1882 at the same place. In an average year, 2.5 billion m^3 of water passes through Serre-Ponçon and 6 billion m^3 at Mirabeau's Bridge. The Serre-Ponçon dam is 780 m above sea level (m.a.s.l.) and Mirabeau's Bridge is 250 m.a.s.l.

Since the completion of the St-Julien Canal in the late 12th century, humans have tried to use the tremendous potential of the Durance: first, by harnessing water to power oil and grain mills, then for irrigation (which saw a considerable increase in adoption in the 16th century under the leadership of Adam Craponne) and, finally, for drinking and industrial water supply in the 19th century, with the realization of the Marseille channel (canal). However, these achievements remained dependent on the vagaries of the river: devastating heavy floods and often-severe droughts. The intakes had to be reconstructed regularly to avoid shortages during drought.

At the end of the 19th century, the idea of a large, regulating reservoir at the head of the Durance valley was gaining ground. The technical services of the French Administration considered the project. The engineer Ivan Wilhelm was part of the team that, in 1897, chose the Serre-Ponçon site—a rocky outcrop located on the territory of the Savines village—for the construction of a large dam to both manage the heavy flooding of the river and address the occasional intermittency of the flow of the Lower Durance.

Despite the technical difficulties, the project was regularly revisited, mainly because of developments in industry linked to hydropower. In 1909, a license application was filed by the Société pour la Régularisation de la Durance (Company for the Regulation of the Durance) to create a dam in the Serre-Ponçon area. Subsequently abandoned for technical reasons, the project resurfaced again in 1919 under Wilhelm. Twelve surveying campaigns were undertaken between 1919 and 1927, when all of the partners agreed that it was not possible to circumvent the technical difficulties associated with the dam foundation (with the layer of alluvium being very thick).

However, although there was overall agreement on agricultural and industrial needs for the project, no one could imagine at the time that it would be possible to destroy Savines village, which occupied part of the planned reservoir site.

7.1.1.3 The Choice of Dam: A Technical Conclusion to Solve the Problem

The development of hydropower between the two world wars encouraged companies to study the thickness of the alluvium foundation. After the wars, surveys were used to define the shape of the rock: two 100 m deep grooves. It was unthinkable in Europe that these could be sealed using existing materials. An embankment dam was therefore needed. This type of dam was perfectly feasible because it was already commonly in use in the United States.

In 1948, EDF opened a “contest of ideas” to select a technically innovative project for a 120 m high embankment dam sited on 100 m of alluvium. The real problem was, of course, the proper sealing of the alluvium. A unique problem needed a unique solution. The large-scale solution was a world first, requiring the improvement of injection technologies: manufacturing of a thin grout capable of not flowing slowly quickly with the use of very-high-pressure injection techniques and reinjection in installments through tube headlines. In total, 10,000 tons of cement and 20,000 tons of clay were used to seal and to solidify 100,000 m^3 of alluvium.

7.1.1.4 The Link to the Overall Development of the Durance

It is essential to keep in mind that the construction of Serre-Ponçon dam in isolation does not make sense. The Durance-Verdon as a whole should be considered: a series of infrastructures from Serre-Ponçon to the sea, with Serre-Ponçon as the centerpiece. Indeed, the flow of the Durance, once the water is stored, relies on the river slope of 2.7 m/km (i.e., a fairly constant slope from Serre-Ponçon to the sea, 750–0 m.a.s.l. along 280 km).

7.1.1.5 The Impact of the Dam on the Socioeconomic Framework

Considering the long-term interests of the whole Durance River, a large dam at Serre-Ponçon was chosen (storing 1.2 billion m^3) rather than the original site 30 m below. This choice had significant economic and social consequences.

The Serre-Ponçon dam flooded 13 communes and covered an area of 2825 ha, including 600 ha of arable land. It had a major impact on two villages, Savines and Ubaye, necessitating the resettlement of more than a thousand people. Many buildings had to be demolished, including two industrial infrastructures in operation. In addition, 60 km of roads and paths and 14 km of railways had to be removed.

However, the development was never challenged, even though there were many critics of EDF’s all-hydropower policy. Under both energy and planning policies, the Durance infrastructures were supported by those who wanted to fight against the “French desert. At the heart of this position was the promotion of the overall development of the region, to ensure the modernization of these valleys, even if this destroyed village communities.

This position was confirmed by debates organized by the consultation process associated with the development of the law. Indeed, given all the conflicting interests involved, it became essential to the French Government and EDF that the Assemblies expressed the general interest of the French Nation for this development by a public vote. All members of Parliament (MPs) agreed to recognize the benefits of the Durance development with the Serre-Ponçon dam as key. While some were concerned about the fate of the people affected, no one considered that this was a sufficient argument to challenge the project, even among those MPs whose constituents were directly affected.

7.1.1.6 The French Law of 1955

Through the law entitled “Development of Serre-Ponçon and Lower Durance” of 5 January 1955 (Loi du 5 janvier 1955), which was adopted almost unanimously by the two Assemblies, the French State declared the development of Durance to be of public interest and licensed the construction and operation to EDF. In addition to regulating the river, the Assemblies assigned the organization two key missions: (1) the supply of water to the Lower Durance for irrigation and drinking and (2) the development of hydropower for electricity generation.

The French State also decided to establish an agricultural water reserve of 200 million m^3 in the reservoir of Serre-Ponçon to address deficits in the natural flow of the Durance during periods of intensive irrigation. Indeed, the Agricultural Ministry contributed to the financing of the Serre-Ponçon dam.

7.1.1.7 Population Answers to the Dam: An Innovative Approach with Engagement and Support

The completion of the Serre-Ponçon dam did not cause tensions within the population affected by the flooded area of the reservoir. Savines village had already experienced a new form of business diversification through the development of an industrial park, started in the earlier 20th century and still active at the time of the dam construction. The dam was endorsed by policymakers and by the public as a means to escape the crisis that was affecting the industrial sector.

The people accepted a development that served their ideals (i.e., the modernization of France), the general benefit, and the best interests of the French Nation. Furthermore, this development offered a future for this population, designed by the elected officials to ensure the well-being of each and access for all to progress.

The consensus was built on willingness to modernize and improve the standard of living and the desire to reach a new era. All of these were achieved by those communities after the dam construction. A majority of people chose to move and resettlement was individualized. By changing site, farmers left a peasant way of life (in mountainous areas) to enter the world of modern agriculture (on large flat areas). This was a tremendous change in terms of working conditions, leading to social advancement for all those involved.

Other elements confirmed the advancements associated with this development. At Savines, the new village rebuilt on the reservoir banks was renamed Savines-le-Lac (i.e., the reservoir became a lake). As a symbol of entering a new era, it was characterized by modern architectural choices for collective housing and the church. Everything in the new village represented a transition to modern living.

7.1.2 *Water Is a Shared Resource in the Durance-Verdon: Water Infrastructures to Support Sustainable Development of the Territories*

The water resources of the Durance, although relatively abundant, are highly variable. Thus, at the confluence of the Durance and Verdon rivers, annual discharge can vary from 3 to 8 billion m^3 across years. These numbers highlight the difficulty in meeting the needs of all users at all times. A dense network of sensors (flow, snow, rain, temperature, etc.) allow, by means of hydrological models developed by EDF, the evaluation of the water stocks of the Durance and Verdon valleys and its fluctuation over time (Figs. 7.1 and 7.2), as well as the discharge volume and its probability of occurrence.

EDF draws upon a high-performance meteorological forecasting system (over 30 measurement points, identifying snow cover, precipitation, and temperatures across the Upper Durance basin), in order to anticipate future water inflows. Infrastructure management is adjusted around the clock to reconcile the needs resulting from all forms of use and the safety of individuals and property.

It is very important to have a step-by-step approach to address the management of multipurpose water uses of hydropower infrastructures: (1) knowledge, (2) understanding, and (3) management, as presented in Fig. 7.3.

The management of seasonal reserves, such as Serre-Ponçon, follows an annual cycle, which must take into account the hydrological characteristics of the river, the various uses and contractual obligations, and environmental requirements. It must also, while satisfying other uses, find the economic optimum in order to generate electricity at the lowest possible cost and at the right moment (i.e., when the electricity grid needs it most).

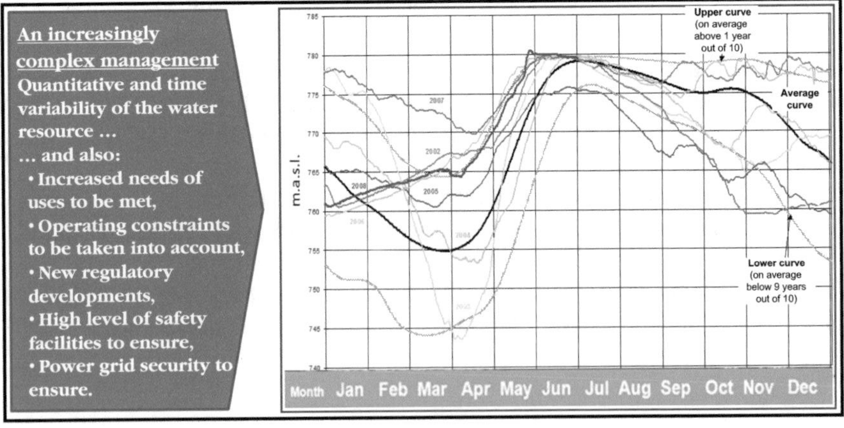

Fig. 7.1 Serre-Ponçon water level from 2000 to 2007

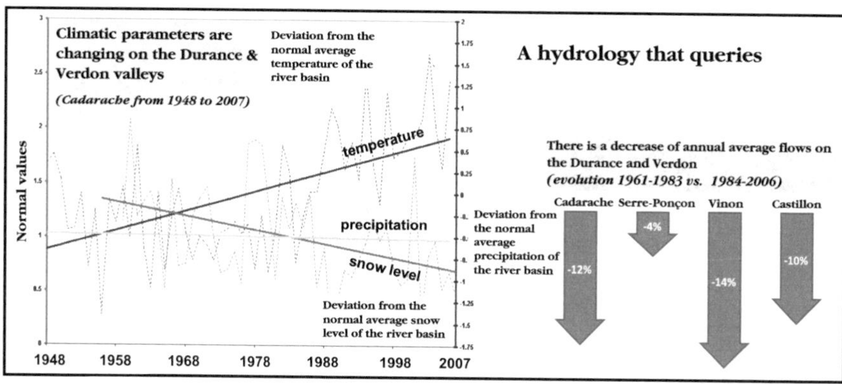

Fig. 7.2 Evolution of the Durance-Verdon inflows from 1948 to 2007

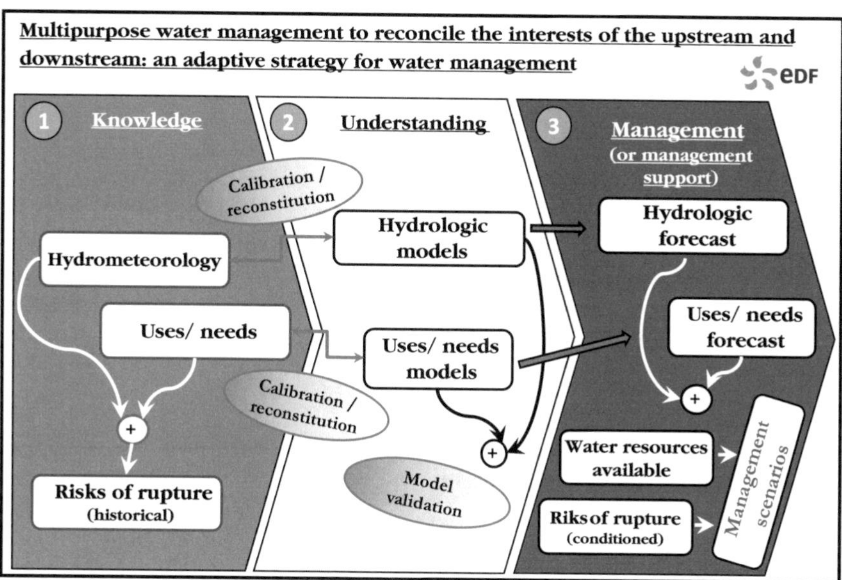

Fig. 7.3 A three-step approach for multipurpose water management

The Serre-Ponçon reservoir is drawn-down during the winter season during peak electrical production. In the spring, power generation is adjusted, according to cumulative inflows, to restore the winter energy stock and agricultural reserves and to reach a water level required for recreational activities on the lake by the beginning of summer. During the summer, after electricity production, discharge in the lower reaches of the Durance provides enough volume for peak irrigation needs, which generally results in a drop in water levels at Serre-Ponçon, but which nevertheless remains compatible with tourism in the majority of cases. The cumulative autumn inflows supplement storage and prepare the stock for a new winter. Drinking water is provided continuously throughout the year.

The management of the Serre-Ponçon reservoir has now acquired a certain maturity and demonstrates the clear will of the stakeholders to show solidarity in the most critical situations (e.g., droughts) to the satisfaction of tourism professionals involved in the reservoir. However, nothing is written in stone in the context of climate change, which has resulted in a sharp rise in environmental, social, and economic issues or new uses.

Figure 7.4 presents a typical filling curve of the Serre-Ponçon reservoir according to its many uses (drinking water, hydropower, agriculture and tourism). Tourism, a new purpose, is not a contractual condition but is also included in current reservoir management objectives.

EDF actively invests in coordinated efforts involving all of the water sector stakeholders in the territory to determine shared action programs for the benefit of all water users, the river, and its ecosystems.

The **SHARE concept** was used in Durance-Verdon (Branche 2015) and is a good example of the concept in action. The following principles were developed for the different water uses of hydropower reservoirs: shared vision, shared resource, shared responsibilities, shared rights and risks, and shared costs and benefits. The SHARE concept also gives guidance for **S**ustainability approach for all users, **H**igher efficiency and equity among sectors, **A**daptability for all solutions, **R**iver basin perspectives for all, and **E**ngaging all stakeholders.

Fifty years after its implementation, the Durance and Verdon program has achieved the economic development objectives that were originally assigned by creating wealth in the entire region, as well as value for all beneficiaries from an economic point of view and in terms of job creation (Fig. 7.5).

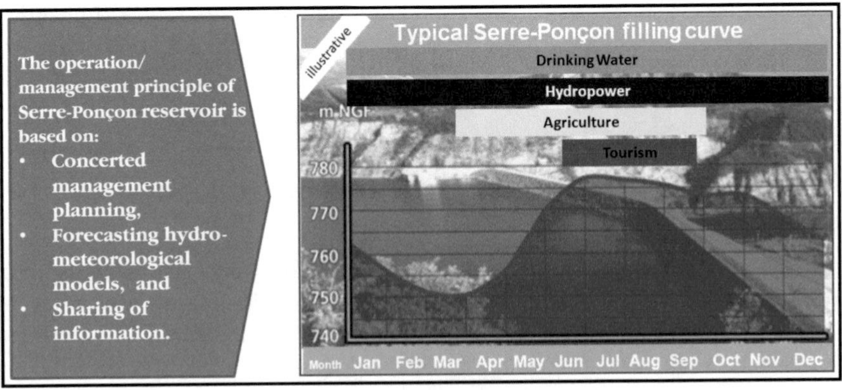

Fig. 7.4 Typical Serre-Ponçon filling curve for different water uses

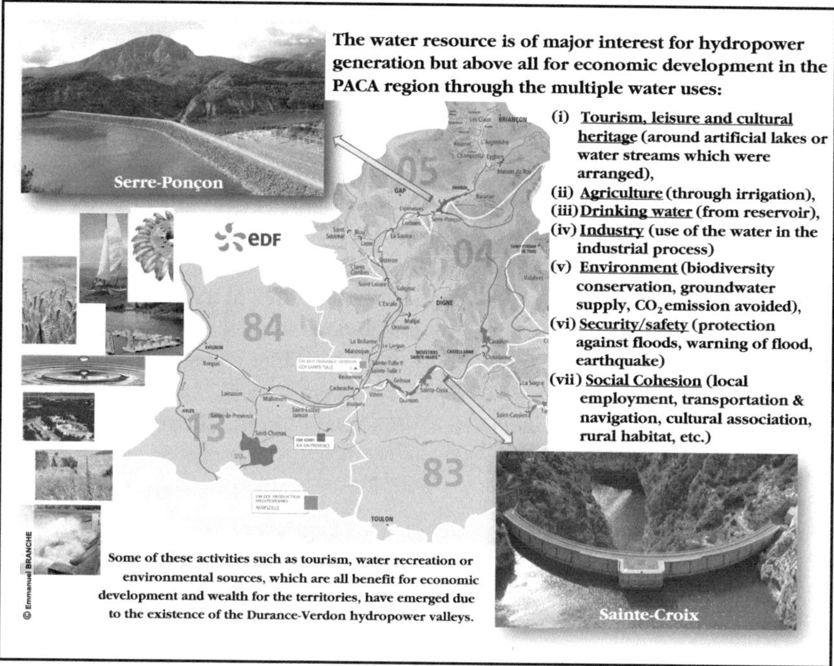

The water resource is of major interest for hydropower generation but above all for economic development in the PACA region through the multiple water uses:
(i) Tourism, leisure and cultural heritage (around artificial lakes or water streams which were arranged),
(ii) Agriculture (through irrigation),
(iii) Drinking water (from reservoir),
(iv) Industry (use of the water in the industrial process)
(v) Environment (biodiversity conservation, groundwater supply, CO_2 emission avoided),
(vi) Security/safety (protection against floods, warning of flood, earthquake)
(vii) Social Cohesion (local employment, transportation & navigation, cultural association, rural habitat, etc.)
Serre-Ponçon
EDF
05
04
84
13
83
Sainte-Croix
Some of these activities such as tourism, water recreation or environmental sources, which are all benefit for economic development and wealth for the territories, have emerged due to the existence of the Durance-Verdon hydropower valleys.
© Emmanuel BRANCHE

Fig. 7.5 Economic development of the territories from Durance-Verdon infrastructures

7.1.3 The River Basin Management Plan: An Appropriate Tool to Address Extreme Climate Events in France

River basin management plans (RBMPs) are effective tools for addressing floods, water scarcity, and droughts. RBMPs required by the European Union Water Framework Directive offer considerable potential to address these extreme events. They become even more useful under a changing climate, in which additional stress is put on water resources.

In France, the purpose of the *Schéma Directeur d'Aménagement et de Gestion des Eaux* (SDAGE; i.e., River Basin Management Plan) is to apply the principles of the French law of 3 January 1992 (Loi du 3 janvier 1992) to major hydrographic basins.

The SDAGE scheme was drawn up by the Basin Committee together with national, regional, and local governments, water users (industry, agriculture, tourism, etc.), and associations (nongovernmental organizations). It is managed by the Agence de l'Eau, the French water agency. SDAGE serves as a general framework for the development and management of the water of each hydrographic unit, system or aquifer. The plan is part institutional, but it is also consulted in crisis situations. It is driven by the aims of the water law of 3 January 1992, inviting stakeholders to work together to seek a balanced sharing of water resources for preserving the environment.

The Basin Committee brings together stakeholders (elected representatives, users, and administrations) to deliberate and make decisions in water resource matters in the interest of all. For the Durance-Verdon valleys, French law provides an excellent framework to initiate dialogue and set up appropriate governance and consultation processes; these valleys have used an innovative consultation process since the 1950s. It should be noted, however, that this is not the case everywhere in France. Figure 7.6 presents a schematic vision of the Basin Committee and authorities in France.

The first Basin Committee was launched in 1996 for the Rhone-Mediterranean Basin. The previous SDAGE produced by the Basin Committee in 2008 covered the period of 2009–2015. For the first time, the new SDAGE for 2016–2021 (SDAGE 2015) sets climate change adaptation as a main priority for the Rhone-Mediterranean basin. Vulnerability maps highlight vulnerable areas in terms of water availability, the drying of soils, biodiversity, and water eutrophication. RBMPs encourage all stakeholders to act according to the following:

1. Mobilizing stakeholders for the implementation of adaptation actions to climate change
2. Developing new facilities and infrastructure: keeping focused on the objectives and planning for the long term
3. Developing prospective vision in supporting the implementation of adaptation strategies

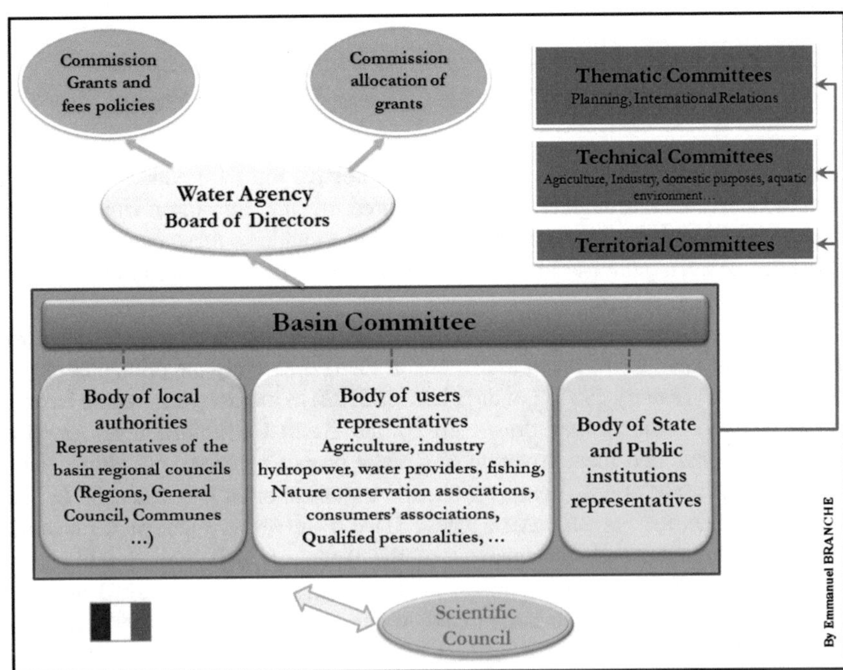

Fig. 7.6 The Basin Committee organization in France

4. Acting with solidarity and in a concerted way
5. Refining knowledge to reduce uncertainty margins and proposing effective adaptation measures.

For infrastructure development (i.e., item 2 above), it is crucial to avoid "bad adaptation," which may have important implications for environmental, economic, and social elements:

- Adaptation is primarily achieved through changes in behavior and practices (urbanization respecting the proper functioning of spaces, choosing crop varieties adapted to climatic conditions, etc.).
- Developments and investments should be reversible whenever possible and consider the long-term changes due to climate change.
- Considering the uncertainties attached to the forecast conditions, stakeholders must be cautious in adopting significant measures that impact sustainable issues.
- Actions and developed activities should not lead to increased vulnerability of the territories and aquatic environments to the vagaries of climate change.
- Adaptation measures should be flexible and progressive to allow their reappraisal and refinement in light of the actual effects of climate change based on the development of scientific knowledge.

Moreover, any development or infrastructure must comply with the objective of not damaging resources as defined in the basic orientation for the resilience of aquatic environments.

7.2 Examples of Adaptation to Climate Change on the Durance-Verdon Basin

This section presents three main examples to highlight the role of infrastructure and governance in the context of adaptation: (1) the R2D2 2050 project, (2) water-saving conventions, and (3) the value creation approach.

Existing infrastructures in the Durance-Verdon have been built under the premise of hydroclimatic stationarity. French engineering standards and dam safety regulation are changing with climate change projections, leading to programs to upgrade and recondition existing dams (e.g., dam spillway upgrades), but there are no plans for new dams to date.

7.2.1 The R2D2 2050 Project: A Partnership Project to Assess the Possible Impacts of Climate Change on the Durance River Basin

The R2D2 2050 program is a partnership project aimed at assessing the possible impacts of climate change on the quantity and quality of water resources, biodiversity, and the changing demand and uses in the Durance-Verdon river basin in 2050 (Roux 2015; Sauquet 2015). The goal is to inform the communities and public authorities about the measures required in order to adapt to climate change. This project sought to engage the stakeholders in a co-construction scenario goals for future water demands and to share the results of the project, increasing ownership of the key findings.

7.2.1.1 Introduction

The evolution of water resources and water demand under climate stress and demographic and socioeconomic changes in territories may require the development of adaptation strategies (water savings are the most obvious lever, followed by crop choice and development scenarios). This will ensure long-term effectiveness of a model to serve the general interest and to find answers to reduce water-use conflicts that could arise if nothing is done. Access to water is a major strategic challenge for all stakeholders. The sound structural design of water infrastructure allows water users time to adjust governance to find and implement solutions to preserve the future supply. In a time of changing needs, EDF is committed, alongside its partners, to preserving water resources and takes an active part in forward-looking studies on the impacts of climate change.

Water management with the Durance-Verdon river basin will have to face global change that may put into question the sustainability of the current water allocation program. Although rarely mentioned, the impact of climate change on energy production is an important issue, as is its impact on other water uses and the environment.

Several EDF teams contribute to a large-scale program dubbed the **R2D2 2050 project** (risk, water resources, and sustainable management in Durance in 2050). This is a partnership project aimed at assessing the possible impacts of climate change on the quantity and quality of water resources, biodiversity, and changing resource demands and uses. EDF works in collaboration with the National Research Institute of Science and Technology for Environment and Agriculture (IRSTEA), which is a public scientific and technical institute directed conjointly by the Ministry of Research and the Ministry of Agriculture and is holder of an agreement with the Ministry of Ecology, the co-signatory of its constitutive decree. Its goal is to inform communities and public authorities about the measures required in order to adapt to one of the 21st century's greatest challenges.

Funded under the *Gestion et Impact du Changement Climatique* (GICC) program of the French Environmental, Sustainable Development and Energy Ministry (MEDDE) and the Rhone-Mediterranean and Corsica Water Agency, the R2D2 2050 project was implemented by seven partners coordinated by IRSTEA: IRSTEA, EDF R&D LNHE Chatou, EDF DTG Grenoble, Pierre and Marie Curie University Paris, LTHE Grenoble, Société du Canal de Provence, and ACTeon. The project lasted 3 years (2012–2015), with the final report published in October 2015.

Research activities were conducted in close collaboration with key stakeholders in the area. They involved targeted interviews and local workshops on themes that complemented the prospective workshops planned.

A Steering Committee for local interaction comprising key stakeholders in the area included the Provence-Alpes-Côte d'Azur (PACA) region, the Rhone-Mediterranean and Corsica Water Agency, and the Regional Directorate for Environment, Physical Planning and Housing (DREAL). This kept the stakeholders regularly informed on the project implementation and results.

A chain of models was developed (Fig. 7.7) to simulate the climate on a regional scale (given by 3300 projections obtained by applying three downscaling methods), water resources (provided by six rainfall-runoff models forced by a subset of 330 climate projections), water demand for agriculture, and water supply for domestic purposes. The models covered different sub-basins of the Durance River basin upstream of Mallemort under present-day and future conditions. A model of water management was developed to simulate reservoir operations for the three main dams (Serre-Ponçon, Castillon, Sainte-Croix) under present-day conditions. This model simulates water releases from the reservoir under constraints imposed by rule curves, ecological flows downstream of dams, and the water levels required in summer for recreational purposes. Four territorial socioeconomic scenarios were also elaborated with the help of stakeholders to project water needs in the 2050 s for the areas supplied with water diverted from the Durance river basin.

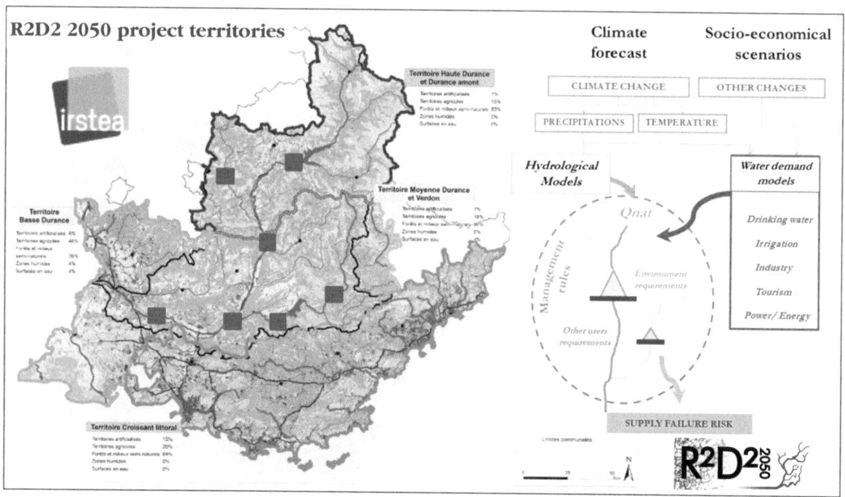

Fig. 7.7 R2D2 project location and chain of models

A multi-model and multi-scenario approach was used to assess uncertainties.

- **Three methods for climate projections** were used for generating statistical downscaling applied to a selection of large-scale model outputs (GCM), all based on the approach by analogy introduced by Lorenz (1969) and already tested in the ANR RIWER 2030 (Hingray et al. 2013; Lafaysse et al. 2014).
- **Seven hydrological models** of different structures and sensitivities under climate change were selected to simulate this resource:
  - CEQUEAU (Morin 2002)
  - GR5 J (Le Moine 2008)
  - Isba-Durance (Lafaysse 2011)
  - CLSM (Ducharne et al. 2000; Magand 2014)
  - MORDOR (Garçon 1996)
  - J2000 (Krause et al. 2006)
  - ORCHIDEE (Krinner et al. 2005).
- **Three irrigation water models** were used by the project partners:
  - FIVE-CoRe (Chopart et al. 2007)
  - MODIC (Sauquet et al. 2010; Braud et al. 2013)
  - SiSPAT (Braud et al. 1995).

This project models the climatic dependence of the Durance-Verdon system: availability of the natural resource, water requirements for nonenergy uses, and energy production (Fig. 7.8).

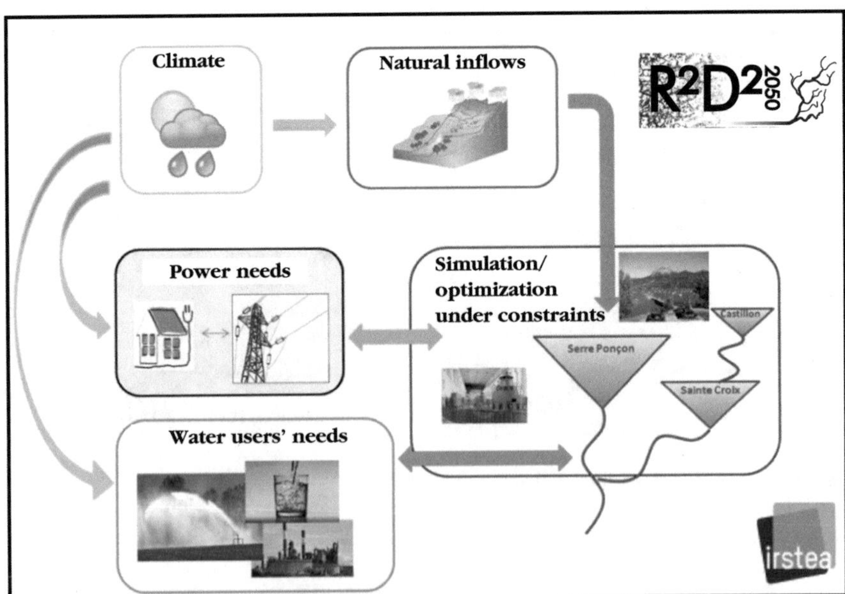

Fig. 7.8 R2D2 simulation/optimization approaches

A hydraulic reserves management model was used. It is based on EDF's operational practices, including anticipation of snow melting. The modelling process was validated using historical data.

A comparison between the present and future was made under the current management arrangements. This comparison was made without capping solicitation reserves but complying with the filling targets. It identified the limits of the integrated management model of the Durance and Verdon under the concession and current commitments and provided lessons and work areas for reducing water conflicts and tensions among users.

This multi-stakeholder research project R2D2 2050 provides insights on the following:

- Changes in the hydrological regime of the major rivers of the watershed and the water supply
- Applications to current and future water use (hydropower, agriculture, tourism, etc.), including the main aquatic ecosystems, local or external, which put pressure on the Durance-Verdon water resource
- Potential future imbalances resulting from the confrontation supply/demand under scenarios of climate change and socioeconomic development
- Leeway and management alternatives to ensure "balanced and sustainable" management of water resources in line with the objectives and challenges of planning
- Uncertainties associated with the main results obtained, as well as the relative importance of potential sources of uncertainty.

7.2.1.2 Main Results

Regarding the **evolution of the climate** in the whole basin, based on the 330 forecasted scenarios, it is envisaged for the projection period that the Durance-Verdon basin will face the following:

- An average increase of +1.6 °C over the whole basin: +1.4 °C in winter, +2.2 °C in summer
- An increase of potential evapotranspiration of approximately 50 mm
- An uncertain trend in terms of rainfall with high spatial dispersion.

Regarding the **future of the water resources**, the main results highlight the following:

- Reduced snowpack and an anticipated merger affecting downstream flows
- The most important changes are to be expected in spring flows (melt season and filling)
- A reduction in summer low flows

- No clear signal for winter flows
- An uncertain future flood regime but a dynamic likely to change the current flood regime dominated by snow-melt.

Hydrological models converge on the following key figures:

- A reduced snowpack, as a consequence of rising temperatures
- A snow peak observed earlier in the year, reduced by at least 80 mm—an equivalent loss of 280 million m^3
- Changes in maximum water volume stored range between −250 and 50 mm
- A reduction in summer low flows, on the order of a 20 m^3/s reduction in the average flow in August and an increase of 20 days in time under the current threshold Q95 at Cadarache
- An evolution of the annual resource around −20 m^3/s (between −70 and 40 m^3/s), which is equivalent to an average decrease of −600 million m^3/year (between 1200 and −2200 Mm^3) at Cadarache—that is, a reduction of about 10 % of the current resource
- An increase in "enhanced crisis" days, meaning more frequent use restrictions.

Regarding the **forecasted filling curve of the reservoirs**, it is expected that there will be:

- Fewer opportunities to lower the water level for energy during winter
- A dominant influence of climate on all scenarios
- A requirement to reach the recreational water level earlier
- An increased risk of spillage during flood events.

7.2.1.3 Conclusions

Air temperature plays a predominant role in hydrology in climate change scenarios, resulting in the following:

- Reduced snowpack
- Early melting period
- Increased evapotranspiration.

Changes in rainfall and flood regime are uncertain.

It is also expected that flows will be lower in spring and low water levels will be more severe and longer. A decrease in overall water demand across the territory is forecasted, which is more dependent on territorial changes than climate change. This is especially likely in the Durance River as the demand on the Verdon grows. There will be contrasting effects on uses between Durance and Verdon.

The physical design of the reservoirs can meet water use demand; however, the storage capacity is limited by the available volume of reserves, especially on the Durance River.

Hydropower generation, being directly affected by the reduction of natural inflows, undergoes a significant reduction of its flexible generation capacity due to changes in flow regimes and filling objectives resulting from climate change.

The tourism management of the Verdon River could be deeply impacted. Serre-Ponçon reservoir operation will be highly dependent on water savings, the management operated by EDF, and the capacity of local actors to develop the lake banks. The spill risk during flood events could be increased and will have to be taken into account to control the associated flood risk.

There are some limits to the R2D2 2050 program. Indeed, this project was unable to consider all components of water management. For instance, assessments of future environmental quality (water temperature, dilution, chemical kinetics) needed for ecological analyses were not undertaken. Groundwater recharge was not examined in this study. Climate projections were based solely on the SRES A1B scenario. These limits are all potential future research topics. More information is clearly required to understand the impact of global climate change and the role of multipurpose water infrastructures for climate resilience.

7.2.2 *The EDF Value Creation Methodology: An Effective Tool to Manage Infrastructures in a Sustainable Way*

7.2.2.1 Introduction and Background

Existing and planned hydropower projects present multiple opportunities to create environmental and socioeconomic value for their host communities and regions (Schumann 2015). EDF, the operator of more than 450 hydropower assets and manager of 75 % of France's surface water resources, is currently conducting assessments to integrate stakeholder value into project planning and operation.

Following a commitment taken at the 6th World Water Forum in 2012 to create value around power projects locally, the hydropower division of EDF has launched a project to develop a methodology to identify and evaluate created benefits of hydropower projects and to take them into consideration in operation and planning. This work is of particular interest to EDF in the French context, where some valleys have a very high density of hydropower projects and where the hydropower operator is often the sole or one of the main large industrial stakeholders in the region and therefore carries responsibility for the development of the surrounding communities.

The project pursues the following four objectives:

1. Identification of created and destroyed socioeconomic and environmental values related to hydropower installations
2. Analysis of the contribution of the hydropower operator to the value created
3. Evaluation in qualitative, quantitative, and, if possible, monetary terms of this value

4. Development of didactic ways to present values and to facilitate discussion with stakeholders.

Estimating the value of water and land uses around hydropower projects is a complicated endeavor. A broad range of values linked to hydropower complexes exists, which of course cannot always be found for all hydropower projects: some are related to water uses, some to nonwater uses; some are made possible because of the infrastructure; others depend on the operators and their collaboration with the surrounding stakeholders. To complicate things, no metric is common to all values and the assessment lists values as varied as greenhouse gas reduction, employment creation, flood protection, and/or number of affected fish species. These can neither be added nor meaningfully opposed to one another. Given the sensitivity of many of these topics, a systematic translation into monetary values is not only a methodological challenge but can also become a meaningless expression in numbers of a topic of potential environmental or social significance.

Five overarching dimensions for assessment, which each contain different types of values, were identified as applying to most sites:

- Electricity services
- Socioeconomic values
- Societal values
- Environmental values
- Risk management, including flood and drought.

This methodology is thus important and useful to put a value on these services, but the value creation approach also allows better integration and decision-making processes for water users in the context of climate change adaptation.

7.2.2.2 Results of the Durance-Verdon Value Creation Assessment in 2015

In addition to the initial purposes for construction (i.e., flood management, irrigation, and hydropower production), the Durance-Verdon hydropower system contributes significantly to the socioeconomic development of the southeast of France. The assessment conducted in 2015 shows that these hydropower infrastructures provide a wide range of socioeconomic benefits, including the following:

- The electricity produced represents between 30 and 50 % of the total electricity consumption of the PACA Region. A thousand jobs (direct, indirect, and induced) are related to this production activity.
- The storage infrastructure supports the delivery of 27 % of the total drinking water supply of the PACA region (3 million people). In addition, it supplies water to around 91,000 ha of irrigated land (mainly orchards, vegetables, and forage crops) that support directly or indirectly around 20,000 jobs.

- Tourism also benefits directly or indirectly from water and from the way water-related infrastructures are managed (e.g., stabilizing water levels in storage lakes so recreational activities can take place, controlling river discharges in line with requirements of kayaking or canoeing). Specific agreements exist between EDF and the actors of tourism. Summer tourism alone generates around EUR 400 million/year of turnover and more than 4000 jobs, with tourism activities being particularly important around the two lakes of Serre-Ponçon and Sainte-Croix.
- The system also provides cooling water for two nuclear energy research and development sites (Commissariat à l'énergie Atomique de Cadarache and International Thermonuclear Experimental Reactor). When combined, these two centers directly and indirectly generate more than 6000 jobs and around EUR 250 million in the PACA region.
- The Durance-Verdon hydropower system helps address climate change challenges by avoiding greenhouse gas (GHG) emissions. For example, it is estimated that the emission of 2.1 Mt/year of CO_2 is avoided due to the electricity produced by the hydropower system (instead of thermal power plants). The EDF Group is finalizing its methodology for GHG emission reduction associated with hydropower operation.

The following figures show these created values in their spatial (Fig. 7.9) and temporal (Fig. 7.10) contexts. The value of the spatial representation is that stakeholders and decision-makers can quickly understand where in a given region a certain value was created or destroyed. The temporal figure adds context to the

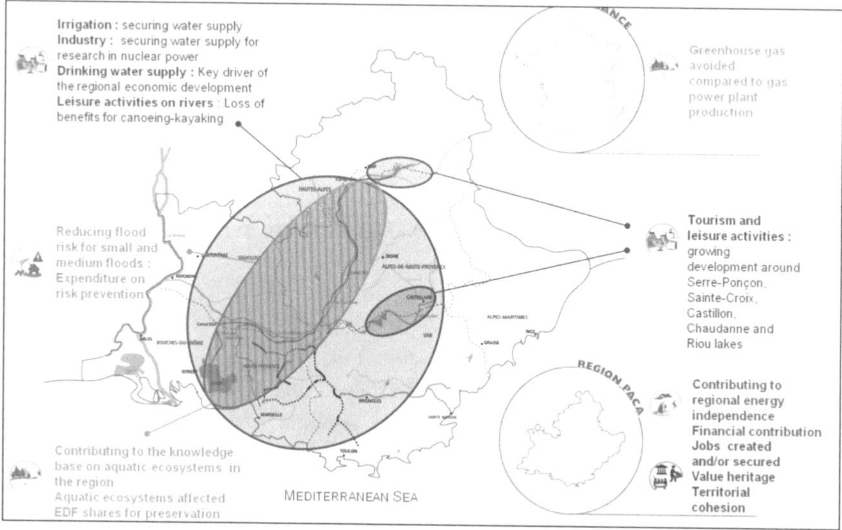

Fig. 7.9 Spatial distribution of values created around Durance-Verdon

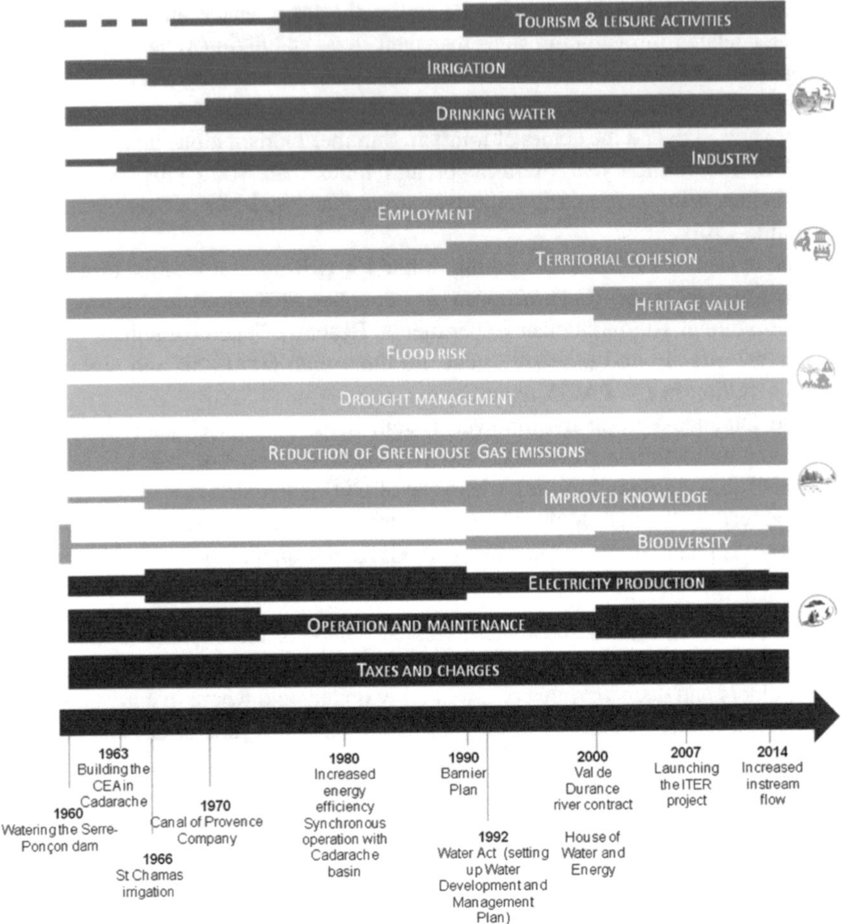

Fig. 7.10 Spatial distribution of values created around Durance-Verdon

temporal development and shows that with the infrastructure created, and also with societal changes, some values only appear over time but can be just as important.

7.2.2.3 Additional Benefits of Using This Methodology

The results show significant socioeconomic values that often exceed the financial and economic contribution of power-related services and tax payments to the region. The Durance-Verdon scheme shows an example where associated economic activities create or secure several hundred million euros of annual turnover and thousands of jobs. When considering investments or changes in water releases, these decisions cannot be solely based on power production but have to be made in concert with regional stakeholders. Moreover, consideration of the significant economic output of some of the associated activities can shed a new light on judgements on whether or not a new hydropower development will be profitable. One of the objectives of this project is the establishment of a systematic assessment tool with easily understandable visualization methods that facilitate dialogue with stakeholders. Finally, the results of this value creation work will be used to engage with stakeholders around the created values and to identify pathways for optimization. This will be particularly important in the context of climate change and climate resilience, as for the R2D2 2050 project.

7.2.3 The Water-Saving Conventions: A Voluntary Approach to Foster a Win–Win Situation Using Existing Infrastructure

As already mentioned, the management of the Serre-Ponçon infrastructure has multiple objectives. EDF is required to deliver 200 Mm^3 of water to irrigators between 1 July and 30 September annually (as the Ministry of Agriculture financed part of the dam construction), and an information bulletin is sent each week to farmers about irrigation flows. A first water-saving convention was initiated in 2002 between EDF and Canaux de Vaucluse. This replicable and innovative voluntary economic approach is based on the following:

1. An obligation to deliver the requested value, with obligation of results
2. A reviewable water-saving target set by the Canaux de Vaucluse
3. Compensation to be paid by EDF if the goal is reached
4. A quick degression to be applied if the target is not reached
5. Incentives to reach the target and to go beyond.

The Water Saving Convention was signed in 2003. The agreement requires EDF to pay back a part of the saving costs if the targeted objectives are reached. EDF encourages farmers to save water by financing modern systems to reduce water consumption. Through this, the agricultural consumption for one partner decreased from 310 Mm^3 in 1997 to 220 Mm^3 in 2012; 60 million m^3 of this has been achieved since signing the agreement.

This is a win–win solution for stakeholders. Indeed, thanks to this convention, significant water savings were reached (30 % reduction of water by the irrigators) and benefits were achieved in several sectors:

- **Irrigators**: attractive compensation and improved control of the Serre-Ponçon reservoir
- **Energy**: better seasonal use with the saved stored water volume (which could be generated during peak hours) but a limited energy gain (as those savings are located downstream end of the chain; the valuation is recovered 20 % on the energy side and 80 % regarding appropriate time generation)

- **Environment**: water savings that were achieved by EDF means more water is released to the environment rather than being consumed
- **Tourism**: maintained to keep the reservoir water level of Serre-Ponçon.

The evaluation of energy loss is based on a method developed jointly by EDF and the Rhône-Méditerranée-Corse Water Agency, who provide part of the financing (Branche 2014). It is a simplified method that provides a reference point in negotiations between EDF and other users for the cost of hydropower generation under new external constraints or uses. It consists of evaluating losses due to lost or shifted energy, with respect to a year divided into five specific periods (Fig. 7.11), and then evaluating them on the basis of a representative price of electricity market futures. The idea is to compare two scenarios (Fig. 7.12): one based on current water withdrawn by the irrigators and another one with x Mm^3/year of water savings.

The first category (peak hours) occurs during the winter period, from December to February. The fifth category, from March to November, covers the least expensive hours of the year. There are never more than four categories during a

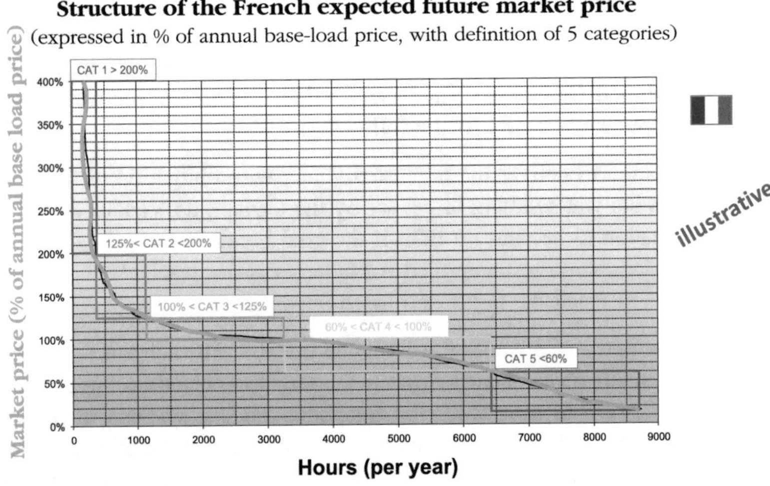

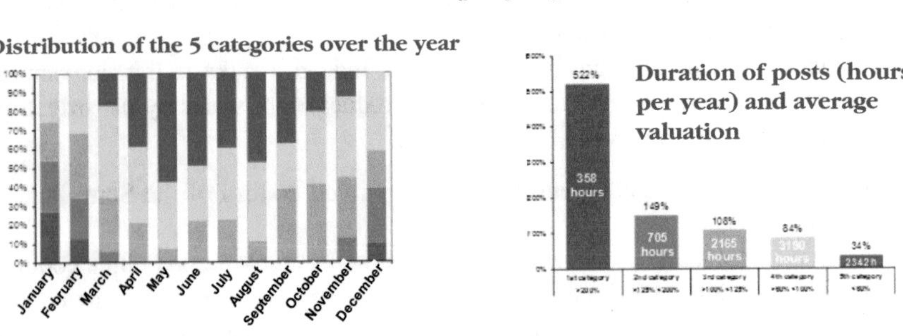

Fig. 7.11 Description of the five categories used in this approach

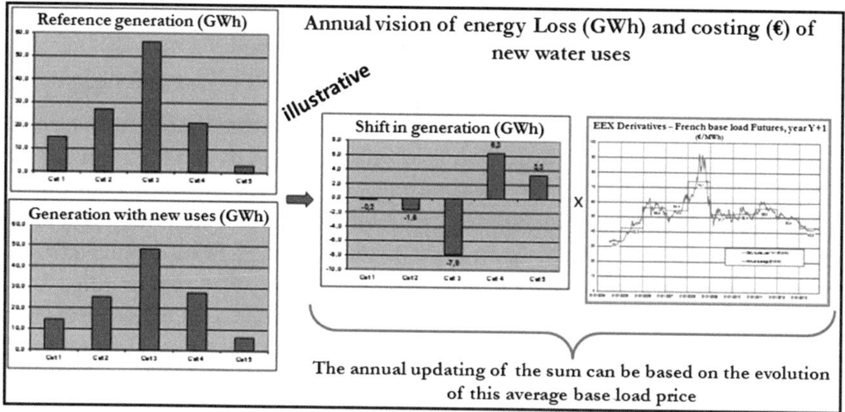

Fig. 7.12 Illustrative representation of the energy loss methodology (Branche 2014)

given month. The power generation profile in relation to market prices results from the optimal management of the reservoir (using EDF software). It is necessary to quantify the losses or the shift of hydroelectric power produced that is induced by the new use or water savings. For example, if a minimum low water level is to be implemented, production over the summer period and during off-peak hours (categories 4 and 5) will be increased and thus the reserves for autumn and the beginning of winter will be lower (generation of category 1, 2, or 3 will also be lower). For any given measures to be implemented, the differences in output (positive or negative) for the five categories are expressed as the percentage of base load price corresponding to each category (e.g., using European Energy Exchange [EEX] power market prices as a reference, which is easily accessible and understandable by all stakeholders). The cost itself is then obtained with reference to a public price for the years to come. This quantification must include all impacted power plants and find a consensus between all stakeholders; for example, in the case of water release in Durance, the Rhône-Méditerranée-Corse Water Agency and EDF decided to co-finance the volumes of water restored to the river.

This successful first convention has created a real dynamic. Indeed, a new convention was signed in 2014 in order to reach an additional water-saving objective of 20 Mm^3 by 2022.

Water saving is becoming an essential issue on the Durance-Verdon basin, and water stakeholders are working together on incentive mechanisms to manage these water efficiencies in a sustainable way. The issue is now focused on the future of these water savings: environmental requirements and water for scarce territories, and a daily operation to meet multipurpose requirements. For instance, EDF in association with the Water Agency and DREAL are currently working towards basin-level water savings. Innovative mechanisms are being discussed so that a part of these water savings could be distributed back to the environment and to water-deficit areas in the basin.

7.3 Conclusion

The Durance-Verdon river basin in France is an excellent example of how a collaborative approach involving all stakeholders can lead to sustainable management of water in the context of adaptation to climate change. It is very important to develop a broad consultation of water management stakeholders in the economic, social, and environmental dimensions. This will allow issues to be shared and the future to be built collectively. A capacity-building strategy is essential. Indeed, it provides visibility to all users on whether the evolution of water resource management will satisfy their needs, thus preventing conflicts. To reach this target, it is essential to provide the means to build this shared future vision. The optimum outcome involves considering every user's interests (agriculture, tourism, hydropower, drinking water production, fishing, etc.) to determine the best balance for the public interest.

In the Durance-Verdon basins, the development of water management and storage infrastructure has played a critical role in water supply and socioeconomic development. The southeast of France has greatly benefited from this in terms of employment, tourism, industrial applications, and pollution reduction. Reservoir management has already been adapted to take into account new purposes, such as recreational activities. This will be even more essential in the future in the context of climate change. In this Mediterranean region, water resources are under high pressure due to significant abstractions for various uses (irrigation, hydropower, drinking water, industries, recreation and ecological services). Water management in the Durance-Verdon river basin will face global changes that may question the sustainability of the current water allocation practices. For the first time, adaptation to climate is set as a key priority in the 2016–2021 River Basin Management Plan.

Although rarely mentioned, the impact of climate change on energy production is a real issue and it also impacts other water uses and the environment. It is essential to communicate a shared vision of this situation with all stakeholders in order to achieve sustainable water use. This will be essential in the future for multipurpose water infrastructures. Hydropower reservoirs are at the heart of the water–energy–food nexus. They are already playing a key role in adaptation to and mitigation of climate change. These multipurpose infrastructures will play an even more critical role in the future and therefore for a cornerstone for increasing resilience to climate change.

It is essential to consider different expectations that are sometimes contradictory for multipurpose water management, in an evolving regulatory, legislative, and climate change context. The objective is to ensure a wide balance between economic, social and environmental issues. However, sharing water is complex: the overall management of this resource involves managing different views and needs. In the case of changes in water use, hydropower generators, for instance, should be amenable to proposals which correspond to sustainable development, priorities between different uses should be well defined, and measures modifying the economic equilibrium of the hydroelectric concession should be co-financed by all stakeholders. Finding the best balance between loss of renewable hydropower and optimal benefit for the environment and/or society must be the principal concern in negotiations between the various stakeholders.

The R2D2 2050 partnership project has assessed the possible impacts on the Durance-Verdon river basin of climate change on the quantity and quality of water resources, biodiversity, and the changing demand and uses in 2050. This project has collected physical, biological, and socioeconomic data. It has helped to improve knowledge of how water infrastructures of the Durance River are connected to the territories. This in turn has enabled the establishment and development of tools dedicated to the modelling of water management and the interaction between local water resources and water volumes stored in large upstream reservoirs. In all cases, it will be necessary to implement water saving policies, reinforced by adaptation strategies to ensure a fair sharing of the water resource in 2050. The objective should also anticipate a potentially even more extreme climate at the end of the 21st century. The physical design of the reservoirs and associated channels of the Durance-Verdon can meet water-use demand in the context of climate change adaptation.

The value creation methodology, evaluating all hydropower services, could be linked to appropriate governance models (e.g., using the SHARE concept) to initiate dialogue among stakeholders and to manage water in the context of climate variability and change. These are effective approaches to foster sustainable use of the water in the context of climate change and climate resilience for multipurpose water infrastructures, as demonstrated by this basin.

Building more water infrastructure does not always increase adaptation or resilience to climate change impacts: this French example highlights the fact that proper governance of water stores are more important than the stored volumes of water themselves. It is essential to customize the governance model to the local context in order to better manage water infrastructures. There are no one-size-fits-all solutions. However, dialogue and a collective approach are essential to achieve a win–win situation and to understand the benefits of bilateral conventions for several parties. It is also important to use robust and relevant data and tools to bring objective information to the negotiation table. Water savings are becoming an essential issue on the Durance-Verdon basin, and water stakeholders are working together on incentive mechanisms to manage these water efficiencies in a sustainable way.

Uncertainty exists. It is up to all stakeholders to collectively manage and prepare for tomorrow's infrastructure development and management with less water and more risks in the context of climate variability and change. There is no single answer to these challenges; they should be addressed on a case-by-case basis. Among the potential solutions, one can promote the following:

- Assessment of adaptation options against a number of climate change scenarios (modelling tools) to anticipate and provide robust visibility to the different stakeholders
- Consultation to understand the issues and share them

- Communication to explain and support guidelines
- Convention development and partnership
- Incentives to save water (for the agriculture sector, for instance)
- Development of a culture of risk prevention and management (for extreme events such as drought, floods).

In a context of change in water resource availability and of significant evolution in water demand, developing adaptation strategies is necessary. These strategies should include water-saving approaches, appropriate development patterns, and crop choices. These adaptation strategies will ensure the long-term effectiveness of the integrated model of the Durance-Verdon valleys, which has been proven to serve the general interest. They will also help to find answers to water-use conflicts that could arise if nothing is done. Access to water is in this context a major strategic challenge for all stakeholders. The good structural design of Durance-Verdon water infrastructures gives all water users time to adjust governance to find and to implement solutions to preserve the future. This is a collective challenge for tomorrow.

References

Branche E (2014) The Durance-Verdon valleys: a model to address the multipurpose water uses of hydropower reservoirs. OECD, Brazil Policy Dialogue on Water Governance, Paris, France, Oct 2014

Branche E (2015) Sharing the water uses of multipurpose hydropower reservoirs: the SHARE concept. Paper presented at the 7th world water forum, Daegu, South-Korea, April 2015. http://www.hydroworld.com/content/dam/hydroworld/online-articles/documents/2015/10/MultipurposeHydroReservoirs-SHAREconcept.pdf. Accessed 3 May 2016

Braud I, Dantas-Antonino AC, Vauclin M, Thony JL, Ruelle P (1995) A simple soil plant atmosphere transfer model (SiSPAT): development and field verification. J Hydrol 166:213–250

Braud I, Tilmant F, Samie R, Le Goff I (2013) Assessment of the SiSPAT SVAT model for irrigation estimation in south-east France. Proc Environ Sci 19:747–756. In: International conference on monitoring and modeling soil–plant–atmosphere processes, Naples, Italy, 19–20 June 2013

Chopart JL, Mezino M, Le Mezo L, Fusillier JL (2007) FIVE-CORE: a simple model for farm irrigation volume estimates according to constraints and requirements. Application to sugarcane in Réunion (France). Proc Int Sugar Cane Technol 26(2):490–493

Ducharne A, Koster RD, Suarez MJ, Stieglitz M, Kumar P (2000) A catchment-based approach to modeling land surface processes in a general circulation model 2. Parameter estimation and model demonstration. J Geophys Res 105:24823–24838

Garçon R (1996) Prévision opérationnelle des apports de la Durance à Serre-Ponçon à l'aide du modèle MORDOR. Bilan de l'année 1994–1995. La Houille Blanche 5:71–76. doi:10.1051/lhb/1996056

Hingray B, Hendrickx F, Bourqui M, Creutin J-D, François B, Gailhard J, Lafaysse M, Le Moine N, Mathevet T, Mezghani A, Monteil C (2013) RIWER2030. Climat Régionaux et Incertitudes, Ressource en Eau et Gestion associée de 1860 à 2100. Report ANR VMCS 2009–2012

Krause P, Bäse F, Bende-Michl U, Fink M, Flügel W, Pfenning B (2006) Multiscale investigations in a mesoscale catchment—hydrological modelling in the Gera catchment. Adv Geosci 9:53–61

Krinner G, Viovy N, de Noblet-Ducoudré N, Ogée J, Friedlingstein P, Ciais P, Sitch S, Polcher J, Prentice IC (2005) A dynamic global vegetation model for studies of the coupled atmosphere-biosphere system. Global Biogeochem Cycles 19:GB1015. doi:10.1029/2003GB002199

Lafaysse M (2011) Changement climatique et régime hydrologique d'un bassin alpin. Génération de scénarios sur la Haute-Durance, méthodologie d'évaluation et incertitudes associées. Doctoral thesis, Université Paul Sabatier, Toulouse

Lafaysse M, Hingray B, Gailhard J, Mezghani A, Terray L (2014) Internal variability and model uncertainty components in a multireplicate multimodel ensemble of hydrometeorological projections. Water Resour Res. doi:10.1002/2013WR014897

Le Moine N (2008) Le bassin versant de surface vu par le souterrain: une voie d'amélioration des performance et du réalisme des modèles pluie-débit? Doctoral thesis, Université Pierre et Marie Curie, Paris, France

Loi du 5 janvier (1955) Aménagement de la Durance. https://www.legifrance.gouv.fr/jo_pdf.do?id=JORFTEXT000000504180. Accessed 4 May 2016

Loi du 3 janvier (1992) Loi n°92–3, loi sur l'eau. https://www.legifrance.gouv.fr/affichTexte.do?cidTexte=JORFTEXT000000173995. Accessed 4 May 2016

Lorenz EN (1969) Atmospheric predictability as revealed by naturally occurring analogues. J Atmos Sci 26:636–646. doi:10.1175/1520-0469(1969)26<636%3AAPARBN>2.0.CO%3B2

Magand C (2014) Influence de la représentation des processus nivaux sur l'hydrologie de la Durance et sa réponse au changement climatique. Doctoral thesis, Université Pierre et Marie Curie, Paris

Morin G (2002) CEQUEAU hydrological model. In: Singh VP, Frevert DK (eds) Mathematical models of small watershed hydrology and applications, Water Res Publ, Chelsea, Michigan (Chapter 13)

Roux D (2015) L'impact du changement climatique sur le bassin de la Durance: résultats du projet R2D2 2050. Ea-Eco Entreprises, Journée techniques, Aix-en-Provence

Sauquet E (2015) Projet R^2D^2 2050: risque, ressource en eau et gestion. Durable de la Durance en 2050. Final Report. IRSTEA, Paris, France. http://cemadoc.irstea.fr/exl-php/util/documents/accede_document.php. Accessed 4 May 2016

Sauquet E, Dupeyrat A, Hendrickx F, Perrin C, Samie R, Vidal JP (2010) Imagine2030. Climat et aménagement de la Garonne: quelles incertitudes sur la ressource en eau en 2030? Programme RDT, Final Report Cemagref, Paris, 149. http://cemadoc.irstea.fr/cemoa/PUB00028876. Accessed 4 May 2016

Schumann K (2015) Value creation at hydropower projects: developing a methodology for systematic assessment and benefit sharing. Hydro2015 event, Bordeaux, France, Oct 2015

SDAGE (2015) Schéma Directeur d'Aménagement et de Gestion des Eaux du basin Rhône-Méditerranée pour les années 2016 à 2021. French Water Agency, Lyon, France, Dec 2015. http://www.rhone-mediterranee.eaufrance.fr/docs/sdage2016/elaboration/cb_20151120/20151106-RAP-SdagePourAdoption-v00.pdf

Chapter 8
Sobradinho Reservoir: Governance and Stakeholders

Antonio Augusto B. Lima and Fernanda Abreu

Abstract The São Francisco drainage basin covers 7.6 % of the Brazilian territory and has tremendous economic, social, and cultural importance for the country. About 60 % of its watershed is in the semi-arid region, which is characterized by having critical periods of drought. To regulate the river flow and to generate power, a cascade of hydropower plants was constructed on the São Francisco River. Among them is the Sobradinho Reservoir, which has an interannual water storage capacity to meet multiple water uses and adds about 4 billion kW of energy per year to the Brazilian northeast. In recent years, the average unregulated flow in the Sobradinho Reservoir has been lower than the historical average. The lowest average annual unregulated flow in the Sobradinho Reservoir occurred in 2014, and the first months of 2015 did not point to better conditions. The current and unfavorable hydrologic situation of the São Francisco Basin represents real governance challenges for Brazil. The minimum outflow from the Sobradinho Reservoir has been reduced several times to ensure that the needs for the multiple uses are met. Today, the Sobradinho Reservoir operates with a minimum outflow of 900 m^3/s. Because of these reductions, there was a cumulative gain of 44.7 % in the active storage of the reservoir. If the mitigation measures that allowed reducing the minimum outflow had not been implemented, this reservoir would have fully depleted of its active storage by November 2014.

Keywords Sobradinho Reservoir · São Francisco River · Outflow · Regulated flow · Mitigation measures

A.A.B. Lima (✉) · F. Abreu
National Water Agency, Brasilia, Brazil
e-mail: antonio.lima@ana.gov.br

C. Tortajada (ed.), *Increasing Resilience to Climate Variability and Change*,
Water Resources Development and Management,
DOI 10.1007/978-981-10-1914-2_8

8.1 Introduction

Construction of the Sobradinho Reservoir in the second half of the twentieth century prompted a huge development in the São Francisco River. Flow regularization allowed an increase in the power generated and the water supply for irrigation. The flow regularization was an important action taken to address the multiple uses during the hydrological year in the São Francisco River Basin, which is a region characterized by having critical periods of drought. The multiple uses are a fundamental element of the National Water Resource Policy in Brazil.

Before the construction of the dams and the anthropic interventions, the minimum flow at the mouth of São Francisco River was about 500 m^3/s; after regularization, the minimum flow reached 1300 m^3/s. A more detailed analysis showed that reservoir construction caused an average decrease of 6–12 % in the greatest stream water levels and an increase of 20–32 % in the smallest stream water levels. It also resulted in a 9 % reduction in the maximum flow (Q_5) and an increase of 27 % in the minimum flow (Q_{95}). Therefore, flow regularization in the Sobradinho Reservoir decreased the effects of large floods and ensured water in the drought periods for the different users downstream of the reservoir. It particularly favored the public water supplies in the cities and some irrigation projects. Despite these benefits, it had some negative consequences for navigation, farming, and fisheries (Martins et al. 2011).

It is important to mention that climate change has probably been acting on the São Francisco River flow. Milliman et al. (2008) observed that the annual average flow in the São Francisco River was reduced by about 20 % between 1951 and 2000. The authors attribute this decrease to climate change and to some anthropic actions in the basin.

Although flow regularization is a positive measure to meet the multiple uses in the São Francisco River Basin, it leads to some specific conditions of conflicting demands for water. This happens mostly in critical situations of water availability (severe drought periods) and in cases that demand increased power generation and result in a reduction of the amount of water stored in the reservoir (Garcia et al. 2007; Collischonn et al. 2006). The main problems in the region of the Sobradinho Reservoir are related to meeting human supply requirements and irrigation and environmental flow demands during drought periods.

The Sobradinho Reservoir has also an important role in flood control. The large multipurpose dams operated by the electricity sector along the São Francisco River are used for flood damping in combination with longitudinal dikes built for the protection of coastal communities. The Sobradinho Reservoir can hold inflow volumes thanks to flood control space designed for this purpose. The use of this space mitigates floods and dampens the large inflows that occur in the reservoir, allowing a smaller flow downstream and respecting determined upstream level limits.

The reservoir operation for floods and the necessary flood control storage space are set at the beginning of each hydrological year based on the expected hydrological behavior of the floods—normal, attention, alert, and emergency. The criteria for characterizing the different river statuses for reservoir operation are set in the operating rules, which are a joint effort between Electric System National Operator (ONS) and operators.

The case of the Sobradinho Reservoir highlights the importance of good water governance in Brazil to meet multiple uses while facing scarcity situations, which are likely to become more frequent and intense with climate change.

8.2 About the Sobradinho Reservoir

On 4 October 1501, Saint Francis' day, Américo Vespúcio discovered on the northeast Brazilian coast a river mouth that would be named in honor of the patron saint of animals. The actual day of the discovery of São Francisco River, however, is under discussion. The São Francisco River starts in the National Park of Serra da Canastra, located in southwest Minas Gerais State. Initially, it flows in a south-to-north direction and then changes to an east-to-west direction. Its watershed drains areas from six states (Minas Gerais, Goiás, Bahia, Pernambuco, Alagoas, and Sergipe) and the Federal District, as well as three biomes (Cerrado, Caatinga, and Mata Atlântica). With an area of 645,000 km^2, its drainage basin covers 7.6 % of the national territory. In the world classification, the São Francisco River has the 34th highest flow (annual medium flow of 2800 m^3/s) and, being 2900 km long, ranks 31st in length (Welcomme 1985). This river has tremendous economic, social, and cultural importance for the states located in its watershed.

The watershed is traditionally divided in four segments—upper, middle, lower middle, and lower. The upper São Francisco River starts at the headwaters and reaches to the municipality of Pirapora—a length of 630 km. The middle segment—1090 km long—runs from Pirapora to the municipality of Remanso. The lower middle segment goes from Remanso to the Paulo Afonso's waterfall—a length of 686 km. The lower stretch is the shortest, being just 274 km (Paiva 1982). The waters in the upper São Francisco River are characterized as fast, cold, and oxygenated. In its middle segment, the São Francisco River is known as a plateau river, with lower velocity and subject to extensive flooding. The lower middle segment is under the effect of dams, while the lowest segment is a plain stretch with slow waters and under marine influence (Sato and Godinho 1999). The river has 168 tributaries—90 of them are located on the right riverbank and 78 on the left.

About 60 % of the São Francisco River watershed is in the semi-arid region of Brazil, which is characterized as having critical periods of drought caused by low precipitation and high evapotranspiration. It is for these reasons that the São Francisco River plays a major role in this region. The annual average precipitation in this region is 1003 mm, in contrast to the national average of 1761 mm. The superficial water availability is 1886 m^3/s, which corresponds to 2.07 % of the national water availability (91,071 m^3/s). The medium flow is 2846 m^3/s, which corresponds to 1.58 % of the national medium flow (179,516 m^3/s). The ratio of the discharge rate to the drainage area (called *specific flow* in Brazil) is 4.5 L/s/km^2.

The per capita maximum storage volume is 5183 m^3/person, higher than the national per capita maximum storage volume of 3596 m^3/person.

To generate energy and regulate the flow in this basin, a cascade of hydroelectric power plants was constructed on the São Francisco River. Among them is the Sobradinho Reservoir. In 2013, the hydroelectric potential installed on the river was 10,708 MW (12 % of national power generation). There are 40 power plants in operation. Of these, 28 are small hydropower plants that contribute 140 MW and 12 are big hydropower plants, responsible for 10,568 MW. Among these latter plants, Xingó (3162 MW), Paulo Afonso IV (2462 MW), Luiz Gonzaga (1479 MW), and Sobradinho (1050 MW) stand out. These plants are the energy supply base for the northeast region of Brazil. Just three hydropower plants have interannual water storage capacity—Três Marias, Sobradinho, and Luiz Gonzaga (Fig. 8.1). These reservoirs serve multiple water uses, such as irrigation, flood control, human and animal supply, navigation, and leisure. Figure 8.2 shows the single-line diagram of the main hydropower plants on the São Francisco River.

The construction of the Sobradinho Reservoir started in 1973 and its filling started in 1978, flooding four cities (Casa Nova, Sento Sé, Remanso, and Pilão Arcado, all in Bahia State). In 1979, a greater flood in the São Francisco River enabled the reservoir to fill in record time. Beyond the multi-year regularization function, the hydropower plant of the Sobradinho Reservoir was projected to add about 4 billion kW of electrical energy per year to the northeast. It also has the function of meeting multiple uses downstream—such as irrigation and water supply—and providing flood control.

The Sobradinho Reservoir is located in Bahia State, about 40 km from the important cities of Juazeiro (Bahia) and Petrolina (Pernambuco). It is located on the lower São Francisco River segment, downstream from the Três Marias Reservoir, which is an important regulator of the inflows into the Sobradinho Reservoir. When compared with the other hydropower plants in the São Francisco River Basin, it is the fourth largest in terms of installed power and first in storage capacity (Correia and Silva Dias 2003). Given its equatorial location, the Sobradinho Reservoir experiences a high evaporation rate that can reach 480 m^3/s during the day (Acioli 2005).

The Sobradinho Reservoir has a regulated flow of 2060 m^3/s, which allows the downstream hydropower plants to achieve a better performance. It has a drainage area of 499,084 km^2, with a storage capacity of 34.116 billion m^3 of water, an active storage area of 28,669 km^2, and a flooded area of 4241 km^2, forming the biggest artificial lake of Latin America. The rainy season starts about November and ends in April. The remaining months constitute the dry season. The greatest inflows are observed in January and March, with March being the month of greater variability.

The construction of the reservoir led to an increase in the amount of water evaporated and the intensity of the anthropic actions in the region. Recently, the flows of the São Francisco River have been regularized for the benefit of hydropower production. This situation creates a high potential for conflict among others water uses, such as human supply, irrigation, navigation, and others.

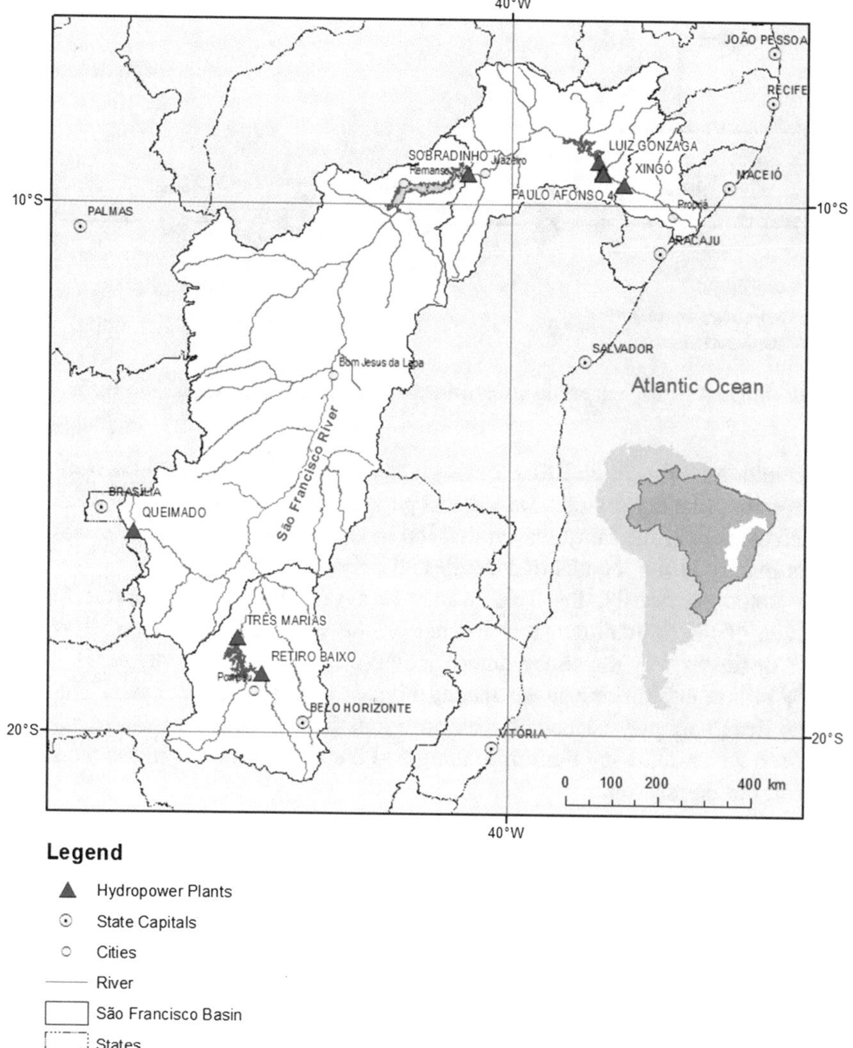

Fig. 8.1 Locations of the main hydropower plants in the São Francisco River Basin and other important benchmarks

The reservoir has changed the natural flow in the São Francisco River by ensuring a minimum outflow of 1300 m^3/s. This rate was defined because lower flows cause navigation problems on the Sobradinho/Juazeiro stretch and cause complications for the catchments for industry, water supply, and irrigation projects in the Sobradinho/Itaparica segment.

The results of an environmental study of the São Francisco River Basin (Nou and Costa 1994) in the period 1977–1989 show that the rate of outflow from and

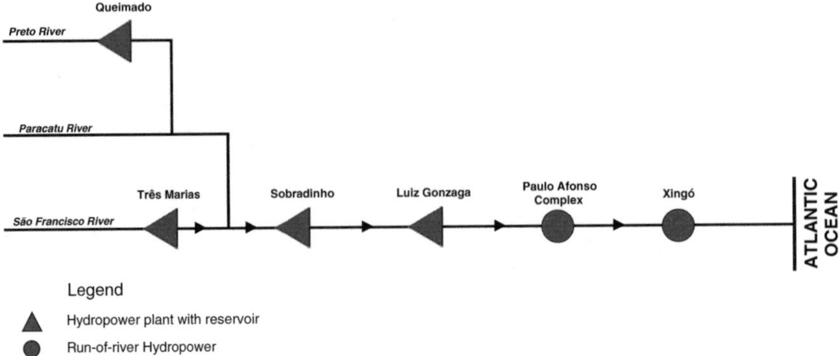

Fig. 8.2 Single-line diagram of the main hydropower plants in the São Francisco Basin

inflow into the Sobradinho Reservoir is greater during the seven months from May to November (the dry season). During this period, the dam releases more water than it receives. It is in the rainy season that the amount of water stored increases; this occurs mainly in the months of January and February.

As noted previously, the Três Marias is a very important reservoir for the operation of the Sobradinho. It is located on the upper São Francisco River segment, upstream of the Sobradinho Reservoir. Its active storage is about 15.278 billion m^3, making it the second biggest reservoir in the basin. The Três Marias Reservoir plays a notable role in regulating the upper waters of the São Francisco River, allowing the surrounding and the upstream areas to use the stored water in the dry season.

8.3 Governance and Stakeholders

Although Brazil owns 12 % of the global volume of fresh water, its scarcity, resulting from severe climatic conditions in the last years, excited a debate about how the water resources can be effectively managed in a country rich in water. Water availability can be monitored and managed locally, and institutions and solid policies are necessary to induce a better use of available water today and in the future.

When the inflows are greatly reduced and a severe climate context persists for a long time, it is difficult to achieve a consensus about a reservoir's operation. In this scenario, it is a challenge to conserve water in the reservoirs while energy production and navigation can be strongly affected if the drought period extends for a long time. Therefore, a decrease in the water level in the reservoir can create a lot of tension. The questions are how it can be adjusted and what is the role of the government.

The National Water Resources Policy, according to Law 9433/1997, has the following foundations:

- Water is a resource in the public domain.
- Water is a natural and limited resource, endowed with economic value.
- The management of water resources should always meet multiple uses.
- In a situation of scarcity, the priority is to supply humans and water livestock.
- The watershed is the territorial unit in which to implement the National Water Resources Policy and to operate the National System of Water Resources Management.
- The management of water resources should be decentralized and should count on the participation of the public power provider, water users, and the community.

ANA, a body of the National Water Resources Management System, was created by Law 9984/2000 as the federal entity responsible for implementing the National Water Resources Policy. According to this law, ANA is responsible for defining and inspecting the conditions of the reservoirs' operation, which is led by public and private agents, aiming to guarantee the multiple water uses, as planned in the Water Resources Plans produced by the respective watersheds. The law also states that the definition of the operating rules for the reservoirs that generate energy should be done in articulation with the National Electric System Operator (ONS). Furthermore, Law 9984/2000 assigns to ANA the responsibility of planning and promoting actions to prevent and minimize the effects of drought and floods, in the context of the National Water Resources Management System, in articulation with the central body of the National Civil Defence System, supporting the States and Municipalities. Because of that, ANA, in conjunction with the ONS, is in charge of establishing the operational rules for the Sobradinho Reservoir and its operation must meet the multiple uses of water resources.

According to Law 9648/1998, ONS is responsible for the coordination and control of the generation of and transmission installations for electrical energy in Brazil's National Interconnected System (SIN), under the control and regulation of the National Electricity Agency (ANEEL). ONS's responsibilities include operation planning, programming, and the centralized command of generation, aiming for optimization of the interconnected system for electrical energy. In the context of the planning and programming activities of SIN, the operative conditions, stated by the agents responsible for operating the hydropower plants and those regulated by ANA, are taken into account. SIN is a large hydrothermal system for electricity production and transmission, the operation of which involves complex simulation models, which are under ONS's coordination and control.

The Sobradinho Reservoir is operated by Companhia Hidro Elétrica do São Francisco (CHESF), a company that produces, transmits, and commercializes electrical power in nine northeastern states and which is looking for sustainability. CHESF operates its reservoirs aiming for flood control, but takes into account the constraints of the other water uses in the watershed. Its operation is based on rules and guidelines established by ANA and ONS, which are certified by ANEEL.

The São Francisco River Basin Committee (CBHSF) is a collective body composed of representatives of the government, community, and private water users. It performs the decentralized and participative management duties for the water resources in the basin in order to protect its sources and to contribute to sustainable development. It has normative, deliberative, and advisory functions.

The Company for the Development of the São Francisco and Parnaíba River Valleys (CODEVASF) is a public sector company linked to the Ministry of National Integration, which promotes the development and revitalization of the São Francisco, Parnaíba, Itapecuru, and Mearim watersheds, with sustainable use of the natural resources and arrangement of the productive activities for social and economic inclusion. This company mobilizes public investments to construct infrastructure works, mainly to implement public irrigation projects and others related with the rational use of water resources.

The Power Company of Minas Gerais (CEMIG) is the agent responsible for the only hydropower plant on the São Francisco River that does not belong to CHESF (Três Marias). As with CHESF, CEMIG plays an important role in the management of the water resources of the São Francisco River.

The current hydrologic situation of the São Francisco River Basin presents real governance challenges for Brazil. It highlights the need for a strategic solution to the conflicts between water uses (human supply, power production, irrigation, navigation, fish-farming, livestock consumption, recreation, and tourism). It is most important in the critical periods when there is insufficient water to meet all the demands of the consumptive and nonconsumptive uses.

8.4 Water Scarcity

During the extended dry season, peculiar to the semi-arid area, water scarcity situations are frequent in the São Francisco Region. In 2013, 276 municipalities (61 % of the municipalities of the semi-arid region) declared emergency situations because of drought. In the region, 206 municipalities have experienced more than 10 drought events between 2003 and 2013. In the São Francisco River Basin, according to ANA's data (2012 and 2013), 21 % of the cities experienced water rationing or were in a state of alert regarding their water supplies in 2013.

Just as the Brazilian northeast region faced a severe drought that affected the rural area and the water supplies in the cities, so now the São Francisco River Basin is in the same situation. The precipitations and inflows into the basin are lower than the observed average, affecting the amount of water stored in the reservoirs. The 2015 precipitation maps show that the adverse situation remains (Fig. 8.3).

In addition to this, meteorologists have identified the formation of an El Niño Southern Oscillation (ENSO) that has a correlation with lower precipitations in the northeast. The Brazilian Weather Forecast Centre and Climate Studies (CPTEC) consensus forecast for November/December/January (2015/2016) shows a 40 % probability of precipitation being lower than average (Fig. 8.4).

Precipitation MAPS

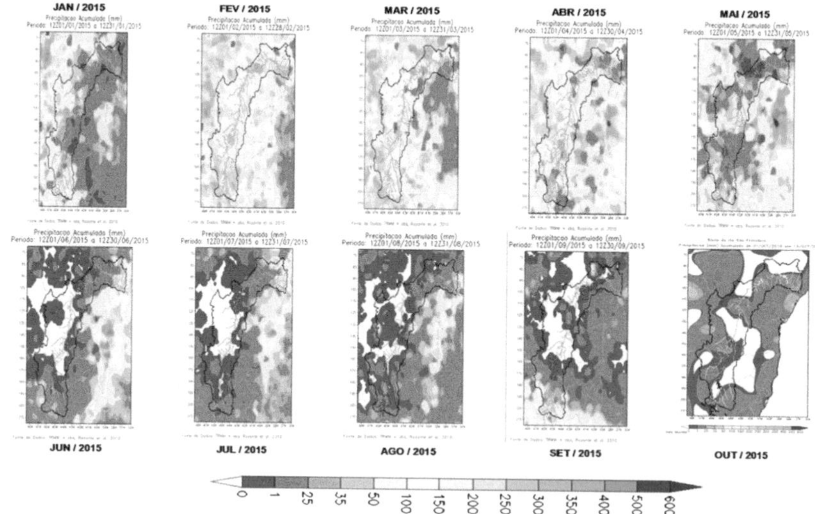

Fig. 8.3 Precipitation maps—São Francisco Basin (reproduced from CPTEC)

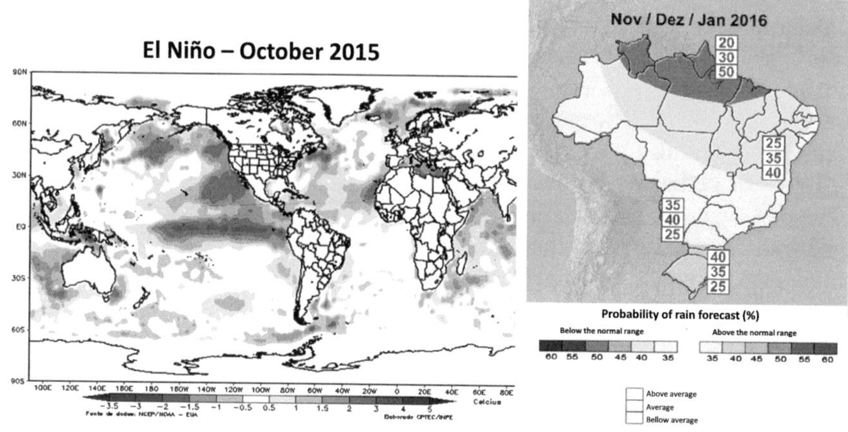

Fig. 8.4 El Niño—October 2015 and consensus forecast (reproduced from CPTEC)

In recent years, the average unregulated flow into the Sobradinho Reservoir has been lower than the historical average, according to the monthly unregulated flow[1] series since 1931 (source: ONS). The lowest average annual unregulated flow into

[1]Unregulated flow is the flow that would occur in a section of the river if there were no anthropic actions, such as reservoirs, diversions, and water catchments.

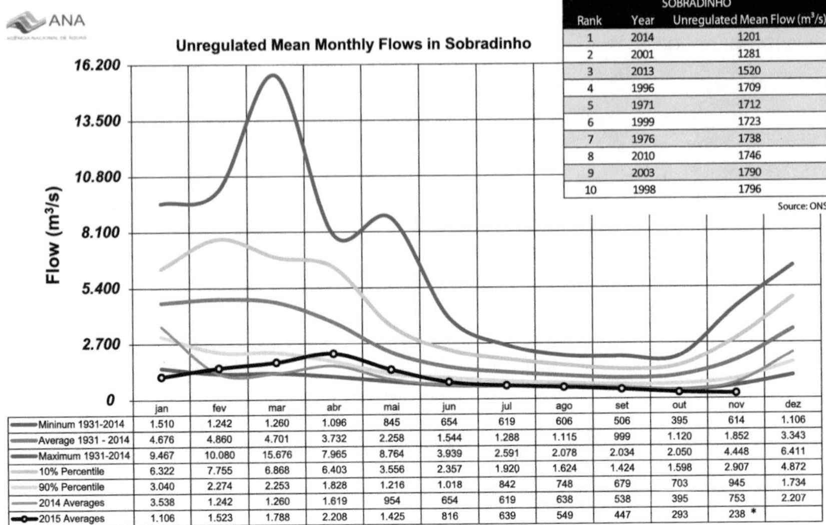

SOBRADINHO		
Rank	Year	Unregulated Mean Flow (m^3/s)
1	2014	1201
2	2001	1281
3	2013	1520
4	1996	1709
5	1971	1712
6	1999	1723
7	1976	1738
8	2010	1746
9	2003	1790
10	1998	1796

	jan	fev	mar	abr	mai	jun	jul	ago	set	out	nov	dez
Mininum 1931-2014	1.510	1.242	1.260	1.096	845	654	619	606	506	395	614	1.106
Average 1931 - 2014	4.676	4.860	4.701	3.732	2.258	1.544	1.288	1.115	999	1.120	1.852	3.343
Maximum 1931-2014	9.467	10.080	15.676	7.965	8.764	3.939	2.591	2.078	2.034	2.050	4.448	6.411
10% Percentile	6.322	7.755	6.868	6.403	3.556	2.357	1.920	1.624	1.424	1.598	2.907	4.872
90% Percentile	3.040	2.274	2.253	1.828	1.216	1.018	842	748	679	703	945	1.734
2014 Averages	3.538	1.242	1.260	1.619	954	654	619	638	538	395	753	2.207
2015 Averages	1.106	1.523	1.788	2.208	1.425	816	639	549	447	293	238 *	

* Average until October 20th

Fig. 8.5 Unregulated mean monthly flows into the Sobradinho Reservoir

the Sobradinho Reservoir occurred in 2014. The first months of 2015 did not seem to point to better conditions and the forecast indicated that the drought conditions would persist. The unregulated flows in January 2015 were the lowest average monthly flow for any January since records began. The unregulated mean monthly flows from January to November 2015 are lower than the 2014 flows for the same period, which probably will make 2015 the worst year in terms of flow (Fig. 8.5).

Table 8.1 compares the medium inflows and outflows and presents the active volume in the Sobradinho Reservoir in 2014 and 2015. It is possible to observe that even with the actions taken to reduce the minimum outflow, the water level in the Sobradinho Reservoir followed a downward trend because the outflows, most of the time, were greater than the inflows.

On 26 September 2015, one of the leading Brazilian newspapers, *Estadão*, published an article with the following statement: "The water level in Sobradinho, main water reserve in the northeast region of Brazil, is falling and can reach, this year, zero percent of the active volume. In a scenario of extreme lack of rain, the reservoir would reach 2–3 % of the active volume at the end of November, according to CHESF. In this case, power production in the Sobradinho plant would be unfeasible and the risk of water shortages for human consumption would increase to worrying levels." The article continues: "Reducing the water level in the reservoir to close to 2 % of its storage capacity would require the shutdown of the Sobradinho plant, which can produce 1050 MW of power, besides representing an adverse situation for keeping the water supply for the regions located along the São Francisco River."

Table 8.1 Medium flows and active volumes in the Sobradinho Reservoir in 2014 and 2015

Month	Inflow 2014 (m^3/s)	Inflow 2015 (m^3/s)	Outflow 2014 (m^3/s)	Outflow 2015 (m^3/s)	Active volume (%) 2014	Active volume (%) 2015	Water level 2015 (m)
Jan	3110	908	1175	1124	50.39	18.88	384.33
Feb	1514	1095	1169	1142	53.44	17.80	384.15
Mar	1278	1256	1182	1153	53.74	18.88	384.33
Apr	1577	1397	1173	1086	57.51	21.91	384.81
May	918	1090	1168	1103	54.44	21.20	384.7
Jun	621	805	1165	958	48.95	19.55	384.44
Jul	524	617	1158	929	43.57	16.74	383.97
Aug	501	531	1140	941	36.66	12.70	383.24
Sep	380	490	1124	938	29.95	8.11	382.34
Oct	297	497	1167	935	21.13	4.33	381.53
Nov	629	524	1150	934	15.78	2.11[a]	381.02[a]
Dec	1816		1147		20.49		

[a]Until November 19, 2015

On 28 October 2015, the Minas Gerais Association of Environmental Defence (AMDA) indicated a huge concern about the Sobradinho Reservoir: "The alarming situation of the Sobradinho Reservoir, in Bahia, which on Monday reached 4.96 % of its active volume, the lowest observed value since its opening in 1979, called our attention to the critical condition of the São Francisco River Basin. In Minas Gerais, a state that contributes 72 % of the São Francisco's water, the drama of the lack of water spreads from the headwaters to the border with Bahia. Important tributaries had drastic volume reductions and it is estimated that most of the small tributaries are dry or intermittent, providing a bleak landscape. There is also the fear that Três Marias Reservoir will reach the critical level because of the difference between the volume that goes in and what comes out of the lake."

Therefore, over the past few years (2013, 2014, and 2015), a set of easing measures to reduce the minimum outflow of the Três Marias and Sobradinho Reservoirs has been implemented to prevent their depletion.

8.5 Intervention and Outcomes

In 2001 and 2002, energy rationing spread across Brazil and, through Presidential Resolution number 39 of August 2001, it was established that ONS and CHESF "should adopt measures to operate the hydropower plant reservoirs, from the Sobradinho Reservoir to the mouth of the São Francisco River, with minimum flows of 1000 m^3/s, having a tolerance margin of 5 %."

The first action from ANA occurred in the transition to the rainy season in 2003/2004, as a result of the unfavorable hydroclimatic situation found at that time.

On this occasion, the Agency considered the ONS study "Easing the minimum outflows constraints in Sobradinho Reservoir," which indicated the critical inflow prospects in the northeast region and maintained the minimum outflow from the Sobradinho Reservoir at 1300 m^3/s. Since then until 2013, on three other occasions, ANA has issued resolutions temporarily authorizing a reduction in the minimum outflow from the Sobradinho Reservoir. These resolutions sought to preserve the volume stored in the reservoir in order to meet the needs of multiple uses.

In April 2013, ANA, after a broad negotiation with the stakeholders involved and taking into account the weather forecasts and the simulations of the reservoir operation, issued Resolution No. 442 authorizing a reduction in the minimum outflow from the Sobradinho Reservoir from 1300 to 1100 m^3/s. From April 2013 to March 2015, this reduction from 1300 to 1100 m^3/s was maintained for the Sobradinho Reservoir. From March until the end of June 2015, the minimum outflow from the Sobradinho Reservoir was established at 1100 m^3/s because of the unfavorable hydro-meteorological situation in the São Francisco River Basin. During specific periods (Saturdays, Sundays, and holidays), the outflow could be reduced to 1000 m^3/s. These reductions were implement by ANA's Resolutions 206/2015, 499/2015 and 602/2015.

However, the critical hydrologic condition in the São Francisco River Basin did not reverse, especially between Três Marias, Queimado, and Sobradinho Reservoir. It was necessary to implement a new reduction in the minimum outflow from the Sobradinho Reservoir to 900 m^3/s. It was authorized through the ANA's Resolutions 713/2015, 852/2015, 1208/2015 and 1307/2015, where the last one was effective until 20 December 2015 (Fig. 8.6).

It is important to point out that all requests made by the electricity sector to reduce the minimum flows were motivated by the occurrence of unfavorable hydro-meteorological conditions combined with low water storage in the reservoirs, which jeopardize the requirement to meet the needs of the multiple uses in the basin.

The lowest level of water storage, considering the historical data of the Sobradinho Reservoir, was reached in 2001 when it registered an active volume of

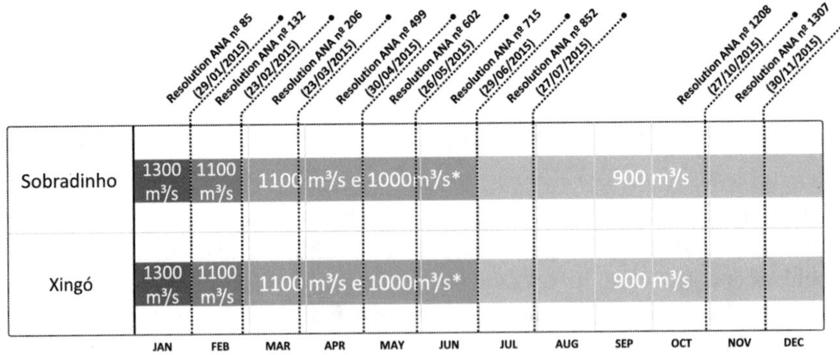

Fig. 8.6 A timeline of resolutions issued by ANA in 2015

5.5 %. Since 23 October 2015, the Sobradinho Reservoir maintained, day after day, negative records in terms of storage. By 1 December 2015, it reached its lowest level, with 1.03 % of the active storage. With the increase in the unregulated flow on 14 December, the active volume reached 1.86 %.

In this context and in anticipation of levels of precipitation lower than the basin's average during the next months, because of the ENSO phenomenon, projections of the water level in the Sobradinho Reservoir indicated a probability of total depletion of the active volume by January 2016. To avoid this situation, on 15 December 2015, ANA coordinated a meeting with the main stakeholders involved in the São Francisco River Basin to discuss the feasibility of implementing a new reduction in the minimum outflow.

For implementing the easing of minimum outflows constraints, the Sobradinho and Três Marias Reservoirs have been operated in a coordinated manner to promote a balance in the reservoirs' water storage in the São Francisco River Basin. Therefore, ANA has adopted easing measures to reduce the minimum outflow from the Três Marias Reservoir as well. At the end of March 2014, because of the low inflows into the Três Marias Reservoir and the low water level in the reservoir, CEMIG, the agent responsible for operating this plant, reduced the minimum outflow from 350 to 220 m^3/s. Until the beginning of July 2014, the outflows from the Três Marias Reservoir were about 250 m^3/s. After some adjustments in Pirapora's city water catchment, which ended on 2 July 2014, CEMIG reduced the minimum outflow from the Três Marias Reservoir from 220 to 180 m^3/s. Since then, CEMIG has been promoting other reductions in the outflow from the Três Marias Reservoir. In the last one, using the incremental flows of the rivers downstream from the Três Marias Reservoir, it was possible to reduce the minimum outflow to 80 m^3/s, with the objective of increasing the volume of water stored in it.

The easing actions of the minimum outflows from the Três Marias Reservoir prevented the total depletion of the active volume of the reservoir in 2014. The situation was different for the Sobradinho Reservoir. For the Três Marias Reservoir, it is not possible under gravity to use the volume of water below its minimum operating level because it does not have an outlet below the active storage level. Considering the amount of water stored in Três Marias Reservoir on 12 October 2015, there was a cumulative gain of 45.7 % in the active storage, resulting from the implemented reductions of the minimum outflow (Fig. 8.7).

The recovery of the water level in Três Marias Reservoir made it possible to progressively increase the minimum value of the outflow from the reservoir (it was about 500 m^3/s). The aim was to raise the amount of water transferred to the Sobradinho Reservoir to reduce its depletion rate, which, following the reduced incremental flows between the Três Marias and Sobradinho Reservoirs in 2015, was the lowest ever observed.

According to ONS simulations and considering the volume of water stored in the Sobradinho Reservoir on 12 October 2015, there was a cumulative gain of 44.7 % in the active storage resulting from the implemented reductions in the minimum outflow (Fig. 8.8). If the easing measures of the minimum outflow had not been

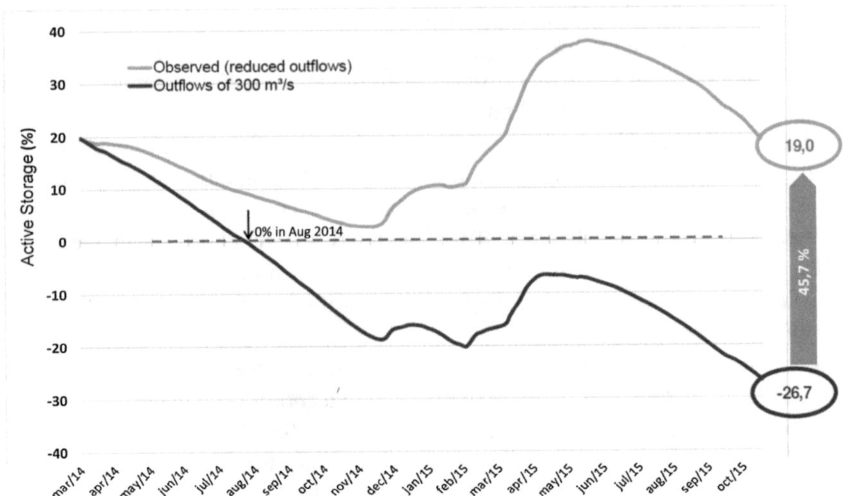

Fig. 8.7 Cumulative gains in the active storage of the Três Marias Reservoir

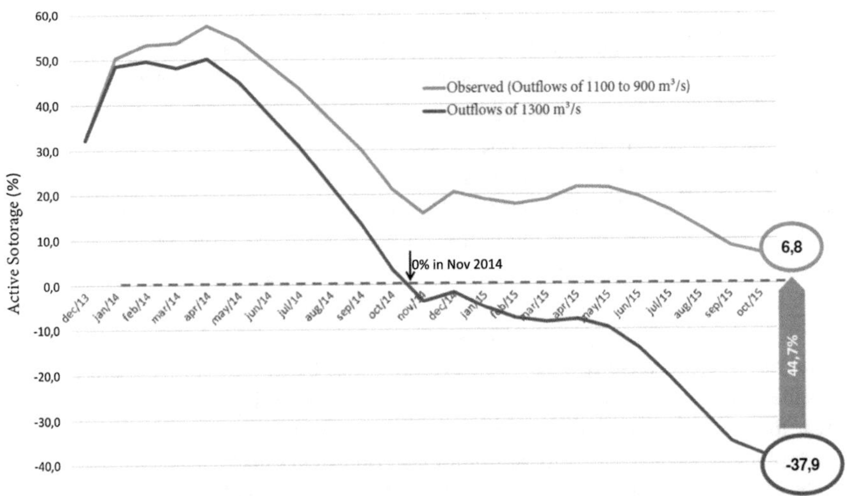

Fig. 8.8 Cumulative gains in the active storage of the Sobradinho Reservoir

implemented, the Sobradinho Reservoir would have fully depleted its active storage by November 2014.

At a meeting on 15 December 2015, ONS presented a study that calculated sustainable outflows from the Sobradinho and Três Marias Reservoirs for the next hydrological year. In the simulations, it was considered that a goal of zero percent of the active volume of the Sobradinho Reservoir and 3 % for the Três Marias Reservoir would be achieved by the end of November 2016. The study showed that,

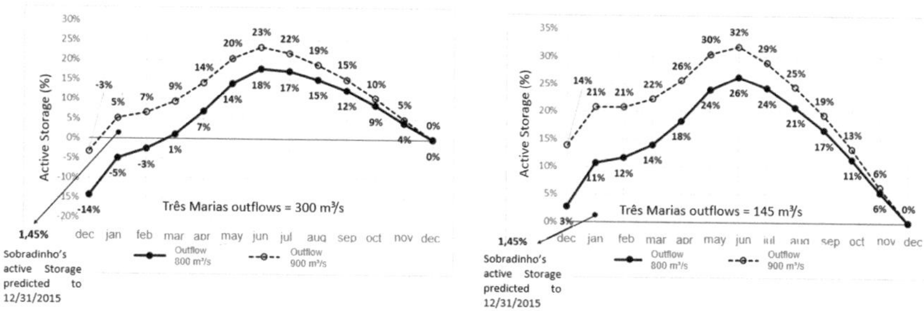

Fig. 8.9 Sustainable outflow for the Sobradinho Reservoir for the next hydrological year

considering the critical inflows observed between December 2013 and November 2014, the Três Marias Reservoir would have to reduce its outflows from 400 to 145 m^3/s in order to reach 3 % of its active storage in the end of November 2016.

For the Sobradinho Reservoir, considering the critical inflows observed between December 2014 and November 2015, ONS presented two scenarios for the Três Marias Reservoir outflows to calculate the Sobradinho Reservoir's sustainable outflow for the next hydrological year. In the first case (145 m^3/s), it would be necessary to decrease the minimum outflow from the Sobradinho Reservoir to a value less than 800 m^3/s to ensure zero percent of the active volume in the Sobradinho Reservoir by the end of November 2016. In the second scenario (300 m^3/s), the minimum outflow from the Sobradinho Reservoir could vary between 800 and 900 m^3/s (Fig. 8.9).

After the ONS presentation, the main stakeholders decided that the current minimum outflow from the Sobradinho Reservoir (900 m^3/s) was not compatible with sustainability of the multiple uses in the São Francisco River Basin. Therefore, they recommended to reduce the minimum outflow from the Sobradinho Reservoir to 800 m^3/s until 31 January 2016 and that of the Três Marias Reservoir to 300 m^3/s.

The main stakeholders are always invited to participate in the discussions involving interventions in the outflow from the Sobradinho Reservoir. National Water Resources Policy establishes that the water management policy should be determined in the presence of representatives from the public sector, community, and water users and with their active participation. However, it is important to note that the final decision belongs to ANA.

If it were not for the Sobradinho Reservoir's regulating effect, nowadays the São Francisco River would have a discharge of below 250 m^3/s and not the currently observed 900 m^3/s. Figure 8.10 shows the difference between the unregulated flows and the outflows in Sobradinho between 1 January 2013 and 19 November 2015. It can be seen that on several occasions the unregulated flows were considerably lower than the reservoir outflows.

Assuming the need to implement these easing measures, the water users should be prepared by the government for the impacts that can occur and the adaptations that they need to implement. The experts are studying the possibility of adopting

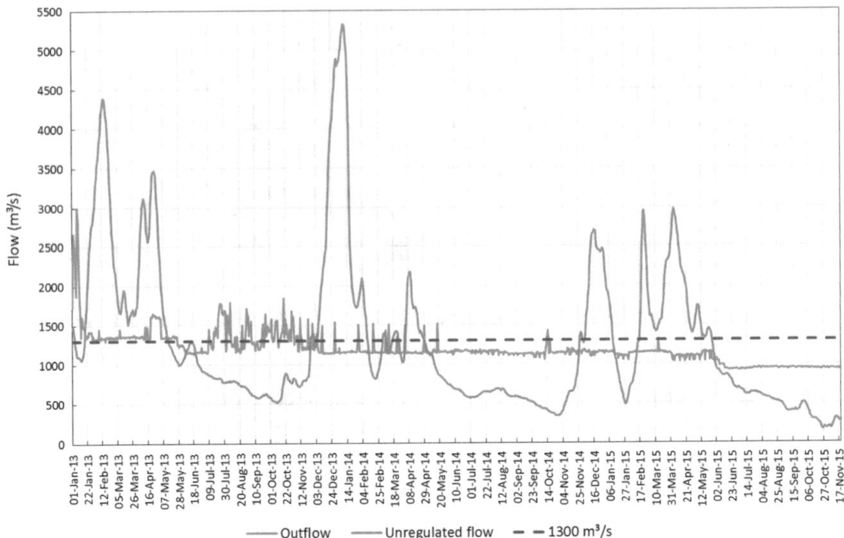

Fig. 8.10 Discharge and unregulated mean flow from the Sobradinho Reservoir

different minimum outflows from the Sobradinho Reservoir according to the season and the likelihood of precipitation.

8.6 Impacts

Before the construction of the dams and the anthropic interventions, the minimum flow at the mouth of the São Francisco River was about 500 m^3/s; after regularization, the minimum flow reached 1300 m^3/s. A more detailed analysis showed that construction of the reservoir caused an average decrease of 6–12 % in the highest stream water levels and an increase of 20–32 % in the lowest stream water levels. Likewise, there was a reduction of 9 % in the maximum flow (Q_5) and an increase of 27 % in the minimum flow (Q_{95}). Therefore, the flow regularization of the Sobradinho Reservoir decreased the effect of big floods and ensured water in drought periods for the different users downstream from the reservoir. In particular, it favored the public water supplies in the cities and some irrigation projects (Martins et al. 2011).

One of the main goals of CODEVASF is to promote the development of irrigated areas in the São Francisco Valley, especially for fruit growing—banana, melon, watermelon, mango, grape—for wine, juices, and seeds. Sampaio (1994) showed that the average net income per hectare from fruit can be up to five times higher than that for other irrigated crops, implying significant income generation, employment, and foreign exchange opportunities for the region.

Preserving the volume of water in the Sobradinho Reservoir is important—not just to keep flow regularization capacity between the reservoir and the mouth of São Francisco River, but also to meet the multiple uses of the basin and maintain productive activities. For example, reduction of the water level in the Sobradinho Reservoir can interrupt the water catchment in Nilo Coelho, which is a very important irrigation center and one of the major fruit producers in the São Francisco Valley.

CODEVASF and EMBRAPA are seeking new options for agriculture to meet the needs of local producers and ensure the sustainability of irrigated production in the valley. The idea is to diversify crops in the irrigated perimeters of the valley and thus enable the supply of other products in different seasons. The estimated investment is around BRL 357,000 annually.

Another important point is that economic production depends on power generation. The São Francisco River is a major source energy production in Brazil and in the northeast region. Without the easing actions to reduce the minimum outflow from the Sobradinho Reservoir, it would have been completely depleted as a result of the drought that began in 2012. Therefore, reducing the minimum outflow has been a necessary measure to maintain power production and thus the productive activities.

It is important to mention that, when faced with adverse hydro-meteorological and storage conditions of the São Francisco River Basin, the operational policy of the energy sector has been to prioritize the use of other hydropower plants (from subsystems in the south, southeast, north, and middle west of Brazil), as well as thermal and wind generation in order to meet the country's demand. This enables the use of hydropower resources from the São Francisco River Basin to be minimized. This measure also contributes to preserving water storage in the São Francisco River reservoirs.

Thus, the positive outcome of adopting reduced minimum outflows from the Sobradinho Reservoir is the increased guarantee of the sustainability of the multiple uses from the reservoir lake to the São Francisco River mouth. This is especially so for the supplies to people in the cities and for agriculture projects. Some of the negative impacts are the reduction in hydropower generation, the inability to capture water using some fixed infrastructure, difficulties in transverse and longitudinal navigation, and problems for tourism activities. Important cities of Brazil, such as Aracaju, Juazeiro and Petrolina, had to make adjustments in their water catchment structures to maintain supplies for human. The same was necessary in some irrigation centers.

8.7 Overall Analysis

The National Water Policy supports an integrated, decentralized, and participatory management approach. The adoption of these principles in the São Francisco River Basin is a challenge because of its size, climatic conditions (50 % of the basin is in the Brazilian semi-arid region) and complexity (the presence of many and active multiple users). The traditional water scarcity in this region resulting from the growing demands of users has generated conflicts because of water use. The main conflict is between reservoir operators who want water for energy generation and other users who require the water for human water supplies, navigation to transport cargo and passengers, fishing, irrigation, agriculture, and tourism.

The São Francisco River Basin census shows that there are approximately 85,000 users in the basin, and of these a just over 1100 are subject to regulation through water-use permits and water-use charges. Of the users subject to regulation, 86 % are associated with irrigation. Insignificant uses that are not subject to these two management mechanisms can take up to 4 L/s. The Sobradinho Reservoir downstream water balance shows that only 8 % of the minimum flow is used by regulated users; of the 800 m^3/s available, the demand is for 66 m^3/s. In this way, the amount of water downstream from the Sobradinho Reservoir is enough to meet the demands of regulated water users; at this time, there is no intention to restrict the use of water from the Sobradinho Reservoir. It can be said that conflicts are related basically to the water level and not to the flow in the water body.

The recent water crisis, occurring since 2013, has been causing more attention to be given to the water allocation issue. The minimum outflow from the Sobradinho Reservoir was defined at that time based on the Xingó power plant's operating needs. Today, it is necessary to include others requirements to define the minimum restriction and maximum flow permitted. One of the ways is to incorporate a cost/benefit analysis, with the users adding more relevant information to the process. The cost/benefit analysis can be a complement to what is recommended by the Water Law (multiple uses and prioritization of human and animal consumption). It would be an input for developing a ranking of users for the prioritization of water use that would take into account the reality in the basin. It could help to determine which activity would have preference in a conflict situation, power generation, or irrigation.

In cases of great complexity, such as the São Francisco River Basin, the use of cost/benefit analyses in the allocation process would allow an economic vision of alternatives, giving technical and economic bases for the decision-making process. Thus, the adoption of the allocation process constitutes an alternative mechanism to the negotiations already being undertaken in the basin to minimize conflicts resulting from water use. However, it is important to highlight that it does represent a challenge for the managers because of the operational complexity of São Francisco River Basin.

The negotiations that are made today to ease the minimum outflow include participation of the user sectors, which means that they are already participatory. However, they cannot be characterized as allocation meetings when they are designed to temporarily approve (or not approve) a reduction in the minimum outflow, without distributing it for each user.

Another mechanism to reduce conflicts is water-use charging. Charging for water use was implemented by the São Francisco River Basin Committee in July 2010 and is raises annually by BRL 20 million (USD 5 million). This measure seeks to educate the user about the indiscriminate use of water. However, according to Santana (2010), the estimated amount raised by charging for water use in the São

Francisco River Basin is equivalent to only 0.041 % of the gross domestic product of the basin. The use of a scarce resource, like water, needs to have a price that leads users to place an economic value on it. This is one possible measure to slow the degradation of water bodies and to reduce conflicts over their use. Entrepreneurs do not complain because they have to pay a price for the inputs obtained in the market. These inputs are needed in manufacturing their products, which are then sold in the market at prices that recover this expense. Thus, the issue about water could be faced in this same way.

The effectiveness of the management of water resources of the São Francisco River Basin is evident when one considers that maintaining the original restrictions (minimum outflow of 1300 m^3/s) would have led to the complete emptying of the Três Marias and Sobradinho Reservoirs by 2014, putting at risk the ability to meet the multiple uses of the water.

The meetings coordinated by ANA during this process made it possible to increase the participation of public authorities, users, and communities in the management of water resources in the São Francisco River Basin.

It is important to remember that until the publication of Law 9433/1997 (the Water Law or National Water Policy), the reservoirs were operated mainly to meet the needs of the energy sector. The water management was not participatory and decentralized, nor was it intended to meet the multiple uses.

Although the National Water Policy has changed the focus of water management in Brazil in practice, some users still complain that reservoir operation imposes restrictions on other users. The arguments are based on the unpredictability of the water level in the river because of the variation in the flows released by the power plants—in particular, the intraday variations that are considered legal, but are harmful to some uses, such as navigation and some pumping equipment. The public managers reject the users' arguments based on ANA's Resolution number 847 of 2011, which stipulates that "all the interferences in the water bodies related to the water permits, including tank-nets, points of water catchments and discharging effluents, should be dimensioned considering the variation of the water level." In this way, the users are responsible for the adaptation of their infrastructure in the water body, counting on the financial support of the Basin Committee, if possible. Because of that, the managers who are responsible for the São Francisco River Basin affirm that most of the difficulties/conflicts come from the reduction of the water level and not because of the flow in the river.

In 2016, ANA will promote the creation of a working group to propose new operating conditions for the main reservoirs of the São Francisco River Basin. This will provide a way to start a longer-term process in order to offer more appropriate conditions for the current reality of the basin. The group will be formed by ANA, the committee, and the management bodies of water resources in the states of Alagoas, Bahia, Minas Gerais, Pernambuco, and Sergipe.

8.8 Conclusions

The Sobradinho Reservoir plays an important role in meeting the needs of water supply in the northeast of Brazil, a region with frequent critical droughts. It is responsible for energy production and flow control throughout the hydrological year. Thanks to flow regularization, it has been possible to achieve multiple uses in the São Francisco River Basin, one of the basic principles of the Brazilian National Water Resources Law. In addition to that, it has helped the regional economy to take a great leap forward and transform the São Francisco Valley into one of the largest centers of fruit production in Brazil. In the past few years, the average uncontrolled flow in the São Francisco River has been lower than the historical average because of extremely low precipitation and high evapotranspiration. This has led to the basin experiencing an unfavorable hydrologic situation, which represents a real governance challenge for Brazil. Therefore, a set of mitigation measures has been implemented to prevent the depletion of the Sobradinho Reservoir to levels that hinder water uses. The minimum outflow from the Sobradinho Reservoir has been reduced several times over the past 3 years in order to meet multiple uses. The Sobradinho Reservoir usually ensures a minimum outflow of 1300 m^3/s, but today it operates with a minimum outflow of 900 m^3/s. Thanks to these reductions, there has been a cumulative gain of 44.7 % in the active storage of the Sobradinho Reservoir. If the mitigation measures had not been implemented, the reservoir would have been fully depleted by November 2014, causing disruptions in all water uses.

References

Acioll GCL (2005) Operação do reservatório de Sobradinho durante o racionamento energético de 2001. In: Third symposium of specialists in the operation of hydropower plants. SP-06-ONS

Collischonn W, Souza GR, Priante GK et al (2006) Da vazão ecológica ao hidrograma ecológico. Congresso da Água, 8, Figueira da Foz. Available at https://www.researchgate.net/publication/237250787_Da_Vazao_Ecologica_ao_Hidrograma_Ecologico. Accessed 05/2016

Correia MF, Silva Dias MAF (2003) Variação do nível do reservatório de Sobradinho e seu impacto sobre o clima da região. Revista Brasileira de Recursos Hídricos 8(1–2):157–168

Garcia AV, de Oliveira ECA, Silva GP et al (2007) Disponibilidade hídrica e volume de água outorgado na micro-bacia do Ribeirão Abóbora, Município de Rio Verde, Estado de Goiás. R Caminhos de Geografia, Uberlândia 8(22):97–107

Martins DMF, Chagas RM, Melo Neto JO et al (2011) Impactos da construção da usina hidrelétrica de Sobradinho no regime de vazões no Baixo São Francisco. R Bras Eng Agríc Ambiental 15(9):1054–1061

Milliman JD, Farnsworth KL, Jones PD et al (2008) Climatic and anthropogenic factors affecting river discharge to the global ocean, 1951–2000. Glob Planet Change 62:187–194

Nou EAV, Costa NL (1994) Diagnóstico da qualidade ambiental da bacia do rio São Francisco: sub-bacias do oeste baiano e Sobradinho. Rio de Janeiro: Instituto Brasileiro de Geografia e Estatística (Série Estudos e Pesquisas em Geociências)

Paiva MP (1982) Grandes represas do Brasil. Editerra, Brasília

Sampaio Y (1994) Experiências de desenvolvimento rural e seus ensinamentos para o Nordeste do Brasil. R Econômica Nordeste, Fortaleza 25(4):529–544
Santana TAR (2010) Estudo econômico dos impactos da cobrança pelo uso da água na bacia do rio são francisco: uma abordagem insumo-produto. Universidade Federal da Bahia, Dissertação de Mestrado 130 p
Sato Y, Godhino HP (1999) Peixes da bacia do São Francisco. In: McConnell RH (ed) Estudos ecológicos de comunidades de peixes tropicais. EDUSP, São Paulo
Welcomme RL (1985) River fisheries. FAO fish technical paper 262. FAO, Rome

Chapter 9
Adaptation and Mitigation Measures for High-Risk Dams, Considering Changes in Their Climate and Basin

Victor H. Alcocer-Yamanaka and Rodrigo Murillo-Fernandez

Abstract Dams in Mexico are periodically assessed to check for hazards to people living downstream. Among the main causes of these hazards are runoff variations due to changes in climate and basin conditions. Consequently, a program of specialized studies and analysis of options for improved structural and functional conditions of dams was launched, enabling the selection of both structural and operational risk-mitigation measures. As an example of dam management under adverse flood conditions, we present the case of a masonry dam from the nineteenth century prone to failure with significant flooding, for which, with the aim of maintaining its safe operation, water management adjustments and structural modification were proposed. Other cases with different conditions are discussed, which have undergone structural rehabilitation or where restriction measures have been imposed in reservoir water storage and downstream discharges, along with an overview of basin and dam management mechanisms focused on water abundance and shortage conditions.

Keywords High-risk dams · Mitigation risk dam · Climate change · ENSO · Operational measure · Legal framework

---

V.H. Alcocer-Yamanaka (✉) · R. Murillo-Fernandez
Conagua (National Water Commission of Mexico), Mexico City, Mexico
e-mail: yamanaka@conagua.gob.mx

R. Murillo-Fernandez
e-mail: rodrigo.murillo@conagua.gob.mx

V.H. Alcocer-Yamanaka · R. Murillo-Fernandez
UNAM (National AutonomusUniversity of Mexico), Mexico City, Mexico

C. Tortajada (ed.), *Increasing Resilience to Climate Variability and Change*,
Water Resources Development and Management,
DOI 10.1007/978-981-10-1914-2_9

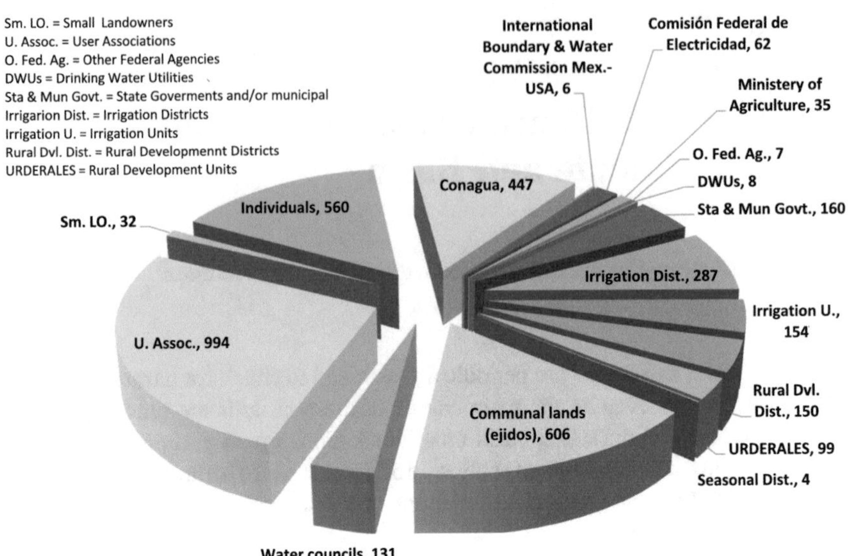

Fig. 9.1 Number of actors responsible for dam management (Conagua NDI 2015)

9.1 Storage Infrastructure and Water Use

Surface water use in Mexico is focused mainly on agricultural, livestock, and industrial activities, and dams play a crucial role in these economic activities. The actors responsible for dam management (Fig. 9.1) are mainly farmers associated in various ways, as well as governmental agencies in charge of water supply for irrigation, power generation, and human consumption, for which Mexico relies on an infrastructure of over 5400 dams distributed across the territory (Conagua SISP 2016). Storage infrastructure consists mostly of earth and rock-fill dams as well as concrete and masonry structures of all kinds. The frequency by dam height is shown in Fig. 9.2, from dams over 200 m high to smaller ones used primarily for livestock. About 41.2 billion cubic meters of surface water are used annually for agriculture, equivalent to 80 % of the available resources (Conagua 2014). This is possible only through adequate water management and conservation, and the operation of storage dams. Sixty-eight percent of the storage infrastructure is dedicated to irrigation (Fig. 9.3).

Most of the dams and their distribution systems were built in the twentieth century, but some date from colonial times (sixteenth to eighteenth centuries). The persistence of these colonial dams shows that their designs were well conceived under the original hydrological conditions, and they have properly operated up to

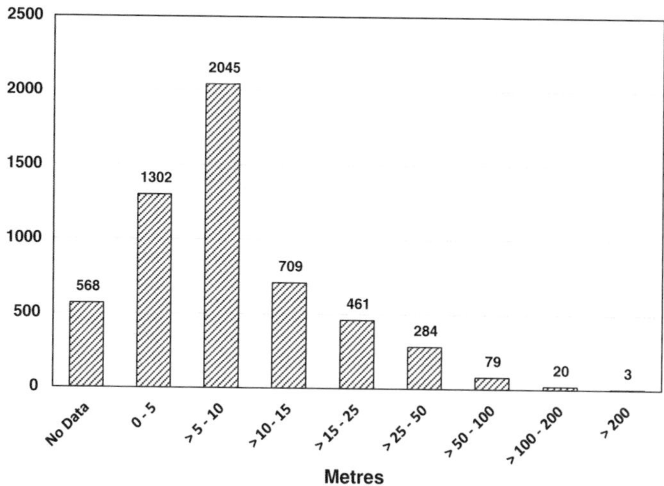

Fig. 9.2 Histogram of height of the 5471 dams in Mexico (Conagua NDI 2015)

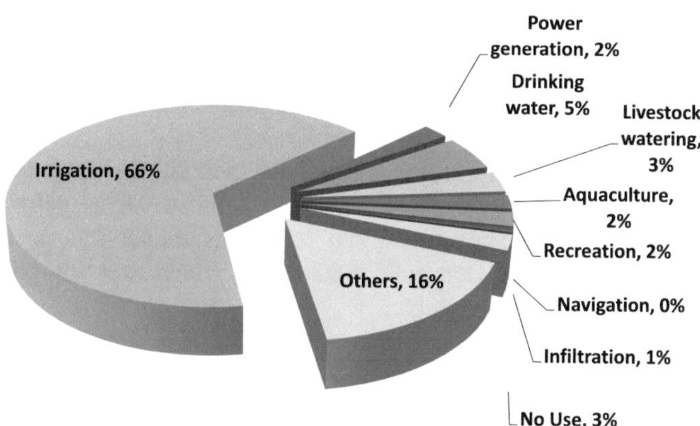

Fig. 9.3 Purposes of dams in Mexico (Conagua NDI 2015)

the present (Fig. 9.4). For decades, however, some dams—some successfully operating in earlier centuries and others recently built with modern designs—have posed risks to populations living downstream.

9.2 Dam Safety Program

Since the early 1990s, dams and flood protection works have received systematic inspections, which led to formalized dam review methodologies: first of the hydraulic works under the responsibility of Mexico's National Water Commission

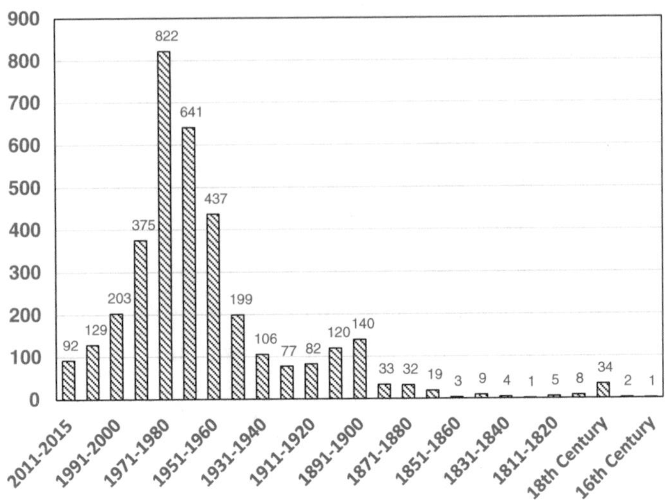

Fig. 9.4 Histogram of dams by year of completion (Conagua NDI 2015)

(Conagua) and subsequently of all remaining dams. In the early twenty-first century, the National Dam Safety Program (Programa Nacional de Seguridad de Presas) was created, through which periodic inspections of dams are carried out (Murillo-Fernandez 2010).

In the first stage, dams are visually assessed and their dimensions, modifications, and overall functional status of components are recorded. Dam, spillway, intake works, reservoir, and downstream riverbed conditions are checked in accordance with established procedures (Conagua-USBR 2000), which were adapted from US Bureau of Reclamation methodologies.

Basic inspection, or level I, involves verifying the dam's characteristics, developing an initial diagnosis of its conditions and classifying it as representing zero, low, medium, or high risk, depending on its physical condition and possible consequences for the population and infrastructure downstream in case of normal or unexpected water discharge. When a high risk is diagnosed, specialized personnel perform a second and more detailed inspection (level II) of the identified anomalies. If the infrastructure requires special attention because of the threat posed, immediate restrictive operational measures are implemented to reduce risks, providing recommendations to improve functional and safety conditions. In the most unfavorable cases, the reservoir is emptied or its filling is restricted. In cases in which the risk is confirmed, a comprehensive program of study is initiated to assess possible operational and structural modifications and establish risk mitigation measures.

In the inspection of over 5000 dams (Murillo-Fernandez 2012), a group of 115 were found to require special attention due to their conditions. Of these, 33 % were at risk due to spillway insufficiency, 11 % had intake structures that were not operating properly, and 7 % could pose significant risks to people living

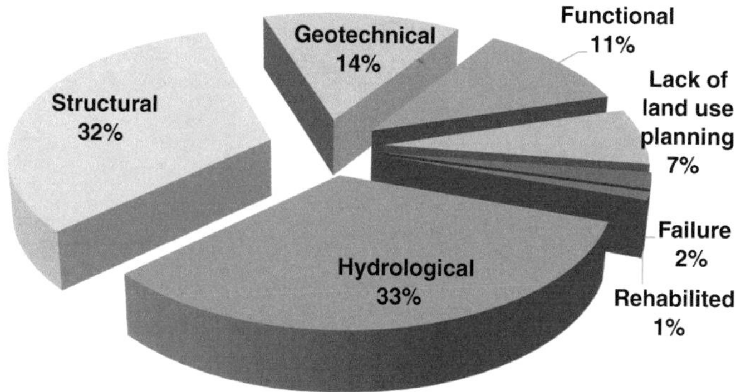

Fig. 9.5 Initial diagnosis of dams by type of hazard (Conagua 2013)

downstream of or adjacent to the reservoir during flood seasons (Fig. 9.5). Furthermore, the hydrological conditions of the basins have changed and rainfall/runoff information has improved since they were designed; thus, it was decided to carry out a hydrologic analysis of most dams. This analysis found that 32 % had possible structural damage and 14 % had potential embankment instability, in addition to the case of some ageing dams with no design information, requiring their stability to be assessed. Some were rigid dams, and some were earthen dams.

The recommendations that followed these reviews were that users or those in charge of such high-risk dams must take operational or structural measures to reduce the threat, which could include taking them out of service permanently. Among those dams classified as high-risk due to their location or size, several were of major importance in their region. Thus, in 2013, Conagua initiated detailed studies of their conditions and sought risk mitigation options, since after several decades of operation their setting and climate conditions had changed.

9.3 Hydrological Conditions in Mexico

Mexico is between the Atlantic and Pacific Oceans, with both having decisive effects on climate, particularly with respect to surface runoff. Tropical waves and cyclones produce most of the summer precipitation, in which the El Niño–Southern Oscillation (ENSO) causes variations. The most severe events bring severe floods in the south and intense droughts in the northeastern border and center of the country during the summer rainy season, while weak or moderate events produce diverse alterations.

Average annual rainfall in Mexico is 760 mm (Conagua-SMN 2015); however, the varied climate, atmospheric circulation, and topography significantly influence the distribution. Precipitation in the southern tropical zones reaches 1846 mm/year, while in the northwestern semi-arid regions it is only 169 mm/year. Annual rainfall

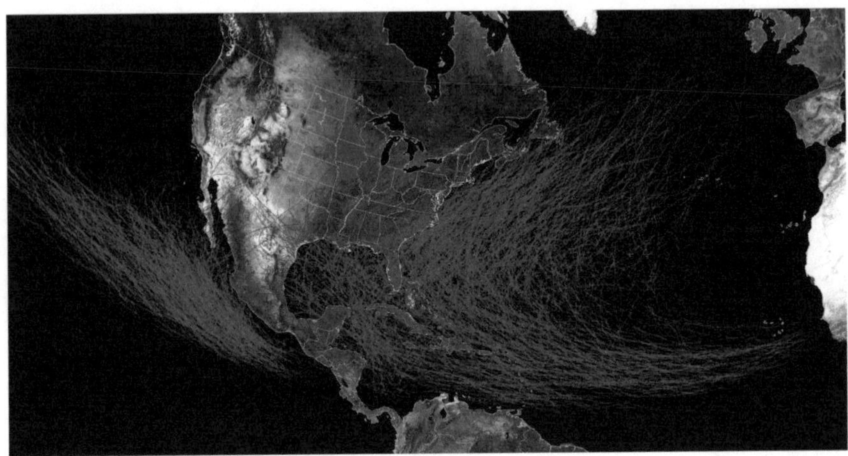

Fig. 9.6 Paths of all North Atlantic (1851–2013) and Northeastern Pacific (1949–2013) hurricanes (at least Category 1 on the Saffir-Simpson Hurricane Wind Scale). From NOAA, NWS, NHC (NHC 2013)

has decreased by 2.2 % over the past 15 years, likely due to climate change, although this figure could be distorted by changes in the location and number of stations.

Tropical cyclones strongly influence variations in regional precipitation and surface runoff. The average annual number of tropical cyclones (named storms) in the Northeast Pacific Ocean over the 1949–2014 period was 13.3; in the North Atlantic Ocean, it was 11.1. The 1992 Pacific hurricane season was the most active on record, with 24 named storms; for the Atlantic Ocean it was 2005, with 27 named storms. The least active hurricane season was 1953 in the Pacific—only four tropical storms formed—and 1983 in the Atlantic. Between 1970 and 2014, 439 tropical cyclones made landfall at Mexican coasts, an average of 9.7 per year (Conagua-Boletines 2015). The paths of these storms are shown in Fig. 9.6.

Evidence of climate change and its effects include the oceans' rising temperature and the substantially higher frequency of tropical cyclones, particularly over the last 20 years. These new conditions have made several dams unsafe from the hydrological point of view.

ENSO effects are of greater relevance in precipitation and runoff. In intense events, they cause a decrease in waves and tropical cyclones in the Atlantic and changes in atmospheric circulation patterns that have an impact in decreased rainfall in the Yucatan Peninsula and the central region of the country. In the Pacific, greater development of tropical cyclones occurs; some make landfall in Mexico, resulting in heavy rain and significant flooding.

Sea surface temperature anomalies for ENSO regions 3 and 4, illustrated in Fig. 9.7, are considered as those affecting climate variations in Mexico. The 1983, 1998, and 2016 intense events were pronounced, and the first two caused widespread droughts. When a strong La Niña event occurs after an El Niño event, a

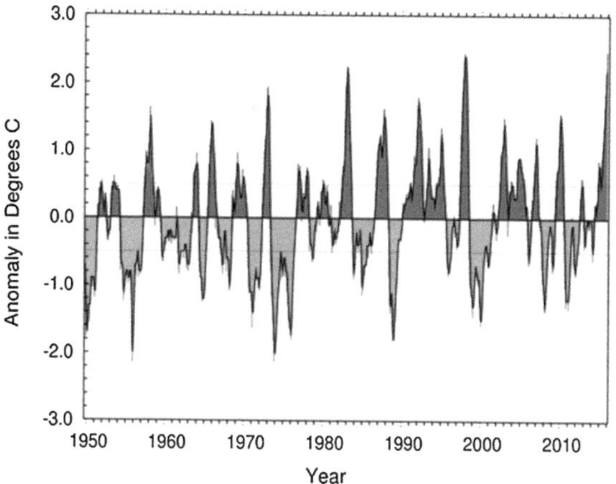

Fig. 9.7 Sea surface temperature anomalies for ENSO regions 3 and 4 (5°N–5°S, 120–170°W) (NOAA 2015)

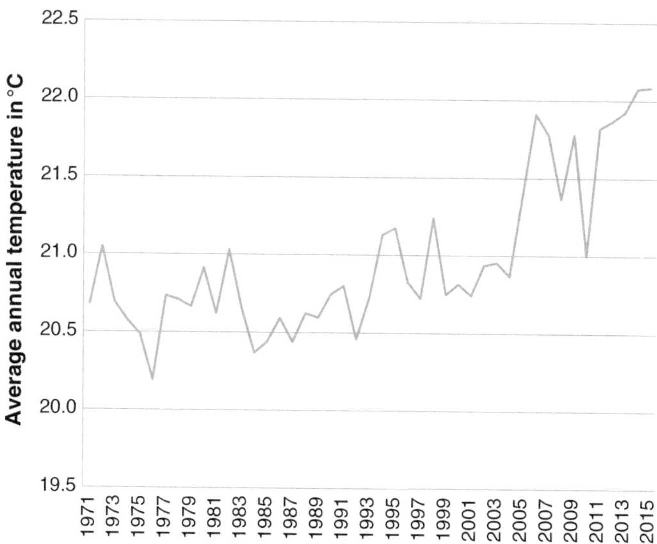

Fig. 9.8 Average annual temperature in Mexico, 1971–2015 (Conagua-SMN 2015)

decrease in precipitation also occurs (e.g. 1973 and 1998). There are also regional variations (Triana 2016).

Average temperatures in Mexico during 1971–2015 show a clear rising tendency over the last 22 years (Fig. 9.8). However, there is no correlation with the average values of annual precipitation (Fig. 9.9), which do not show a significant trend for 1950–2015, possibly due to multiple factors influencing rainfall.

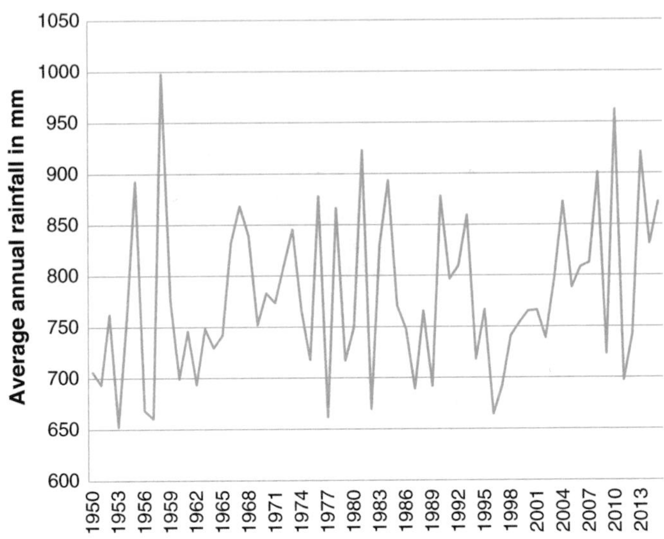

Fig. 9.9 Average annual rainfall (mm) in Mexico, 1950–2015 (Conagua-SMN 2015)

9.4 Regulatory Framework and Governance for Dams

The National Water Law and Regulations (Conagua LAN 2004) stipulate that the management of national surface water, streams, and groundwater is to be provided by the federal government through its National Water Commission, which undertakes various supervision and evaluation tasks for water resources and infrastructure.

The articles of the LAN contain specific guidelines for building in the zones of the federation. These require that water concessionaires, and those constructing dams and other facilities in rivers on federal property, apply for water concessions and occupancy permits for federal land and construction works, for which they must submit a technical plan and related hydrological studies showing that the works are feasible and will not damage the environment. Regulations also oblige concessionaires to prevent damage from inappropriate infrastructure use or water management; in the event of such effects, they are obligated to provide redress.

Major dams (those higher than 60 m) are designed taking into account the most adverse hydrological effects, using a 10,000-year return period or local storm maximization and storm transposition. Smaller dams in rural areas upstream of cropland or pastureland are designed for a 500-year return period or, if populations live downstream, a 1000-year return period.

Dam construction was traditionally carried out by the federal government at the request of water irrigation users for smaller dams, and through planning mechanisms for regional development in the case of large dams. Today, state and municipal governments mainly construct them, with direct involvement of users.

For major storage works, in addition to the above, the federal government participates through Conagua or the Federal Electricity Commission.

Overflows of small dams (mainly earthen dams) have occurred over the last 30 years, due to cyclonic effects and other atmospheric events such as severe storms, which are attributable both to climate change and to runoff basin modifications, especially deforestation and land use changes. Unfortunately, these events have caused the loss of human lives and considerable damage, which led to reinforcing the existing regulations regarding water users' responsibilities and particularly of those operating the infrastructure. Thus, in May 2010, Conagua issued the "Agreement to Identify Those in Charge of Dams in Operation," which provides greater clarity on the responsibility involved in managing water resources and infrastructure (Diario Oficial de la Federacion 2010).

The agreement states that those in charge of dams, whether owners or not, who manage or operate them to use water or the land where dams are located, are responsible for their maintenance, monitoring, and structural and functional safety. It covers any storage with a capacity of 0.25 hm^3 or more, measured up to the ordinary high water level. Dam failure due to inadequate operation or lack of maintenance or care, for administrative, civil, or criminal purposes, causing damage to people or property, infrastructure, the environment, ecosystems, or the artistic, archaeological, or historical heritage of the country, will be attributable to those responsible.

This agreement prevents legal loopholes and sets out liabilities for repairing damage caused by mismanagement or neglect of dams and other reservoirs. These aspects are covered by the LAN, but they need to be clarified.

Among Conagua's attributions is that of establishing water management measures for proper resource management for all uses and simultaneously preventing excesses or deficiencies in surface waters from becoming hazards or disasters to the population. For this, Conagua relies on an interinstitutional collegial body, the Technical Committee on Hydraulic Works Operation (Comité Técnico de Operación de Obras Hidráulicas, CTOOH), which involves the following:

- Conagua's technical areas
- Mexican National Weather Service (SMN)
- Ministry of Agriculture, Livestock, Rural Development, Fisheries and Food (SAGARPA)
- Ministry of Health
- National Autonomous University of Mexico (UNAM)
- National Polytechnic Institute (IPN)
- Federal Electricity Commission (CFE)
- National Centre for Disasters (CENAPRED)
- Mexican Institute of Water Technology and other institutions related to water management and use.

CTOOH holds weekly meetings and establishes the general operating rules for agricultural cycles, water extraction programs for power generation, and operational

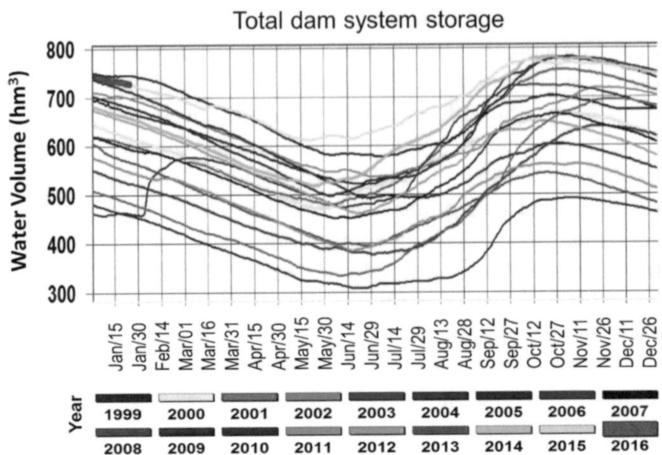

Fig. 9.10 Historical storage evolution in the dam system (Conagua SIH 2015)

aspects for the nearly 200 major dams in the country (Fig. 9.10). When the situation warrants, the hydrological conditions of some basins and dams are analyzed in order to take special hydraulic management measures to prevent flood threats or shortage risk due to possible hydro-climatological extreme events such as cyclones, floods, and droughts.

The National Program Against Drought (PRONACOSE 2015) provides monitoring, mitigation, and prevention of this recurrent phenomenon. Its objective is to develop tools to enable integrated management of basin councils (organizations involving users and authorities) regarding water resources management under extreme shortage conditions, with a proactive and preventive approach. The program provides information on drought monitoring, interinstitutional action policies, and prevention and mitigation for drought periods. It also provides regional programs on hydraulic contingencies for rainfall excess and deficit. In both cases, actions on dams and other infrastructure are considered.

With the aim of providing clear operational rules to those responsible for this infrastructure, voluntary safety standards for dams are developed under consensus with participation of experts and are disseminated for public consultation to allow the multiple stakeholders involved in dam management to express their opinions. The proposed standards include in their technical conception the good practices for design, construction, operation, and maintenance that have been used by the Mexican government agencies for over 85 years. To date, the first standard (*Risk Analysis and Classification of Dams*) has been approved and will soon be published for implementation; a second (*Safety Inspections*) has also been developed and will be submitted for consultation in the coming months; and a third (*Emergency Action Plan*) is being developed (Grupo de trabajo 2015).

9.5 Dam Condition Assessment

Conagua conducts systematic assessments of the quantity and quality conditions of surface water and groundwater, as well as of hydraulic infrastructure status, through the National Program for Dam Safety. Through this, the operational conditions, physical status, and the preliminary classification of possible risks are assessed. The results of assessments enable identification of those works that pose threats to people, infrastructure, or the environment, for which measures to reverse this condition are undertaken.

Preliminary assessments found 115 dams had high potential hazard conditions due to the following causes:

- Hydraulic and hydrological deficiencies caused by flood overtopping failure during spillway and reservoir operation
- Structural failure from dam instability or deterioration from natural causes or negligence
- Functional failure from intake structures and bottom drainage not properly operating, which lead to critical dam conditions
- Lack of land use planning, where flood zones downstream are inhabited under the assumption that the dam controls all flooding and no spillway discharges will occur. This false assumption is likely to hinder proper spillway and reservoir operation.

9.5.1 Methodology for the Inspection and Evaluation of Dams

When specialized studies are conducted to verify current conditions, the procedures are as follows:

- Gathering of historical information on design, construction and operation in the records of those in charge of their operation from the Historical Water Archive (AHA 2015) and the Sistema Informático de Seguridad de Presas (Conagua SISP 2016).
- Physical structure inspection to verify their conditions and actual dimensions, and complement the observations made where appropriate.
- Determination of probable floods using from 5-year to 10,000-year return periods
- Reservoir topography where applicable and study of the reservoir-spillway operation and hydraulic conditions downstream.
- Geotechnical sampling to corroborate the properties established in the original design and construction of flow networks with normal and maximum water level and under sudden drawdown conditions.

- Slope stability analysis of earth dams (geotechnics and geology) under filling and seismic conditions.
- Stability and deformability study of rigid dams if deficiencies, structural damage, or collapse likelihood were reported.
- Operation assessment of intake structures and bottom drainage.
- Determination of the flood zone for normal events (100-year return period) and for dam rupture conditions, with damage assessment.
- Proposal for risk mitigation options.

In the hydrologic and hydraulic performance evaluation, the current basin conditions are checked. Other checks include any natural or anthropogenic modifications due to soil and water use changes, previous floodplains being inhabited and runoff variability effects due to climate change or basin modification by land surface modification (deforestation, urbanization and other causes).

The case of a nineteenth-century masonry dam was selected from the works studied. The dam is upstream of a large city, and there was no background information on whether its design or construction was likely to cause flooding.

9.5.2 Case Study: El Cajón, a Nineteenth-Century Masonry Dam

9.5.2.1 Historical Condition

El Cajón Dam was built in 1880, in what was then the capital of a province in the central region, in which the cyclonic effects of both oceans are present. It receives water from the El Arenal and San Isidro streams, among others, in La Laja River basin, which flows into the Lerma River, and is located in the Lerma–Santiago hydrologic region. The dam location is show in Fig. 9.11. It was built by private individuals and originally used exclusively for irrigation of 200 ha.

It is a masonry dam bonded with hydraulic lime mortar (*calicanto*), including hand-carved basalt pieces. The maximum height is 15.3 m from the foundation; it is 187.6 m long, in three straight sections, with a stepped cross-section (Fig. 9.12). The dam frequently operates as a spillway along its length, because the right bank, with its 22-m crest, is sufficient for floods of only 48 m^3/s (Fig. 9.13). The dam was built on vesicular basalt of good quality. For nearly a century, it operated without presenting deficiencies or posing danger to the population. There is no information on its original design. At the time it was built, there were no regulations establishing periodic inspections of spillway capacity or stability of the curtain under different conditions.

The Dolores and Santa Catarina Dams, upstream of El Cajón, also date from the nineteenth century and operated under similar conditions for most of the twentieth century, providing water for irrigation and regulating floods in the basin.

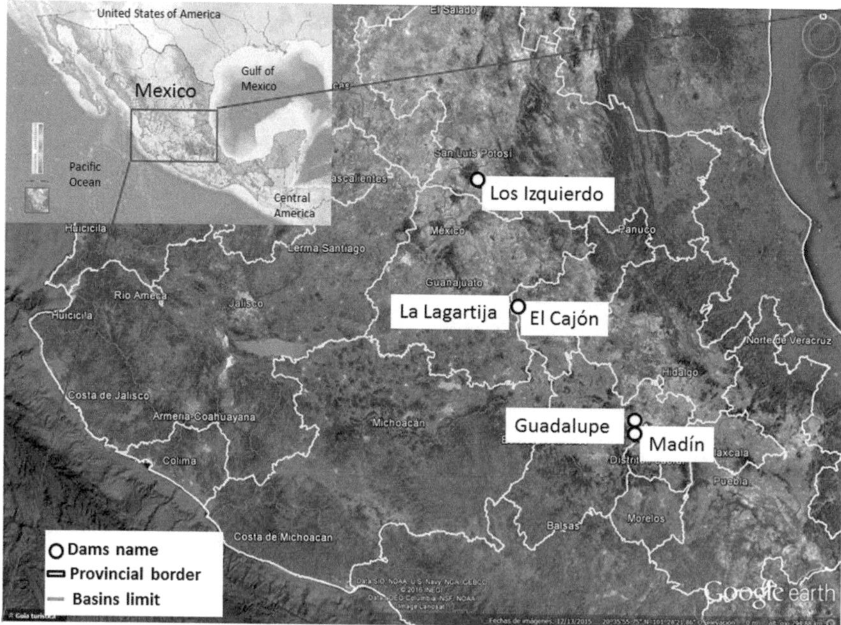

Fig. 9.11 Location of dams

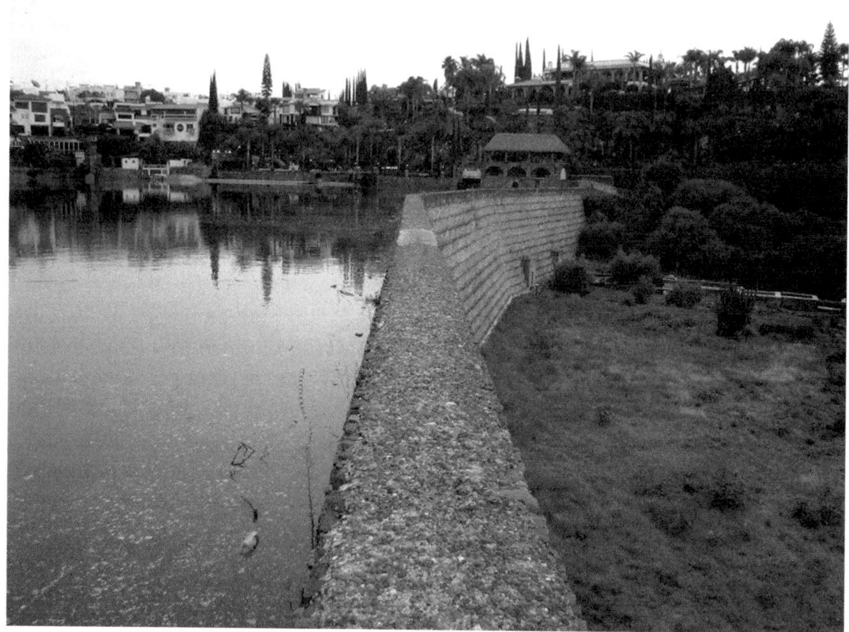

Fig. 9.12 El Cajón Dam in 2013 (Conagua SISP 2016)

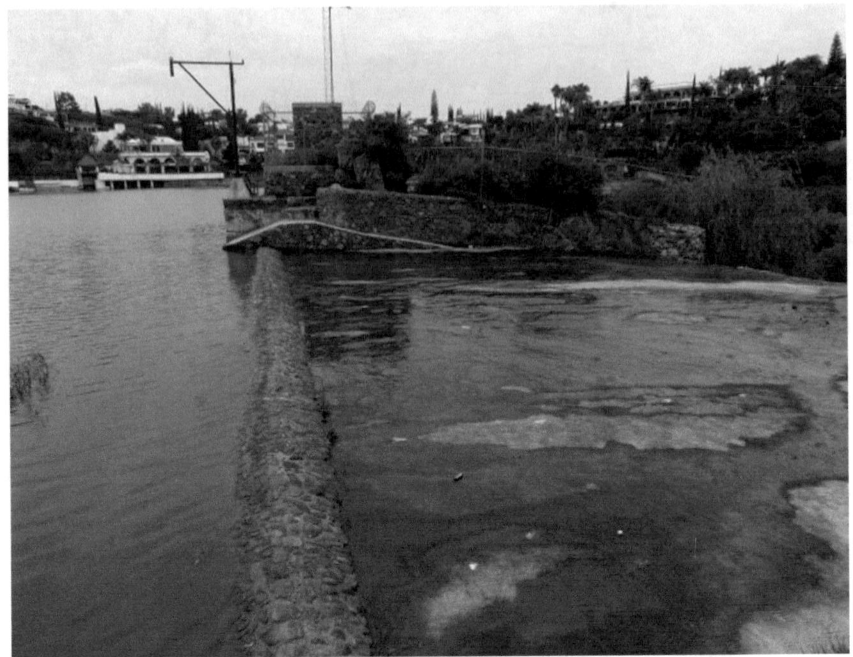

Fig. 9.13 El Cajón Dam spillway in 2013 (Conagua SISP 2016)

9.5.2.2 Current Status

The dam is now almost completely surrounded by urban areas, and its irrigation zone has virtually disappeared. A heavy rainfall event in October 1986 caused water overtopping. Conagua had conducted operational and stability studies confirming that for floodwaters producing flows higher than 1.0 m over the crest, equivalent to a 10,000-year return period flood, the structure's stability was precarious (Nieto 1992). According to current criteria, dams upstream of major cities must have a stable dam and weirs with sufficient capacity to discharge and withstand flood flows.

Conagua encouraged the owners to build a spillway controlled by gates to allow flood discharges as well as to reinforce the dam. However, the proposal was not accepted. A prestigious national engineering company then carried out a flood and stability assessment of the dam (CIEPS 1998). CTOOH issued a finding in late 1997 informing the owners that they should maintain low levels in the reservoir during rainfall season to reduce the likelihood that the dam wall would overturn.

In 1998, studies concluded that a 10,000-year flood was likely, discharging across the dam crest at a height of close to 1.0 m and probably causing instability, and therefore that the dam should be reinforced. Given this threat of collapse, Conagua proposed an operation policy of maintaining low water levels during the rainy season (July–October) such that the water would not rise above the crest of

Fig. 9.14 Effects of the 2002 flood, El Arenal River (Conagua SISP 2016)

the dam and regulating flood flows without spilling over the crest. The owners increased the spillway length from 11 to 22 m, without following the recommendation to manage storage at low levels in the rainy season.

The 66.4-mm rainfall event of 16 September 2002 (Conagua Cicese 2015) and the rupture of small flood control works upstream caused the dam to spill over with a hydraulic load of 15 cm, flooding the El Arenal River, which did not have the flood-carrying capacity for the discharge water; the river overflowed (Fig. 9.14).

Due to increasingly frequent flooding downstream, state and municipal governments launched the construction of dams in the basin's upper section. Los Jazmines and La Lagartija Dams and other flood control reservoirs were built during 1996–2002, as described in the *Stormwater Master Plan Update of Queretaro Metropolitan Area 2008–2025* and sponsored by local authorities (UAQ 2015).

Given the dam causing unexpected flooding of populated areas, it was classified in 2012 as high-risk and a comprehensive study of structural and hydrological conditions was carried out of dams in the basin, both old and new, as well as the area that would be affected. New hydrological studies confirmed the basin conditions (UNAM II 2013), with the following findings.

Given the current conditions of the 328.5-km^2 basin (Fig. 9.15), runoffs have been modified by changing land use from a predominantly agricultural and forested area to mostly agricultural and urban. The faster-moving water concentrates floodwaters. Moreover, agricultural water use has decreased, and today the reservoir is also used for recreation, so it is kept filled most of the year. If the two dams

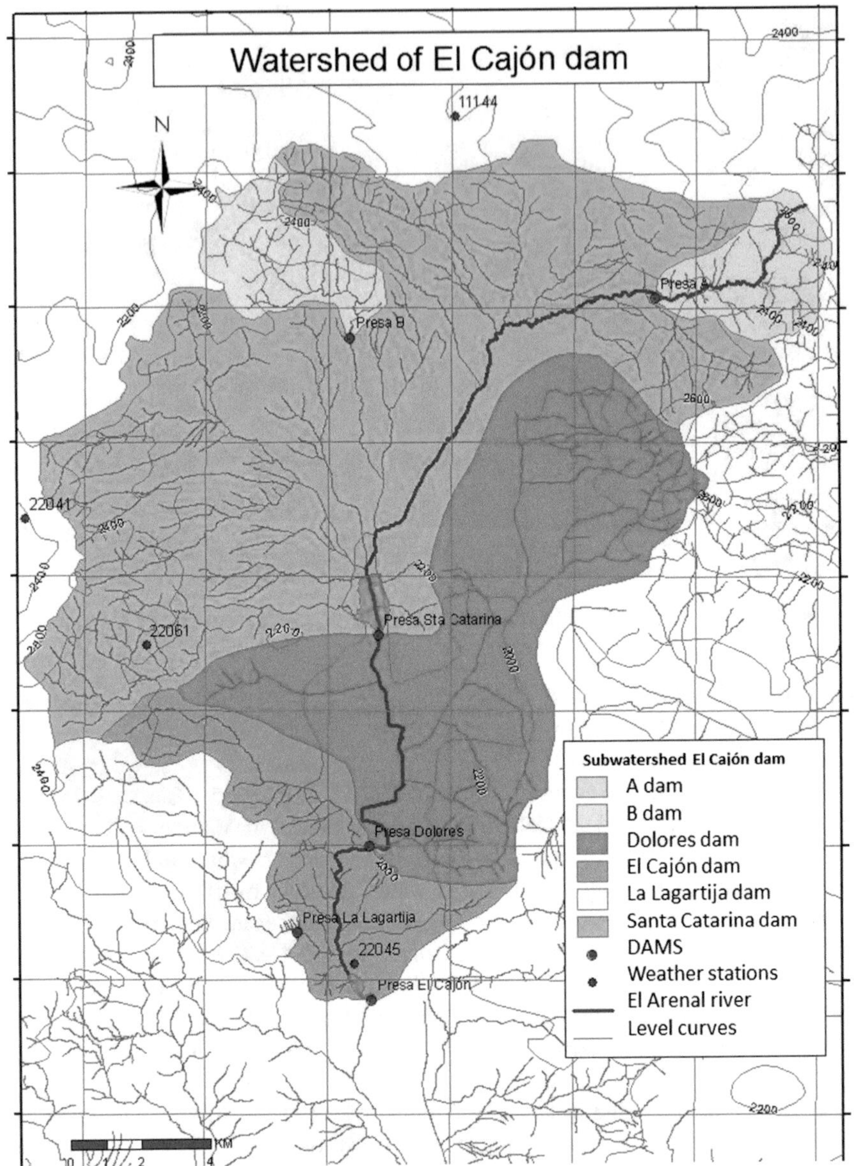

Fig. 9.15 El Cajón Basin and location of other dams (UNAM II 2013)

upstream are also considered, the basin is reduced to 116 km^2, and the maximum floods in the basin bottom are 222.8 and 416.5 m^3/s for 100-year and 10,000-year return periods, respectively. In both cases, floodwater would be discharged over the crest of dam, causing severe flooding in urban and industrial areas downstream

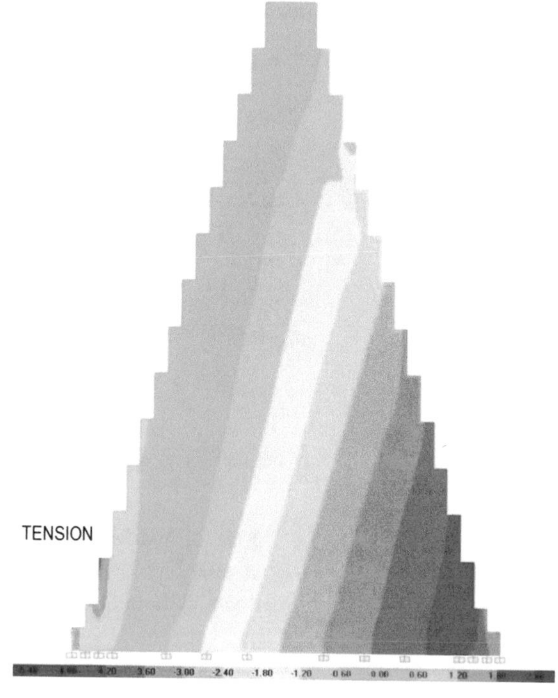

Fig. 9.16 Cross-section and axial stress on the dam in kg/cm^2, under 100-year return period flood conditions (UNAM II 2013)

because the El Arenal River has a flow capacity of only 40 m^3/s, which is insufficient for a greater floodwater flow or a possible infrastructure collapse.

Stability analysis was performed in view of the thinness of the dam's cross-section, presenting potential instability for high flows during floods; thus, finite element method analysis was also performed. This dam is unique in the country, due to its structure and hydraulic performance. Stability studies confirmed that for floodwaters producing flows higher than 1.0 m over the crest, equivalent to a 100-year return period, the structure presented significant stress in the upstream heel (Fig. 9.16). This could cause collapse or overturning, suddenly discharging 650,000 m^3 of water plus 300,000 m^3 of mud into the state capital, as the reservoir is greatly silted.

The owners were warned of the potential instability of the dam. Both corrective and preventive mitigation measures were proposed, such as maintaining the intake and spillway in adequate condition, as well as low water levels in the reservoir during floods, to protect the masonry from strain exceeding its resistance or causing cracking.

These studies confirmed the potential hydraulic–structural hazard conditions. Therefore, in addition to the previous temporary measures, the construction of a side-channel spillway was proposed. A spillway with a capacity of 350 m^3/s would

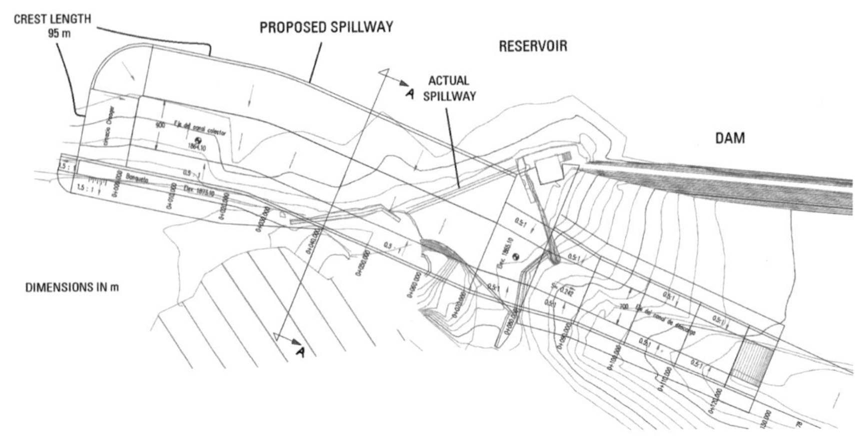

Fig. 9.17 Proposed spillway plan for El Cajón Dam (UNAM II 2013)

be sufficient to discharge floods of a 1000-year return period (Fig. 9.17), while simultaneously maintaining a suitable level in the reservoir for recreation and allowing efficient flood discharges without stressing the dam to the point of rupture, with all its consequences.

A hazard or threat is defined as the occurrence of a given event, whose magnitude can cause damage to susceptible goods and, in the case of flood control, is often estimated based on one or various return periods. In a flooding event, there may be two hazardous components: flood depth and flow velocity. Risk is the likelihood of damage occurring in an event (hazard), depending on the susceptibility of goods exposed (vulnerability). In decision-making, it is common to express risk in terms of economic losses. Typically, flood hazard zones are defined in terms of flood depth and flow velocity (in some settings, the product of depth and speed).

Danger zones from controlled and uncontrolled discharges of the dam were evaluated in this study using topographic data at 1:50,000 scale (INEGI 2013), and the output hydrographs calculated in the hydrological inspection. This information was loaded into a two-dimensional model to determine flood depth and areas, and velocities expected, for both types of discharge.

Although there is no consensus on the limits to be observed in depth and speed variables (Alcocer Yamanaka 2016), this study defined high-risk areas as having depth greater than 1 m, velocity greater than 1 m/s, or depth × velocity greater than 0.5 m^2/s.

For this dam, taking into account flood depth greater than 0.20 m from discharges for a 100-year return period, it was found that the urban area affected would be 490 ha. If a population density of 1162 per km^2 (INEGI 2001; INEGI SEM 2015; INEGI Censo 2010, 2015) adopted, 5690 inhabitants and 182 ha of industrial zones would be affected.

For the worst-case scenario, equivalent to dam rupture from a gap formation (USBR 1987), calculations indicate a flood of at least 0.20 m over an area of

Fig. 9.18 Map of hazard potential for the rupture of El Cajón Dam (Murillo-Fernandez 2014)

1714 ha, of which 688 ha is urban area, with a total affected population of 7993 inhabitants and 388 ha of industrial areas. The rest of the flooded area would be farmland and land for other uses (Fig. 9.18).

Due to the threat it represents, other mitigation measures have been proposed, which include operational aspects of storage, from the extreme case of avoiding storing water to operate only as a flood control dam, to structural measures such as constructing various types of weirs or a bottom drainage. Other options, such as building flood control dams in the upper part of the basin, have already been implemented through the Stormwater Master Plan (UAQ 2015). The option considered most suitable is building a free-operating side-channel spillway, which would maintain reservoir levels to meet current uses of the water body and at the same time permit the rapid discharge of low-level floods in the reservoir to prevent the dam being subjected to strain that could produce structural damage, or rupture in the worst-case scenario.

The owners of the dam are aware of the conditions and options but have not yet decided whether to conduct structural measures (building a new spillway) or operational measures (reducing the reservoir storage in the rainy season).

9.5.2.3 Other Dams in the Basin

During the 2002 event, floodwaters overtopped La Lagartija Dam, which was broken by erosion and caused a sudden floodwater increase at El Cajón Dam. La Lagartija Dam is homogeneous, 6.1 m high, and 377 m long, with a capacity of 0.12 hm^3 for irrigation and flood control (Fig. 9.11). This dam was also ranked as high risk, and studies recommended raising the embankment by 1 m to withstand the probable floods, as well as improvements to its spillway and intake works. These actions were undertaken by the state government in 2015 (GEQ 2014), and it is currently under a diagnosis process to be classified as secure (Fig. 9.19).

Fig. 9.19 La Lagartija Dam rehabilitated, January 2016 (Conagua SISP 2016)

9.6 Other Representative Cases

One of the common situations in dams for agricultural use has been that because of siltation it is not possible to store sufficient water for crops. This leads those responsible for operation to obstruct the spillway in order to store the water volumes needed using various materials, such as soil-filled sacks (which can be dragged by floodwaters). In the worst cases, a masonry or concrete wall has been built on the spillway crest, causing significant freeboard reduction and subjecting earthen dams to overtopping conditions not originally considered in the design, with potential embankment erosion from floodwaters and associated ruptures. These interventions have serious consequences, including loss of life.

Los Izquierdo Dam in San Luis Potosí (Fig. 9.11) was under these conditions. It was a homogeneous earth dam, 10 m high and 355 m long, with a 4-m-wide crest. It had an estimated 0.750 hm^3 of storage capacity and a free-discharge spillway, originally made of masonry, formed by a straight cornice 1.5 m high and 80 m long, in which a reinforced concrete slab 20 cm thick had been placed, thus reducing its low freeboard. The difference between the spillway crest and the dam crest was 1.2 m, which was insufficient to discharge frequent floods. The crest was uneven, with height differences of up to 0.40 m; therefore, the maximum depth that the spillway could discharge was 0.80 m before overtopping the crest. Furthermore, the embankment had been damaged by waves in the upstream slope and by runoff in the crest and the downstream slope. The intake lacked maintenance and was incomplete and out of service.

In 2010, floods caused by Hurricane Alex caused the reservoir water level to reach 0.40 m below the embankment crest, and the spillway discharged 0.80 m of floodwaters, with imminent danger of discharging over the crest of the dam and causing its erosion and collapse (Fig. 9.20). The embankment near the spillway was

Fig. 9.20 Los Izquierdo Dam in 2010, during flooding caused by Hurricane Alex (Conagua SISP 2016)

undermined by vortex effects, causing cracks and fractures in the spillway and cracking of the discharge channel due to loss of supporting material in its foundation. Given the critical conditions, rapid modification was called for. In 2013, the embankment was raised 1.84 m; the crest was raised 0.20 m with reinforced concrete, and a buffer tank was built in its discharge; the discharge channel was protected; and the intake works were modified. Figure 9.21 shows the dam after the remediation measures.

Other typical cases are those dams that over time have become surrounded by urban growth, both in small towns and mega-cities, whose hydraulic performance and dam stability have been assessed. In some cases where their reservoirs cannot regulate floodwaters, because the intake flow is virtually the same as the outflow, the solutions implemented were to eliminate those dams that accumulate a water volume only temporarily; in case of rupture, thus may increase damage downstream. Alternatively, flood control dams were built in the upper section of the basin.

Some dams with good reservoir capacity have proven to be structurally safe, and their hydraulic operating conditions for recommended return periods have shown that their spillways have adequate flood discharge capacity. However, due to urban

Fig. 9.21 Los Izquierdo Dam in 2014, after rehabilitation (Conagua SISP 2016)

growth in downstream areas, it was decided to limit spillway discharges. This has been through reservoir operating policies that reduce the potential storage in the rainy season and allow the free evolution of the reservoir from late October to the beginning of the rainy season, in May of the following year. Such is the case of the Guadalupe and Madín Dams in Mexico City (Fig. 9.11): their filling is restricted to discharge flows according to their intake works, without flooding occurring downstream.

9.7 Role of Dams During Droughts

When it is evident that a drought has begun, recommendations are established for users to optimally use stored water. Actions mainly consist of limiting crop areas, changing to less water-consumptive crops, and using groundwater sources to complete the necessary amount of water for irrigation. Surface water is mainly used for agriculture but is essential for livestock, which is why use priorities are established in the LAN as follows, in decreasing order: domestic, urban public, livestock, agriculture, ecological conservation, electric generation for public service, industry, aquaculture, and others of lesser importance.

9.8 Resilience to Flooding from Dam Failure

Disasters caused in recent years by the unexpected discharge of small dams have called for greater attention to their general condition and operation, inspection programs that provide greater certainty of their conditions, and preventive actions for dams of all sizes.

To reduce the likelihood of events affecting the population, various types of measures have been established, such as the following:

- Reservoir operating policies, limiting storage when they pose a danger until corrective measures are implemented or their security is improved.
- Permanent monitoring programs for dams not in good condition, to reduce the likelihood of unsafe operation.
- Rehabilitation of structures for better water performance and use.
- Dam safety laws, regulations, and standards that define conservation, monitoring, and supervision procedures, as well as measures to reduce the likelihood of harm to the population, productive areas, infrastructure, environment, and other goods.
- Engagement of water users in infrastructure care and management of volumes stored under both abundance and shortage conditions.
- Greater knowledge in society of the vulnerability of the population, dam risk, and emergency care plans through dissemination of civil protection measures implemented in various hurricane, heavy rainfall, and flooding programs, as well as drought indicators and mitigation programs.
- Attention to dams with respect to their likelihood of causing social, environmental, or economic damage, allowing exploration of different ways to reduce flooding hazards, with the participation of water users, the general public, and government authorities at local, regional, and national levels.

9.9 Key Lessons Learned

Knowledge of hydroclimatological variables, their historical evolution, and forecasting, coupled with determining the current conditions of storage infrastructure and flood control, allow the prevention of threats associated with water abundance or shortage and implementation of mitigation measures for extreme events such as cyclones, floods, and droughts.

Adequacy of reservoir management regulations carried out with the involvement of society, experts, and authorities has reduced damage due to variations of surface water extent, and consensus technical measures have been established to address these hazards.

The participation of water users and of those in charge of infrastructure operation allows improvements in dam functional conditions, reducing the adverse effects of floods and enabling better use of available surface water.

The detailed analysis of current basin hydrological conditions, stability, conservation, and functional conditions of dam structures, and the study of flood influence zones, provides essential tools for decision-making that enable better water and infrastructure management.

It is necessary to further analyze the variables affecting surface runoff, update technical studies of dams, seek consensus with users on water management, and update the legal framework for consensus standards for their use, conservation, and operation. These measures will enable better water use and safe and efficient operation of dams.

9.10 Conclusions

There is no evidence that global warming directly affects precipitation or surface runoff. Furthermore, the geographical location of Mexico is significantly influenced by the development of ENSO, producing increased rainfall in some regions and drought in others, when there is a positive anomaly event. When an intense La Niña follows an intense El Niño, precipitation decreases (1972 and 1998). The negative anomaly event following the 1983 El Niño was accompanied by increased rainfall.

Deforestation and changes in land use modify floods, which subject previously secure dams to hydrological conditions different from those they were designed for, making it necessary to review the functioning of reservoirs facing new runoff conditions. The cases presented in this chapter illustrate how it is possible to improve the operability and safety of storage infrastructure to withstand the hydrological and structural stresses to which they are currently subjected—when the conditions considered in the original design have been modified—whether by changes in the basin environment, by anthropogenic impacts or climate variations, or by structural modifications.

Of the group of dams identified to date, 23 have been rehabilitated, 3 are out of service, 6 will be closed down, 17 were determined to present lower risk than was preliminarily estimated, 48 have been assessed to verify their conditions and implement improvement measures, and 8 are under rehabilitation. For the remainder, provisional recommendations have been issued to improve safety conditions while operational or permanent structural actions are determined, in coordination with water users and those in charge of their management or operation.

In all cases, nonstructural measures, such as limiting reservoir fill depth or keeping it empty, enable the search for alternatives and reduce hazards, while structural options improve the operability and safety of dams. In some cases, measures will be drastic, such as closing down the dam or building new infrastructure to replace it. To reduce the risk posed by these dams, the aim is to complete the recommended remediation by 2018. Nonstructural measures to reduce vulnerability and risk are permanent and are periodically reviewed.

In addition to reducing harm, these structural and nonstructural mitigation measures provide experience in safety, efficiency, and operational improvement, which may be applied to other dams in the country with similar conditions.

References

AHA (2015) Archivo Historico del Agua. Retrieved from http://www.conagua.gob.mx/archivohistoricoybiblioteca/

Alcocer Yamanaka VH (2016, June) Metodologia para la generacion de mapas de riesgo por inundacion en zonas urbanas. Tecnologia y Ciencias del Agua VII(3):28

CIEPS, S.A. (1998) Presa El Cajon, Arroyo Jurica, Qro. Informe sobre la modificacion propuesta. Mexico City

Conagua (2013) Diagnostico de presas por tipo de peligro. Subdireccion General Tecnica. Mexico City (unpublished)

Conagua (2014) Estadisticas del Agua en Mexico, 2014 edn, vol 1. Conagua, Mexico City, Mexico

Conagua Cicese (2015, Oct 8) CICESE. Retrieved from CLICOM http://clicom-mex.cicese.mx/

Conagua LAN (2004) Ley de Aguas Nacionales y su Reglamento, 2004 edn, vol 1. Conagua, Mexico City, Mexico

Conagua NDI (2015) National Dams Inventory 2015 (unpublished)

Conagua SIH (2015, Oct 8) Sistema de Información Hidroclimatológica—SIH. Mexico City, Ciudad de Mexico, Mexico

Conagua SISP (2016, Feb 14) CONAGUA, V2. Retrieved from Tramites y Servicios/Aguas Nacionales/Sistema de Seguridad de Presas: http://172.29.151.136/sisp_v2/hinicio.aspx

Conagua-Boletines. (2015, Aug 08). CONAGUA. Retrieved from Comunicados de prensa: http://www.conagua.gob.mx/CONAGUA07/Comunicados/Comunicadodeprensa525-15.pdf

Conagua-SMN (2015) Precipitacion y Temperatura Media anual 1941–2015. Servicio Meteorologico Nacional. Mexico City (unpublished)

Conagua-USBR (2000) Manual para la capacitación en seguridad de presas, 2000 edn. Conagua, SGT, Mexico City, Mexico

Diario Oficial de la Federacion (2010, May 27) Acuerdo mediante el cual se identifica a los responsables de la presas en operación. DOF, Primera seccion, pp 59–61

GEQ (2014) Proyecto para la rehabilitación del bordo La Lagartija. Design, Gobierno del estado de Queretaro, Secretaria de Desarrollo Agropecuario, Queretaro

Grupo de trabajo (2015, Apr 21) Diario Oficial de la Federacion. Retrieved from www.dof.gob.mx/nota_detalle_popup.php?codigo=5391698

INEGI (2001) Superficies de las areas geoestadisticas estatales y municipales, vol 1. Mexico

INEGI (2013, Oct 10) Continuo de elevaciones mexicano CEM v 3.0. Mexico

INEGI Censo 2010 (2015) Instituto Nacional de Geografia y Estadistica. Retrieved 28 Sept 2015, from Censos y conteo: www3.inegi.org.mx/sistemas/ageburbana/consultar_info.aspx

INEGI SEM (2015) Instituto Nacional de Geografia y Estadistica. Retrieved 28 Sept 2015, from Sistema Estatal y Municipal de Base de Datos: sc.inegi.org.mx/cobdem/

Murillo-Fernandez R (2010, 12) PNSP (Colegio de Ingenieros Civiles de Mexico, Ed.). Ingeniería Civil LXI(500):30–34

Murillo-Fernandez R (2012) Sistema Informatico de Seguridad de Presas (SISP). In Asociacion Mexicana de Hidraulica, Memorias del XXII Congreso Nacional de Hidraulica. Acapulco, Guerrero, Mexico, p 6

Murillo-Fernandez R (2014) Medidas de mitigacion para presas en alto riesgo. In Asociacion Mexicana de Hidraulica, Memorias del XXIII Congreso Nacional de Hidraulica (p 6). Puerto Vallarta, Jalisco, Mexico

NHC (2013) National Hurricane Center. Retrieved 7 Sept 2015, from www.nhc.noaa.gov/climo/images/1851_2013_hurr.jpg

Nieto SF (1992) Presa El Cajon, Queretaro Revision de la estabilidad de la cortina. Study, Queretaro

NOAA (2015) National Oceanic and Atmospheric Administration. Retrieved 04 Jan 2016, from Equatorial Pacific Sea Surface Temperatures: www.ncdc.noaa.gov/teleconnections/enso/indicators/sst.php

PRONACOSE (2015) PRONACOSE. Retrieved 08 Oct 2015, from www.pronacose.gob.mx/

Triana RC (2016) Manejo del Sistema Hidroeléctrico Grijalva, Escenarios 2016. Study, CONAGUA, Organismo de Cuenca Frontera Sur, Tuxtla Gutierrez

UAQ (2015) IMPLAN Queretaro. Retrieved from http://implanqueretaro.gob.mx/web/inicio/programas/plan-pluvial

UNAM II (2013) Revision hidrologica, presa El Cajon. Universidad Nacional Autónoma de México, Instituto de Ingenieria, Mexico City

USBR (1987). In: Department of the Interior (ed) Design of Small Dams, 3 edn. USA

Chapter 10
Seyhan Dam, Turkey, and Climate Change Adaptation Strategies

Bülent Selek, Dilek Demirel Yazici, Hakan Aksu and A. Deniz Özdemir

Abstract Climate change adaptation is the adjustment in natural or human systems in response to actual or expected climatic stimuli or their effects, which moderates harm or exploits beneficial opportunities (IPCC 2007). Water resources are significantly affected by climate change. Many studies show that the Mediterranean region of Turkey will be affected by decreasing precipitation and increasing temperatures. In this study, Seyhan Dam, located in this region, was studied as a tool for climate change adaptation strategies. This chapter provides a brief overview of the requirements that led to the construction of the Seyhan Dam, the characteristics of the basin and the dam, the results of climate change studies for the basin, and an evaluation of the adaptation strategies and policies with a broader perspective. The roles of the Seyhan Dam in increasing resilience to climate change are detailed in the first part of the chapter. The second part of the chapter explores the recent legislative and institutional changes in the water sector and assesses the impacts of these developments in the context of climate change.

Keywords Seyhan dam · Climate change · Adaptation · Flood · Drought · Infrastructure

10.1 Introduction

In recent years, many institutional and legislative developments have been seen within the Turkish water policy agenda. The majority of these developments are closely related to adaptation strategies for climate change impacts on water resources. A newly established institution, the General Directorate of Water Management (DGWM), was tasked with conducting an assessment of climate change impacts on water resources, which, in turn, has boosted the number of

B. Selek · D. Demirel Yazici (✉) · H. Aksu · A.D. Özdemir
Investigation, Planning and Allocation Department, General Directorate of State Hydraulic Works (DSİ), Ankara, Turkey
e-mail: dilekdemirel@dsi.gov.tr

C. Tortajada (ed.), *Increasing Resilience to Climate Variability and Change*, Water Resources Development and Management,
DOI 10.1007/978-981-10-1914-2_10

studies in this area. Furthermore, management strategies for extreme hydro-meteorological events—the severity and frequency of which will be the most affected by climate change—have been fundamentally modified through legislative changes.

As the effects of climate change increase, water scarcity and the efficient use of water assumes the highest priority in the adaptation strategy. Many parts of Turkey are projected to face significant changes in water availability, including the Seyhan Dam Basin in South Central Turkey. The dam was one of the first dams constructed by the Republic of Turkey. It began operation in 1956 with the joint aims of producing energy, providing water for irrigation, and flood control.

10.2 Need for the Seyhan Dam

The Seyhan Dam is part of a wider project that includes a range of water management infrastructure constructed throughout the catchment. The project was designed to increase agricultural production, supply energy, and provide flood protection as well as environmental benefits for the fertile Adana Plain in South Central Turkey (Fig. 10.1). Its development was based on the control and use of the Seyhan River, which has a catchment area of 20,731 km^2. The catchment upstream of the dam is defined as the Upper Seyhan River Basin, while the Lower Seyhan River Basin consists of the Tarsus and Yüreğir plains (President Demirel 2004).

The Adana Plain is a delta formed by three rivers—the Seyhan, Ceyhan, and Berdan, of which the Seyhan is the largest and most important. It contains some of the most productive agricultural land in Turkey and is particularly well suited for growing cotton, oilseeds, and citrus fruits. Adana, Tarsus, and Mersin are important industrial centers in the plain, and the latter is Turkey's second largest Mediterranean port. The development of the region has been stunted during times of flood because of insufficient electrical power and water for irrigation (IBRD 1952).

Aside from loss of life, the average annual losses resulting from flooding were estimated at about USD 2800,000 in the 1950s when Turkey was an agricultural country, with about 78 % of its population engaged in farming. There were only three dams in Turkey at the time and very little irrigation for a country with about 8.5 million ha of arable land (IBRD 1952).

Public and private industries and utilities in the area suffered serious power shortages; most of the power was supplied by numerous small hydropower facilities, thermal plants fed with coal, oil and oilcake, and expensive mechanical power. Total power consumption, both electric and mechanical in 1951, was just 39 million kW. Only 789 million kWh of energy were generated (with just 4 % from hydropower)

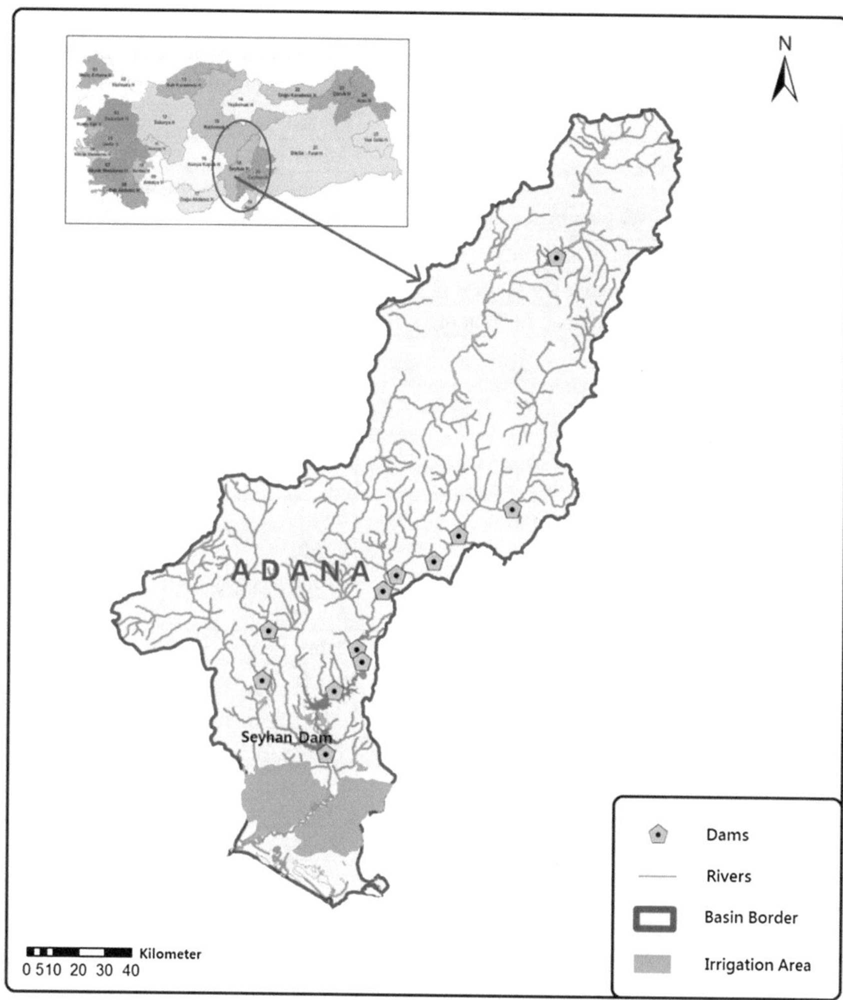

Fig. 10.1 Seyhan Basin (DSİ)

throughout the country and there was no interconnected power system. The lack of capacity restricted the use of power by existing consumers and discouraged the establishment of new enterprises (IBRD 1952).

Given these summarized shortcomings, the Seyhan Project was constructed. The purpose of the project was to prevent flood damage on the Adana Plain and provide water for irrigation, thus increasing agricultural production (part of which could be

Table 10.1 Characteristics of Seyhan Dam (DSİ 2014a)

General	Owner	DSİ, State hydraulic works of Turkey
	Year of completion	1956
	River	Seyhan
	Location	Adana
	Purpose	Irrigation + energy + flood control
	Catchment area (million m^2)	19.254
Reservoir	Reservoir area (million m^2)	79
	Total reservoir capacity (million m^3)	865.42
	Active storage volume (million m^3)	799.00
Dam	Dam type	Clay core earth fill
	Dam volume (million m^3)	7.500
	Crest level (m)	72.20
	Crest length (m)	1955
	Height from foundation (m)	77
	Height from river bed (m)	53.20
	Normal water level (m)	67.00
	Spillway type	Radial gated spillway
	Number of spillway gates	6
	Discharge capacity of the spillway (m^3/s)	2860
	Annual mean flow (million m^3)	6183
	Catastrophic flood peak (m^3/s)	7000
Benefits	Installed capacity (MW)	54
	Annual power generation (GWh)	350
	Irrigation area (ha)	148,616
	Area protected from flooding	85,000 ha

exported) and helping to expand industries and public facilities by raising power production. The project included construction of a system of levees along the Seyhan River south of Adana and along the Ceyhan and Berdan Rivers. Collection canals were added at the foothills of the Taurus Mountains to catch the runoff from small streams. The Seyhan Dam, with 865.42 million m^3 of storage, provides irrigation water and assists in controlling floods. New irrigation canals on both sides of the river increase the area under irrigation. It also supplies 54 MW to the power-generating facilities. Table 10.1 provides more details of the dam. Dams located in the Seyhan River basin are shown schematically in Fig. 10.2.

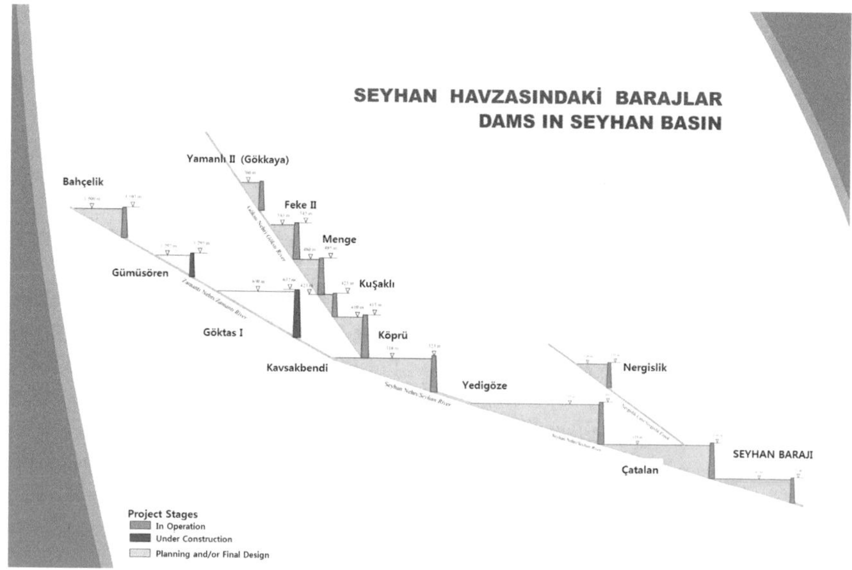

Fig. 10.2 Dams in the Seyhan Basin (DSİ 2014a)

10.3 Main Benefits of the Construction of the Seyhan Dam

10.3.1 Environmental Benefits

Agricultural activities in the Seyhan Plain (also called Çukurova) started as a result of the draining of the flood plain of the Seyhan River by the Ottoman Empire in the middle of the 19th century (Sönmez 2012). Çukurova, a low-lying land flooded in wet seasons, was the habitat of mosquitoes that were causing malaria during that time. The Seyhan Dam was declared a Wildlife Development Area in 2006 by the Ministerial Council. Since then, the Seyhan Dam has become a biological conservation area (Official Gazette No. 26,310, 5 October 2006).

The Seyhan Dam also serves as a recreational center because it lies near the city of Adana. The Balcalı Campus of Çukurova University, established on 2000 ha of land on the east coast of the Seyhan Dam Reservoir Lake, is a natural park area with an impressive view. Commercial and amateur fishermen benefit from the existing aquaculture of lake stock. An aquaculture cooperative has also been established by the local communities living in the surrounding villages.

10.3.2 Economic Benefits

Adana is in the center of Çukurova, which is surrounded by the Ceyhan and Seyhan Rivers. The rivers have brought the Seyhan, Kozan and Çatalan dams to the province and the rivers have provided very fertile alluvium lands. Not only the geographic location but also the convenient climate brought significant advantages for agriculture. In terms of socioeconomic development, Adana is ranked as the eighth province in Turkey.

In Adana, fertile land makes up 39 % of the total area. In this regard, agriculture—cotton cultivation in particular—is the main economic activity in the region, with 15 % of the cotton production of the whole country.

The welfare of the people in the basin has been increased by the development of agriculture and industry as a result of construction of the Seyhan Dam. New settlements were established. The minimum increase in net income is USD 2000/ha. The irrigation ratio of the Seyhan Dam Irrigation Facility is 80–90 %. Agricultural gross national income increased 7–9 times after the dam construction (DSİ 2014c).

Adana was one of the first industrialized cities, as well as one of the most economically developed cities of Turkey. The city was a mid-sized trading city until the mid-1800s. It came to the attention of European traders when the major cotton suppliers in the United States suffered export turmoil during the Civil War. Çukurova farmers exported agricultural products for the first time and thus started accumulating capital (Özcan 2010).

With the construction of Seyhan Dam and improvements in agricultural techniques, there was a boom in agricultural production during the 1950s. The Seyhan Dam project made it possible to irrigate a gross area of 186,000 ha; it produces 343.7 million kWh of energy annually with an installed capacity of 54 MW. Since the dam construction, approximately 36 % of the Adana territory has been devoted to agriculture. Commercial agricultural products (for domestic and foreign trade)—particularly cotton—are grown in Adana. Approximately 17,145 billion kWh of energy has been generated between 1956 and 2014. The Seyhan Dam meets all the electricity requirements (residential, industrial, underground transport, government offices, ambient lighting, etc.) for about 75,510 people (www.enerjiatlasi.com).

10.3.3 Flood Control

Flood History of the Seyhan River Basin

The Seyhan and Ceyhan Rivers have caused numerous floods in the city of Adana and the Çukurova Region in the past, resulting in significant loss of life and economic damage. Historical flood events caused serious damage in the Seyhan Plain until 1956. In the flood events of 1937, 1946, 1948, and 1950, the city of Adana was flooded, farmlands in the Seyhan Plain were inundated, agricultural yields were damaged, and many settlements were destroyed with loss of lives and property (DSİ 2014b).

River embankments, totaling 100 km in length, were built around the southern part of Adana between 1949 and 1953 in order to protect the Seyhan Plain from floods. The construction work of the dam began in 1953; it was completed and put

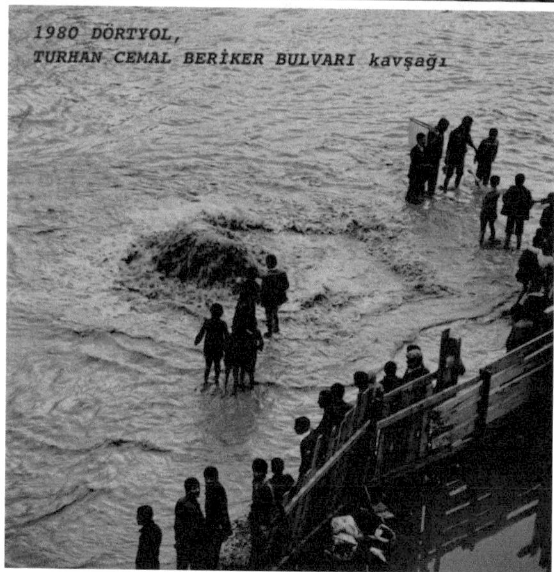

Fig. 10.3 Photographs from the 1980 flood event

into operation in 1956. The dam protected 85,000 ha of land from floods as a first direct benefit.

After the beginning of the Seyhan Dam operations, floods of the dam were controlled to a greater extent and flood damages have been reduced. Major floods in the period from 1956 to 1980 were controlled by the dam until major flooding in 1980 (Fig. 10.3). Since the capacity of the Seyhan Dam was not enough on its own to control floods with a return period of 500 years, a decision was taken after the

Table 10.2 Historical flood damages before the Seyhan Dam (Verbundplan et al. 1980)

Date of the flood event	Flood damage
05.12.1937	Adana and 4000 ha of farmland were inundated, 15 villages remained under water
10.05.1946	Seyhan Plain flooded, 15 village houses were damaged, animals were drowned
09.11.1948	Adana and 50,000 ha of the Seyhan Plain were flooded, 277 houses were damaged, 234 animals were lost, 11 people died
28.11.1948	Adana and 50,000 ha of the Seyhan Plain flooded
05.12.1948	Adana flooded
15.02.1948	The city of Adana, 70,000 ha of the Seyhan Plain and 115 villages flooded, 170 houses were damaged, 26 people died
15.05.1950	The city of Adana and the Seyhan Plain flooded, agricultural products were damaged, people were confined by the flood

Table 10.3 Historical flood damage after the Seyhan Dam (Verbundplan et al. 1980)

Date of the flood event	Flood damage
02.12.1958	Dam controlled the flood, otherwise immeasurable amounts of property would have been flooded and loss of life would have occurred
28.01.1959	Seyhan Dam controlled the flood
26.12.1969	Floods were controlled by discharging 1186 m^3/s
30.04.1975	Floods were controlled by discharging 1146 m^3/s flow downstream
03.01.1979	Since the reservoir was not full, floods were controlled easily
28.03.1980	The reservoir elevation was 61.00 m when the flood started. Although the capacity of the dikes was 1200 m^3/s, the flood was controlled by discharging 2830 m^3/s downstream. Adana survived through critical days (peak discharges 6079 m^3/s).

1980 event to build the Çatalan Dam upstream to improve flood control in the basin.

After the Çatalan Dam was put into operation, major flood damage to the Seyhan Plain and Adana city were prevented by improved operational practices. The main reasons for the loss of lives and property were unplanned urbanization or the use of areas subject to flooding in urban areas in and around the basin. Historical flood damage in the Seyhan Plain prior to the construction of the Seyhan Dam is listed in Table 10.2. Historical flood damage in the Seyhan Plain after the construction of the Seyhan Dam is listed in Table 10.3.

The annual flood control benefits of the project are both direct and indirect. One measure of the direct benefit is the average annual loss that occurred because of floods before the 1950s. This was estimated by the government in 1950 by taking a 100-year period and calculating from known records how many floods would have occurred, how large they would have been, and the damage caused by each. It was

Table 10.4 Estimated flood damage (IBRD 1952)

Seyhan discharge rate (m^3/s)	Frequency in 100 years	Damage at each discharge rate (USD)	Total damage in 100 years (USD)
950–1550	80	2,142,857	171,428,571
1550–1800	10	2,750,000	27,500,000
1800–2100	6	3,000,000	18,000,000
2100–2300	2	3,535,714	7,071,429
2300–2500	1	3,535,714	3,535,714
>2500	1	3,750,000	3,750,000
Total			231,285,714

calculated that in 100 years, four floods of 2100 m^3/s or more may be expected. Two of these would be 2300 m^3/s or more; of these two, one would be 2500 m^3/s or more. Table 10.4 was prepared for proceeding in this manner.

The average annual direct flood damage was calculated as 1 % of this total—approximately USD 2,300,000. The indirect damages (which cannot be accurately assessed), such as disruption of traffic, sickness, and occasional loss of life, were estimated at 20 % of the direct losses or USD 460,000. Thus, the total annual flood control benefits to the Turkish economy could be estimated at approximately USD 2,760,000.

10.3.4 Drought Control and Operational Studies

Analysis of the annual inflows into the Seyhan Dam indicated that the Seyhan Basin is prone to drought conditions when annual inflow is below about 3750 million m^3. The drought periods in the basin, after the Seyhan Dam was put into operation, were 1973–1974, 2001, 2007–2008, and 2014 (DSİ 2014b). In these periods, the negative impacts of drought were eliminated by a series of operational procedures put in place by DSİ, including restricted irrigation, night watering, and second crop prevention.

The General Directorate of State Hydraulic Works is responsible for the operation of the Seyhan Dam. The operation of the dam is rather complicated because of its competing targets, such as maximum energy production, maximum irrigation supply, maximum flood protection, and maximum drought production. The main issues for operation studies are as follows:

- The Seyhan Dam cannot be operated below the 61.50 m (minimum) elevation because of pumped irrigation.
- The maximum operation level of the Seyhan Dam is up to 67.50 m during the flood season (November–March) and excess water is used to produce energy.
- Depending on the amount of water inflow, the dam is filled as of February and energy is produced at full capacity at the power plant.

10.4 Assessment of the Results of Climate Change Studies on the Seyhan Basin

The Fourth Assessment Report of the IPCC (2007) predicts that a 1–2 °C increase in temperature will be observed in the Mediterranean Basin, aridity will be felt over an even wider area, and heat waves and the number of very hot days will increase, especially in inland regions. For Turkey, the average increase in temperatures is estimated at around 2.5–4 °C, reaching up to 5 °C in inner regions and up to 4 °C in the Aegean and Eastern Anatolia. The IPCC report and other national and international scientific modelling studies demonstrate that in the near future Turkey will become hotter and more arid and unstable in terms of precipitation patterns (Sen et al. 2010; Onol and Semazzi 2009; Bozkurt and Sen 2011). Global warming will likely worsen the already existing water scarcity and water allocation problems in Turkey, especially in the western and southern regions (Yilmaz and Yazicigil 2011).

According to the results of previous studies focused on the Seyhan River Basin (RIHN 2007), it is predicted that, by the year 2070, average air temperatures will increase by 2–3.5 °C and precipitation will decrease by 25–35 %, with changes in seasonality leading to earlier snowmelt in the mountains. The resulting decrease in water available for irrigation is predicted to place additional pressure on groundwater resources, increasing the risk of pollution and salt water intrusion (up to 10 km inland) along the coastal regions. All of these changes will be reflected by a changing agricultural product pattern in the basin.

According to the results of RIHN (2007), the water scarcity index (I_{ws}) is considered to be highly water stressed if it exceeds 0.4. It is found to be less than 0.4 at present, but it ranges from 0.6 to 0.8 in the future period under Scenario 1, and from 0.6 to 1.2 under Scenario 2 in any region. An I_{ws} higher than 0.4 has about a once in 6 years' probability under present conditions. However, the decreasing precipitation scenario indicates that droughts will occur more often at a once in 2 years probability.

In another study conducted under the research project *Impact of Climate Changes on Agricultural Production System in Arid Areas* (RIHN 2007), the potential impact of climate change on the hydrology and water resources of the Seyhan River Basin in Turkey has been explored. A dynamic downscaling method, referred to as pseudo warming, was used to connect the output of the raw general circulation models and the river basin hydrologic models. The downscaled data covered two subset periods (1993–2004 and 2070–2080) and were used to drive hydrologic models to assess the impact of climate change on the water resources of the Seyhan River Basin. The results showed an average annual temperature change of +2.0 °C using the Meteorological Research Institute (MRI) of Japan model and +2.7 °C using the Centers for Climate System Research (CCSR), Columbia University projections (Fig. 10.4). The average annual precipitation change is projected to be −159 mm (MRI) and −161 mm (CCSR), with marked changes in seasonal distributions (Fig. 10.5). Compared with the present data, reductions in

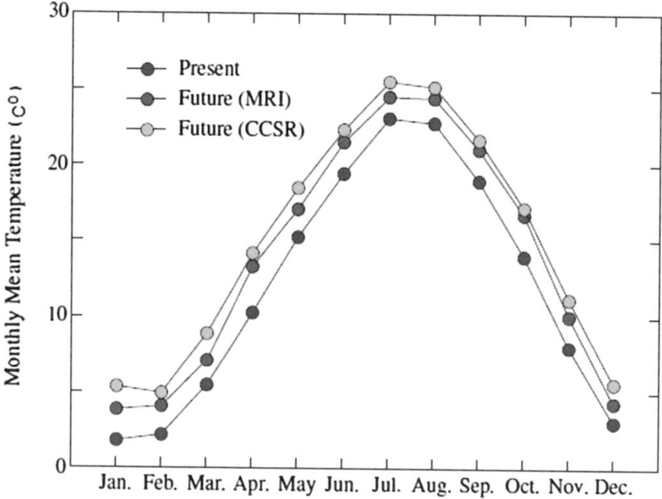

Fig. 10.4 Temperature changes predicted under different scenarios (RIHN 2007)

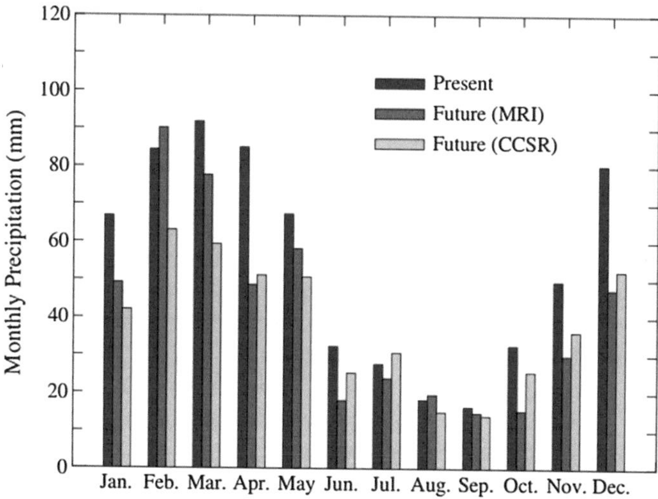

Fig. 10.5 Precipitation changes predicted under different scenarios (RIHN 2007)

precipitation will considerably decrease the flow, with the peak monthly inflow occurring earlier than at present. The ratio of water withdrawal to discharge will increase because of the effects of global warming on the decreased discharge (Fig. 10.6).

A further study of the effects of climate changes on surface water management in the Seyhan Basin was conducted by Selek and Tunçok (2013) and supported by the Seyhan River Basin Grants Programme, under the United Nations Joint Programme

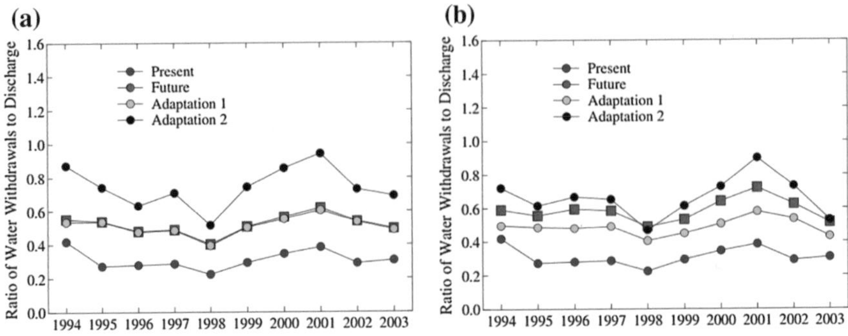

Fig. 10.6 Ratio of water withdrawal to discharge (RIHN 2007). **a** MRI. **b** CCSR

(UNJP) on Enhancing the Capacity of Turkey to Adapt to Climate Change Project. In the study, water resources planning and management policies, specifically irrigation scenarios and associated reservoir operations, were evaluated within the climate change perspective. The study aimed to set up the basis for climate change adaptation of water resources management policies in the Seyhan River Basin. The first priority was to identify the balances between water resources and water users with respect to existing and planned projects. Various aspects of the Seyhan Basin were evaluated, including existing water resources, determination of water demand for existing/planned projects, and water resources' supply-demand characteristics. The global climate change model was downscaled to the basin scale and demands were determined for specific projection years. Water resources management scenarios were developed to evaluate adaptation alternatives to climate change scenarios at the basin level. The results were associated with the hydro-meteorological monitoring network and finally the impact of climate change on surface water resources. It was determined that even though there was no water stress in the Seyhan Basin in 2010, many parts of the basin were expected to suffer from significant water shortages over the coming years.

Under Scenario 1, potential deficits and surpluses along the Seyhan Basin were identified by considering existing water resources in the year 2010. As per the operational records, no deficits are present for this base condition along the Seyhan Basin. Basin-scale water balance was also investigated through the use of the MIKE BASIN model (DHI 2011). Scenario 2 was developed primarily to evaluate the effects of climate change on the availability of water resources along the Seyhan Basin and the associated impacts on irrigation systems. As shown in Fig. 10.7, water budget deficits are expected to take place in the upper basin from 2020 and show a significant increase in 2040. Scenario 3 was developed to evaluate the effects of climate change and associated adaptation measures and practices (i.e., use of drip irrigation) to minimize water deficits along the Seyhan Basin. As shown in Fig. 10.8, changes in the irrigation systems and the associated optimal use of water through the reservoir network resulted in significant savings, especially within the upper segments of the Seyhan Basin.

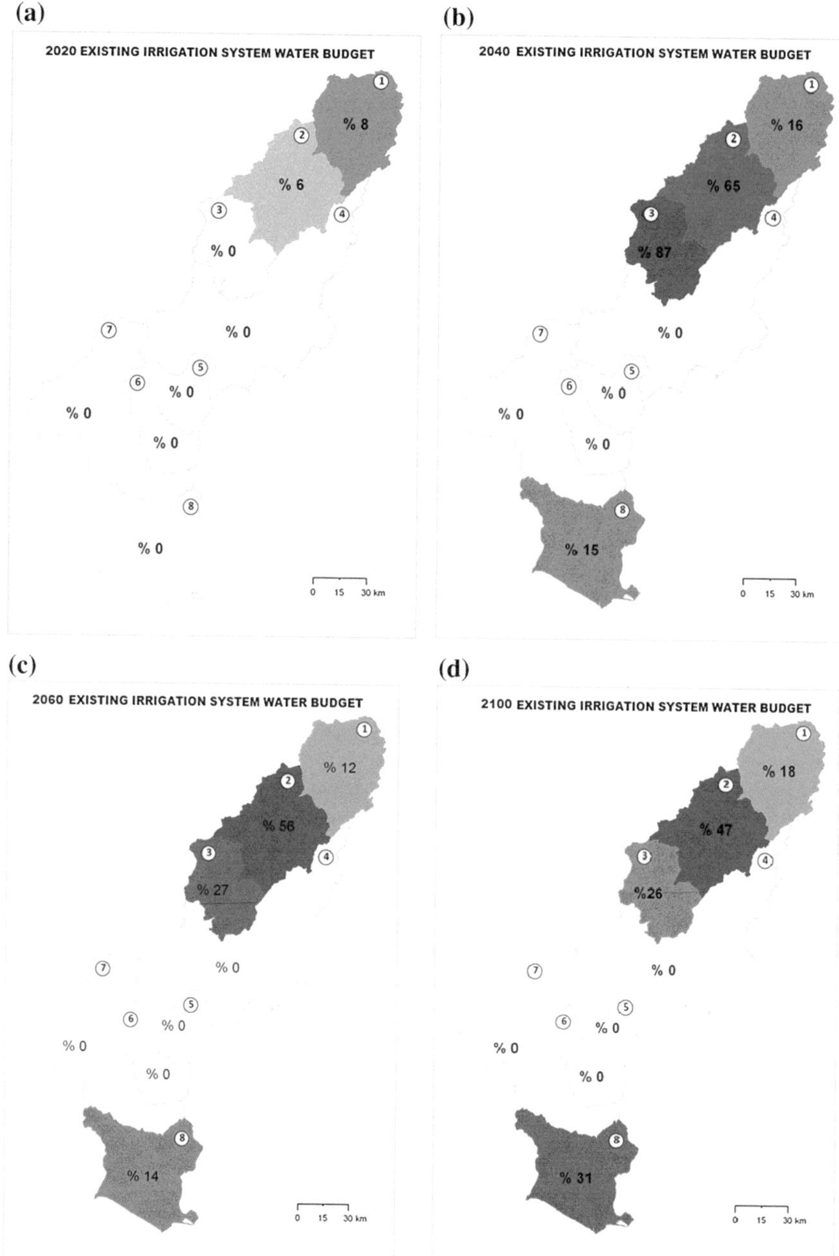

Fig. 10.7 Surface water deficits for scenario 2. **a** Expected future conditions for 2020. **b** Expected future conditions for 2040. **c** Expected future conditions for 2060. **d** Expected future conditions for 2100 (Selek and Tunçok 2013)

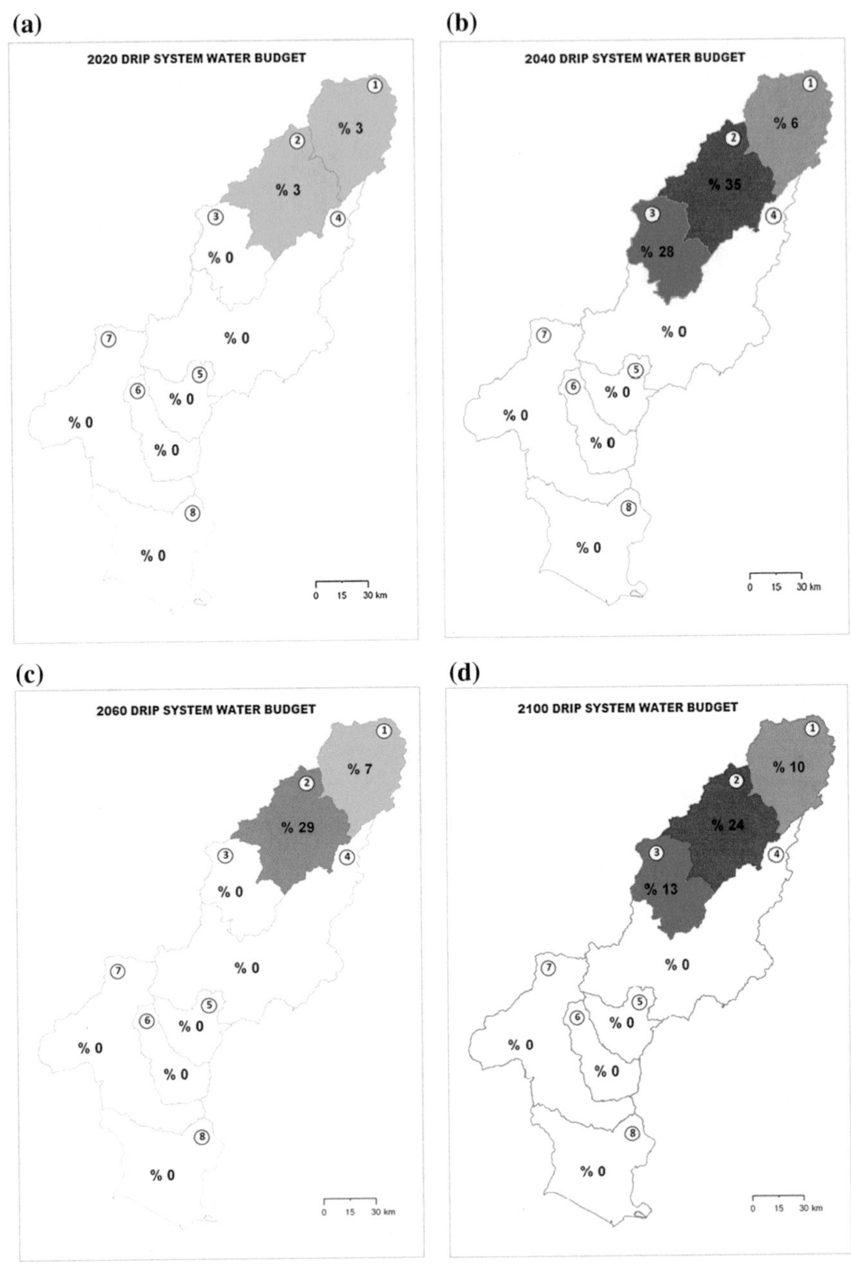

Fig. 10.8 Surface water deficits for scenario 3. **a** Expected future conditions for 2020. **b** Expected future conditions for 2040. **c** Expected future conditions for 2060. **d** Expected future conditions for 2100

According to the climate change model developed as part of the study supported by the Seyhan River Basin Grants Programme, under the UNJP, it is projected that average temperatures within the Seyhan Basin will increase by 5 °C within the 2010–2100 timeframe. While there was no water stress in the Seyhan Basin in 2010, it is apparent that many parts of the basin will suffer from significant shortages over the coming years. This will be further exacerbated by the rising water demand in many sectors, particularly in agriculture. The fact that the average total precipitation over the basin is expected to drop by 35 % in the lower segments, 23.5 % in the middle segments, and 17 % in the upper segments within the 2010–2100 timeframe will make the consequences for local communities and wildlife even worse.

The Seyhan River Basin has important agricultural potential, not only for the region's agricultural industry and high population density but also for Turkey more widely. The projected increases in temperatures and decreases in effective rainfall amounts found in previous studies (Selek and Tunçok 2013) suggest that surface and groundwater resources will likely suffer from high evaporation loss as well as low recharge rates. To avoid significant impacts on irrigated agriculture as a consequence of increased water stress, new water resources have to be developed. Furthermore, while the reduction in rainfall amount will likely intensify droughts, the increase in the intensity of rainfall will likely increase the occurrence of floods.

It has been demonstrated through the use of climate change and basin management models that shortages in the upper segments of the basin are inevitable. Therefore, best management practices in irrigation alone will not be sufficient to ensure delivery of water resources to future planned projects. Effective water conservation initiatives will have to be implemented across all sectors to adapt to the predicted changes in water supply. Technological changes in irrigation and water distribution systems will also be required. In this context, industry should apply appropriate environmental measures to assist in the conservation of natural resources.

10.5 Climate Change Policy in Turkey

The foundation of Turkey's policies regarding climate change was laid with the Eighth Five-Year Development Plan. In 2000, the Climate Change Special Expertise Commission Report was published within the context of the Eighth Development Plan. The Ninth and Tenth Five-Year Development Plans prepared after that have added objectives for the development of the process. While it was stated that the studies for the process of being a party to the United Nations Framework Convention on Climate Change (UNFCCC) will be carried out in the Eighth Five-Year Development Plan, it was also stated that regulations will be carried out in the energy efficiency field for the reduction of greenhouse gases. As foreseen in the Ninth Five-Year Development Plan, another step has been taken toward the fight against climate change and a Climate Change National Action Plan showing the policies and measures for greenhouse gas reduction in accordance with Turkey's conditions was prepared. The Tenth Five-Year Development Plan, which was prepared last and which is still in effect, states that the "green growth" concept is to be taken as a basis in order to reach sustainable development targets (MoEU 2016).

In 2001, the Coordination Board on Climate Change (CBCC) was established. The CBCC was restructured in 2004 after Turkey became a party to the UNFCCC; in 2010, its remit was expanded with the participation of new members. In 2013, the CBCC was merged with the Coordination Board on Air Emissions and renamed the Coordination Board on Climate Change and Air Management. The Board, composed of relevant ministries and industry representatives, determines the policies, measures, and activities to be pursued related to climate change. Turkey has put into effect policies and measures to tackle climate change in the energy, agriculture, forestry, transportation, industry, and waste sectors (GRI-CCE 2016).

Turkey became a member to the Convention in 2004 and ratified the UNFCCC by law in 2003 after a decision that invited all member countries to recognize the special circumstances of Turkey relative to other Annex-I countries. Parliament endorsed the ratification of the Kyoto Protocol and officially became a party to the Protocol in 2009 (GRI-CCE 2016).

After Turkey ratified the climate change convention, it was required to produce a National Communication—a document detailing emissions trends and steps to comply with the goals of the convention. Within the scope of its commitments, Turkey completed a First National Communication in 2007, a Fifth National Communication (as the second-to-fifth National Communication) in 2013, and a Sixth National Communication in 2016.

The main document used to generate policies regarding climate change studies is the "National Climate Change Strategy Document" covering the years from 2010 to 2020. The document was prepared in coordination with the former Ministry of Environment and Forestry using a widely participative study containing CBCC members, related public and private sector representatives, universities, and non-governmental organizations (NGOs). The document was approved by the Higher Planning Council on 3 May 2010 (MoEU 2016).

National Climate Change Strategy Document (2010–2020)

This document was prepared to guide the studies to be carried out on climate change between 2010 and 2020 and to determine the key policies in this field. The document contains the mitigation, compliance, financing, and technology policies that Turkey can realize, with the help of national and international resources, based on the "common but differentiated responsibilities" principle (MoEU 2016).

National Climate Change Action Plan (2011–2023)

The National Climate Change Action Plan was prepared in accordance with the Ninth Development Plan and National Climate Change Strategy Document in coordination with the Ministry of Environment and Urbanization. The group preparing the plan was widely representative, containing CBCC members and other stakeholders, and was published in July 2011. The plan provides actions for the control of greenhouse gas emissions and adaptation studies within the context of the

National Climate Change Strategy Document and defines the responsibilities and timing for applying these actions (MoEU 2016).

Turkey is located in the Mediterranean Basin, which, according to the Intergovernmental Panel on Climate Change (IPCC) Fourth Assessment Report, is one of the regions most negatively affected by global climate change (MoEU 2013). Because of its climatological features, Turkey does not have an abundance of water resources. For this reason, there is a need to take into consideration the ongoing desertification trend in the Mediterranean Basin and Turkey since the early 1970s. In order to prevent serious water problems in the future, an integrated basin management approach and water policies are being pursued (MoEU 2013).

The most important tools used in the preparation stage of the climate change policy documents and National Communications of Turkey are the climate change projections, which have been produced according to climate change scenarios in line with the scientific studies of the IPCC. Climate change scenarios are being used to design climate change policy at all levels (international, national, regional, and local) and have formed the scientific base in Turkey since 2000 (Climate Change Special Commission—Eighth Five-Year Development Plan). Several projects have been carried out to understand and forecast climate change and its effects for Turkey by related institutions and universities. More detailed information about climate change scenarios and projections for Turkey can be accessed from the National Communications of Turkey.

It is widely acknowledged that climate change will have a broad-ranging impact on the social, economic, and environmental developments of countries. For this reason, climate change scenarios provide the basis for policy decision-making and strategy development for climate change mitigation and adaptation.

Already, it is known that uncertainty is inherent in climate predictions and in the rate and magnitude of climate impacts. The scientific studies on climate change projections provide decision-makers—from small-scale farmers, to city mayors, to government ministers—with actionable information to adapt to shifting climate risks (WRI no date).

10.6 Adaptation Policies

To integrate policy measures into the technical initiatives established in the study (Selek and Tunçok 2013), the recommendations of the Food and Agriculture Organization of the United Nations (FAO 2007) with respect to climate adaptation strategies were evaluated to reflect the speed of response and choice of options. This was achieved with the support of the Seyhan River Basin Grants Programme, under the UNJP. As such, it was agreed by the Regional Directorates of DSİ (State Hydraulic Works of Turkey) that a framework for climate change adaptation needs to be directed simultaneously along several interrelated lines:

- Legal and institutional elements need to integrate various levels of institutional mechanisms, such as legislation, human rights norms, tenure and ownership, regulatory tools, legal principles, governance, and coordination arrangements.
- Policy and planning elements need to structure risk assessment and monitoring mechanisms to reflect analysis, strategy formulation, and sectoral measures.
- Livelihood elements need to be planned to ensure sustainability by incorporating the strategic components of food security, hunger and poverty alleviation, and nondiscriminatory access.
- Ecosystem elements need to define the composition of and evaluate biodiversity, resilience, ecosystem goods, and services.

In addition to the practices defined above, the underlying factors in the regulatory framework need to be in place to support and implement adaptation policies and programs. Then, these have to be further supported through capacity-building and technology transfer initiatives, in which skills at the rural community level are improved at the basin scale. The duties and responsibilities of institutions and ministries related to the adaptation policies of the water sector are given below.

The goals of DSİ in line with the DSİ Strategic Plan (2015–2019) (DSİ 2015), which are related directly or indirectly to climate change, are as follows:

- Basin master plan studies for 25 river basins will be completed.
- Protection, improvement, and monitoring of the quality and the quantity of the water will be undertaken.
- Flood hazard maps will be prepared and early warning systems will be set up.
- Drinking, domestic, and industrial water needs of municipalities will be met.
- River basins will be protected against domestic and livestock pollution.
- Stream reclamation and flood protection facilities will be built and the services of existing facilities will be maintained.
- The irrigation network area will be increased to 4.85 million ha by DSİ.
- The irrigation rate will be increased from 62 to 68 %, and irrigation efficiency will be increased from 42 to 50 % in irrigation areas developed by DSİ.
- Hydraulic energy supply will be increased.

Moreover, in the 10th Five-Year Development Plan (2014–2018) (Ministry of Development 2013), the "Effective Use of Water in the Agriculture Programme" coordinated by Ministry of Forestry and Water Affairs (MoFWA) is currently being implemented. The following will be achieved by this program:

1. Increasing the share of the total irrigation area developed by DSİ with modern water-saving irrigation methods (drip and sprinkler) from 20 to 25 % in the plan period.
2. Increasing the irrigation ratio from 62 to 68 % and irrigation efficiency from 42 to 50 % in DSİ irrigation during the plan period.
3. Increasing the number of modern water-saving irrigation systems by 10 % each year in the course of the plan period.
4. Reducing groundwater use by 5 % during the plan period.

The transfer of irrigation management from DSİ to Water Users' Associations can be regarded as a kind of adaptation activity. Until the early 1980s, operations and maintenance (O&M) for irrigation systems were highly centralized, imposing an increasing institutional and financial burden on the government. A very low ratio of billing and collection rates (or no collection at all); very high water consumption; no cost recovery for investment; and no local interest by the farmers in protecting the infrastructure were all contributing factors.

Although some small irrigation proposals were transferred to users, the pace of change was very slow. By 1993, DSİ was still operating 95.2 % of all irrigation schemes and only 72,000 ha of the irrigation command areas had been transferred to Water Users' Associations. After 1993, with advice from the World Bank, an accelerated process of handing over irrigation O&M to Water Users' Associations was undertaken. The Accelerated Transfer Programme took off very quickly, with areas operated by Water Users' Associations increasing to about 96 % of the total irrigated land in 2005. The recovery rate for O&M costs increased from less than 40 % to more than 80 % after the facilities were handed over to water users' organizations (Water and DSİ 2013).

In addition, water overuse and its consequent negative environmental impacts (e.g., salinity) have gradually decreased. The irrigation program that was a "government program with the assistance of the farmers" became "a farmer program with the assistance of the government" (GWP 2012).

The agricultural drought dimension that stems from climate change in water basins in Turkey is of the utmost importance. In the future, given the population growth and increasing demand for food, reduction in agricultural production because of agricultural drought will be more significant. Hence, combating agricultural drought will have prominence. Protection of land resources, sustaining an ecological balance, development of plants resistant to drought, and minimizing hydrological drought as adaptations to the impacts of climate change will help in reducing the negative social and economic consequences of drought. This situation is important in meeting an increased food demand in the future. Strategies and policies concerning agricultural drought in Turkey have been implemented since 2008 and the impacts of climate change are being taken into consideration in this interval.

In the Turkish Strategy for Combating Agricultural Drought and Action Plan (TAKEP) (MoFAL 2013), carried out by the Ministry of Food, Agriculture and Livestock, a number of goals and actions were determined in an attempt to reduce the pressure of climate change impacts on water resources in Turkey. These precautions should also be taken into account in the activities undertaken to fight against drought on the provincial scale. In order to combat drought, a Provincial Drought Action Plan (MoFAL 2013) has been prepared in every province, taking each province's dynamics and special circumstances into consideration. Provincial agricultural drought crises centers (MoFAL 2013) have been founded in the cities.

There has been a restructuring of the Ministry of Food, Agriculture and Livestock and other ministries which that responsibilities under the Agricultural

Drought Management Coordination Board. The Five-Year Agricultural Drought Combating Strategy and Action Plan for 2008–2012 has expired. These two situations have been taken into account in the drafting of new legislation and a Cabinet Decision on Activities for Combating Agricultural Drought and Drought Management has been published (MoFAL 2013). Corresponding with this Cabinet Decision, Regulations on Agricultural Drought Management Duties and Working Procedures and Principles have been prepared as well (MoFAL 2013). In accordance with these regulations, the Agricultural Drought Management Coordination Board has been established under the leadership of the Undersecretary of the Ministry of Food, Agriculture and Livestock and with the participation of representatives from related Directorates, NGOs and universities, and has begun its activities.

Since drought is a natural disaster and can occur at any moment, a Five-Year Agricultural Drought Combating Strategy and Action Plan for the 2013–2017 period (MoFAL 2013) was prepared. Under the coordination of the Ministry of Food, Agriculture and Livestock and with the participation of ministries, universities, governorships, local administrations and NGOs, as determined by the Cabinet Decision, the action plan has been put into force according to country conditions.

The main objective in combatting agricultural drought is to take all necessary measures when there is no drought for sustainable agricultural water usage planning. This includes taking supply and demand management into consideration by including all shareholders to the process and increasing public awareness. Also, it entails minimizing drought effects by applying an effective program to combat these during the crisis.

Activities under the Action Plan have been prepared by grouping, on the basis of a determined strategy, the main development centers and priorities. The titles of measures for which related institutions are responsible have been prepared and duties have been distributed. These activities are as follows:

(a) ***Drought Risk Estimation and Crisis Management***

- Crisis management of estimates of the extent of the agricultural drought will be applied.

(b) ***Provision of Sustainable Water Supply***

- Potential water holding capacity will be increased.
- Water delivery channels will be modernized, investments in maintenance and renewal of water storage and delivery channels will be realized in a timely manner.
- Measures will be taken for the collection of waste water and ensuring the use of treated waste water in the agricultural and industrial sectors.
- Effective management of groundwater will be ensured.
- Land use techniques increasing the preservation of water in the soil will be developed and land use plans for protecting and developing soils that are the most important for natural water storage will be prepared.

(c) ***Effective Management of Agricultural Water Demand***

- The most suitable cultivation areas for agricultural products in the selected agricultural basins will be determined by taking water availability into consideration and effective water usage will be ensured.
- Water delivery systems will be modernized.
- Effective use of underground water for agriculture will be ensured.
- Incentive programs directed to sectors and productions that are most affected by drought, including plant production, animal production, bee-keeping and domestic aquaculture, will be established. Plant production and livestock production policies will be implemented by taking drought risk into consideration.

(d) ***Accelerating Research and Development and Training Activities***

- Research and development activities supporting the fight against drought will be accelerated.
- Training activities, especially for farmers, will be increased.

(e) ***Developing Institutional Capacity***

- Necessary legal regulations for effectively combatting agricultural drought will be prepared and institutional structuring will be strengthened.
- Necessary institutional capacity for fighting forest fires will be developed.

In line with these titles, the duties of related institutions are determined together with their activities addressing the priorities/measure for which they are responsible. At the end of the year, these institutions will report their findings for assessment.

Four steps have been determined for activities on irrigated and dry agricultural lands under the prepared Provincial Drought Action Plan (MoFAL 2013). These are:

- Step 1: Preparation for drought
- Step 2: Drought alert
- Step 3: Immediate action
- Step 4: Limitation

There is coordinated action with the Provincial Directorates in order to follow these steps regularly. Every April and May, an evaluation is made by considering appropriate data to determine which province is at which level. The agricultural drought provincial crisis center prepares its report by making an annual evaluation, whether or not a drought has occurred, and sends its decisions on the matter to the General Directorate of Agricultural Reform. This institution then carries out the necessary coordination with the secretariat.

Almost all of the duties of the General Directorate of Water Management (one of the service units of MoFWA) are directly related to water resources management in basins. These duties include:

- Determining policies related to the protection, improvement, and use of water resources.
- Providing coordination of water management at national and international levels.
- Preparing and making river basin management plans and carrying out the relevant legislative studies concerning integrated river basin management plans. This is done to protect and improve the ecological and chemical quality of the aquatic environment by taking protective actions and using/taking balance into account.
- Determining, assessing, and updating the precautions on the basis of basins, together with related institutions and associations, and following up their implementations.
- Determining, together with related institutions and associations, objectives, principles, and receiving environment standards for the protection of surface waters and groundwaters, and monitoring water quality.
- Determining and monitoring sensitive areas in terms of water quality and nitrate sensitivity.
- Determining strategies and policies related to floods and preparing related legislation and flood management plans.
- Making the necessary coordination related to water allocations on the basis of sectors, according to river basin management plans.
- Following up processes arising from international agreements and other legislation related to the protection of water resources and management and carrying out works related to transboundary and frontier waters in coordination with related institutions.
- Building a national water data-based information system.
- Performing studies on climate change impacts on water resources.

The main achievements of General Directorate of Water Management to date are preparation of National River Basin Management Plans, water quality monitoring in line with EU Water Framework Directive, and drought and flood studies. The studies carried out by the General Directorate of Water Management related to climate change and adaptations are as follows:

- In order to analyze how much the water resources of the country would be affected by climate change and to determine the adaptation measures required, the Climate Change Impacts on Water Resources Project (DGWM) was started in 2013 and is planned to end by 2016.
- Flood management plans and drought management plans for 25 river basins will be prepared by the General Directorate of Water Management by 2023 (DGWM).

- The National Flood Management Strategy Paper, Regulation about Preparation, Implementation and Monitoring of Flood Management Plans, and Drought Management Strategy Paper and Action Plan studies are being carried out (DGWM).

The General Directorate of Forestry, affiliated with MoFWA, is responsible for protecting forestry and forestry resources against dangers of all sorts. Its mandate is to develop forestry and forestry resources using a nature-friendly approach and manage them within the integrity of the ecosystem and in a manner that will provide society with multipurpose, sustainable outcomes.

In 2011, the General Directorate of Combating Desertification and Erosion (ÇEM), affiliated with the MoFWA, was established. Its mandate is "protection of soil, improvement of natural resources, combating desertification and erosion, setting policies and strategies related to avalanche and flood control, and providing cooperation and coordination among related agents and agencies."

Also established in July 2011 was, the Ministry of Environment and Urbanizations. Its mandate is, "To determine the principles and policies for the protection and improvement of the environment and for the prevention of environmental pollution, to develop standards and scales, to prepare programmes; to organize and develop training, research, projects, basin protection plans and contamination maps, to determine and to monitor application principles for these and to carry out all the related tasks and duties concerning climate change." This ministry is the principal institution responsible for combating climate change (MoEU 2013).

The General Directorate of Environmental Management of the Ministry of Environment and Urbanization has the following duties regarding protection of water resources. "To protect surface and ground waters, seas and land, to determine objectives, principles and contaminants with a view to prevent or eliminate pollution, to determine the methods and principles regarding removal and control of pollution, to carry out emergency response plans and to have them executed." (Talu and Özüt 2011).

As the responsibilities in the field of water management, within the context of adaptation to climate change, have been allocated to many ministries and their subdivisions, the need for coordination among these institutions has become a significant part of the current issue. An example of proper coordination within this complicated structure is in regards to legal grounds. The MoFWA is authorized for integrated management of water resources and for the issues relevant to climate change and water resources, while the Ministry of Environment and Urbanization is authorized for the adaptation to the impacts of climate change.

In conclusion, determination of the water potential and consumption in all the basins throughout Turkey and the reflection of long-term water resource planning are equally important. Modelling the impacts of climate change on the water resources plus determining adaptation needs and costs under various scenarios are the key issues. As explained above, the institutional structure shows once again the importance of cooperation and coordination.

10.7 Recent Developments in Water Policy in Turkey

10.7.1 Flood Management and Flood Control

The General Directorate of State Hydraulic Works (DSİ) was established by Law No. 6200, 18 December 1953, as a legal entity and brought under the aegis of the Ministry of Energy and Natural Resources. It is charged with "single and multiple utilization of surface and ground waters and prevention of soil erosion and flood damage." For that reason, DSİ is empowered to plan, design, construct and operate dams, hydroelectric power plants, domestic water, and irrigation schemes.

Pursuant to the duties and responsibilities defined by law, the DSİ General Directorate generally implements structural works projects to mitigate the effects of torrents. Hence, under to the provisions detailed in Laws No. 4373 and No. 7269, DSİ executes diverse works in every phase of torrent disaster. The Prime Ministry Circular, "Creek Courses and Torrents," 2006/27 came into force on the date of publication (Official Gazette No. 26,284, 9 September 2006). Another Prime Ministry Circular, "Improvements of Rivers and Creek Courses," 2010/5 (DSİ 2013) came into force on the date of publication (Official Gazette No. 27,499, 20 February 2010).

DSİ involves assorted works in the whole process of torrent disaster by taking the necessary precautions and warning the relevant organizations of flood emergencies. Building and operating protective structures against floods is one of the basic duties of DSİ.

MoFWA, previously structured as the Ministry of Environment and Forestry, was established through the Decree Law No. 645 on the Organization and Duties of the MoFWA (Official Gazette No. 27,984, 4 July 2011). The General Directorate of Water Management is responsible for determining strategies and policies related to floods and preparing related legislation and flood management plans according to Decree Law No. 645.

10.7.2 Drought and Drought Control

The General Directorate of Water Management is responsible for preparing drought management plans on the basis of river basins or in having them prepared and following up on their implementation according to Decree Law No. 645. DSİ is responsible for the operation of dams (for irrigation purposes). In drought periods, the negative impacts of the drought were eliminated by some operational studies, such as restricted irrigation, night watering and preventing a second crop.

The Ministry of Food, Agriculture and Livestock is also responsible for agricultural drought. Agricultural drought activities are based on:

- The Decision of the Council of Ministers dated 7 August 2007 (UNW-AIS).
- The Decision of the Council of Ministers dated 6 June 2012 (UNW-AIS).
- Regulations on the Duties of Agricultural Drought Management, Working Procedures and Principles dated 18 August 2012 (UNW-AIS).

10.7.3 Irrigation Management Transfer

Until 2004, Irrigation Unions were founded according to Articles 133–138 of the Municipality Law No. 1580. After that, arrangements related to Irrigation Unions have been involved in the Municipality Law (Decree Law No. 5393) and Local Administrative Unions Law (Decree Law No. 5355) (Water and DSİ 2013). These arrangements did not meet the requirements regarding arranging the facilities of Irrigation Unions. To solve the problems regarding Irrigation Unions, 15 years of research and analysis was carried out beginning in 1993. The need emerged for a separate Irrigation Unions Law in order to address the sustainable irrigation industry along with the controllable and transparent structure of the union. Intensive studies were held between 2008 and 2011 and the Irrigation Unions Law No. 6172 was published and put into action (Official Gazette No. 27,882, 22 March 2011).

10.8 Outcomes and Conclusions

One way to manage the impacts of climate variability on water resources is through "hard engineering" to capture and control river flow. Storage dams are built to retain and store flows in excess of user requirements and to release these during periods of low flows—a practice that can also serve to maintain aquatic ecosystems. Alternatively, during floods, peak flows can be stored for later release, avoiding flood damage by reducing the maximum flows. Both functions are important to sustain urban settlements and to avert disasters caused by floods and droughts.

Dams also harness water as a form of potential energy to generate electricity, without which healthy urban life would be difficult to sustain as settlements increase in size. Nineteen percent of the world's electricity is currently generated from hydropower and there is a substantial potential to expand this, particularly in Turkey. A specific benefit of hydropower is that it does not usually generate significant quantities of greenhouse gases and thus allows economic and social development to occur without aggravating global warming.

Water managers need to address solutions in climate variability and extreme events. These can be achieved not only by construction of infrastructure, but also through institutional mechanisms. These are important means of helping to deal with climate variability and achieving goals, such as water supply for people, industries and farms, flood protection, and ecosystem maintenance. Water allocation, conservation, use efficiency, and land use planning are soft tools to manage demand as well as increase supply. As a long-term policy problem, the governance of adaptation to climate changes relies on knowledge about long-term climate change impacts, and this knowledge is riddled with uncertainties (Vink et al. 2013).

During the construction period of the Seyhan Dam, concepts such as climate change and adaptation to climate change were not on Turkey's agenda to the same extent as they were globally. The Seyhan Dam was built with a "no-regret" approach in which the dam would generate net benefits whether or not climate change occurred. The Seyhan Dam is an excellent example of adaptation to climate change impacts by storing water for irrigation and enabling farmers to cope with the variability of rainfall while adapting to the reduced water resources available. As the frequency of floods in the coming future may significantly increase, the dam will control and regulate these floods. In addition, being a clean energy project, it will help mitigate climate change by avoiding the emission of greenhouse gases.

References

Bozkurt D, Sen OL (2011) Hydrological response of past and future climate changes in the Euphrates-Tigris Basin, GRA 13:EGU2011-11072

Danish Hydraulic Institute (DHI) (2011) MIKE 11 GIS floodplain mapping and analysis. DHI, Horsholm

FAO (2007) A framework for climate adaptation in agriculture, forestry and fisheries, p 32

Directorate General for Water Management (DGWM) Available via http://suyonetimi.ormansu.gov.tr/AnaSayfa/Yayinlarimiz/sygmsunumarsiv/taskinkuraklik_sunumlari.aspx?sflang=tr. Accessed 16 Mar 2016

General Directorate of State Hydraulic Works (DSİ) (2013) Water and DSİ (1954–2013). DSİ, Ankara

General Directorate of State Hydraulic Works (DSİ) (2014a) Dams of Turkey, international commission on large dams Turkish national committee (TRCOLD) Dams of Turkey: Seyhan Dam. DSİ, Ankara

General Directorate of State Hydraulic Works (DSİ) (2014b) Seyhan Basin master plan report. DSİ, Ankara

General Directorate of State Hydraulic Works (DSİ) (2014c) Assessment report on operations by DSİ and transferred irrigation facilities, 2013. DSİ, Ankara

General Directorate of State Hydraulic Works DSİ (2015). DSİ strategic plan 2015–2019 (DSİ stratejik planı 2015–2019). Available via http://www.dsi.gov.tr/docs/stratejik-plan/dsi-sp-2015-2019.pdf. Accessed 16 Mar 2016

General Directorate of Water Management (DGWM). Available via http://iklim.ormansu.gov.tr/Eng/. Accessed 16 Mar 2016

Global Water Partnership (GWP) (2012) Case studies, Turkey: transfer of irrigation management to water users associations (#57). Accessed 16 Mar 2016

Grantham Research Institute on Climate Change and the Environment (GRI-CCE) (2016) Available via http://www.lse.ac.uk/GranthamInstitute/legislation/countries/turkey/. Accessed 16 Mar 2016

International Bank for Reconstruction and Development (IBRD) (1952) The Seyhan river multipurpose project in Turkey. DC, World Bank Group, Washington

Intergovernmental Panel on Climate Change (IPCC) (2007) Appendix I: glossary. In Parry ML, Canziani OF, Palutikof JP, et al. (eds) Climate change 2007: impacts, adaptation and vulnerability. Contribution of working group ii to the fourth assessment report. Cambridge University Press, New York

Ministry of Development (2013) 10th five-year development plan (2014–2018). Ministry of development, Ankara. Available via http://www.kalkinma.gov.tr/Lists/Kalknma%20Planlar/Attachments/12/Onuncu%20Kalknma%20Plan1.pdf. Accessed 16 March 2016

Ministry of Environment and Urbanization (MoEU) (2013) Turkey's fifth national communication under the UNFCCC. MoEU, Ankara

Ministry of Environment and Urbanization (MoEU) (2016) Turkey's sixth national communication under the UNFCCC. MoEU, Ankara

Ministry of Food, Agriculture and Livestock (MoFAL) (2013) Agricultural drought combating strategy and action plan 2013–2017. MoFAL, Ankara. Available via http://www.tarim.gov.tr/TRGM/Belgeler/Duyurular/2013_2017_Kuraklik_Eylem_Plani.pdf

Ministry of Forestry and Water Affairs (MoFWA). website, www.ormansu.gov.tr

Önol B, Semazzi FHM (2009) Regionalization of climate change simulations over the eastern Mediterranean. J. Climate 22:1944–1961

Özcan Z (2010) Pending the giant: Adana (Bekleyen dev: Adana). Aksiyon

President Demirel S (2004) International Workshop on impacts of large dams Istanbul: impacts of large dams in Turkey with a special reference to Seyhan Dam. Turkey

Research Institute for Humanity and Nature (RIHN) (2007) The final report of the research project on the impact of climate changes on agricultural production system in arid areas (ICCAP). RIHN, Kyoto. Available via http://www.chikyu.ac.jp/iccap/finalreport.htm. Accessed 17 Feb 2016

Sönmez ME (2012) The urban expansion in Adana, Turkey, and the relationship between land use change in its surroundings. Turk J Geogr Sci [Türk Coğrafya Dergisi] 57:55–69

Selek B, Tunçok IK (2013) Effects of climate change on surface water management of Seyhan Basin, Turkey. Environ Ecol Stat 21:391–409

Şen Z, Uyumaz A, Öztopal A et al (2010) Final report on the impacts of climate change on İstanbul and Turkey water resources. İSKİ Project, Su Vakfı

Talu N, Özüt H (2011) Strategic steps to adapt to climate change in Seyhan River Basin. T.R. Ministry of Environment and Urbanization (MoEU), Ankara

UN-Water Activity Information System (UNW-AIS). Retrieved from http://www.ais.unwater.org/ais/pluginfile.php/548/mod_page/content/75/Turkey_CountryReport.pdf. Accessed 06 January 2016

Verbundplan GmbH, Romconsult SA, Temelsu (1980) Lower Seyhan Basin master plan report. DSİ, Ankara

Vink MJ, Dewulf A, Termeer C (2013) The role of knowledge and power in climate change adaptation governance: a systematic literature review

World Research Institute (WRI) (no date) Decision making and climate change uncertainty; setting the foundations for informed and consistent strategic decisions. Available via http://www.wri.org/our-work/project/world-resources-report/decision-making-and-climate-change-uncertainty-setting. Accessed 06 Jan 2016

Yılmaz KK, Yazıcıgil H (2011) Potential impacts of climate change on Turkish water resources: a review. In: Baba A, Tayfur A, Gündüz O et al (eds) Climate change and its effects on water resources. Springer, Netherlands, Rotterdam

Chapter 11
Lake Nasser: Alleviating the Impacts of Climate Fluctuations and Change

Asit K. Biswas

Abstract Egypt as a country may not have existed without the Nile River. Its population was 84 million in 2015 and is estimated to reach 124 million by 2050. It is already the largest global importer of wheat. Even after Lake Nasser became operational in 1970, its arable land is still less than 4 % of its land area. The major constraint to expand agriculture is water, not land. Agriculture currently accounts for more than 80 % of total water use. Throughout history, Egypt has suffered from frequent droughts and floods. Construction of a storage reservoir at Aswan allowed it to stabilize significant interannual and intra-annual fluctuations of the Nile. Lake Nasser enabled Egypt to increase its cultivable land by 30 % and allowed cropping intensities to reach 180–200 % in old lands. When it became operational in 1970, it accounted for over half the country's electricity generation. While this reservoir has effectively taken care of climatic fluctuations, a main issue now is how to seamlessly integrate future climate change considerations with climatic fluctuations. At the present state of knowledge, it is impossible to predict even what will be the direction of river flow changes, let alone their real magnitudes. Three well-respected global circulation models indicate flow increases of 12 and 18 % and catastrophic decline of 77 %. Under such uncertainties, Egypt needs to increase all types of water storage, improve significantly water use efficiencies in all sectors, and monitor river flow changes over years very carefully.

Keywords Egypt · Lake Nasser · High Aswan Dam · Climate fluctuations · Climate change · Benefits of storage reservoirs · Climate change adaptation

---

A.K. Biswas (✉)
Lee Kuan Yew School of Public Policy, National University of Singapore,
469C Bukit Timah Road, Singapore 259772, Singapore
e-mail: prof.asit.k.biswas@gmail.com

A.K. Biswas
Third World Centre for Water Management, Atizapan, Mexico

C. Tortajada (ed.), *Increasing Resilience to Climate Variability and Change*,
Water Resources Development and Management,
DOI 10.1007/978-981-10-1914-2_11

11.1 Background

"Egypt," observed the Greek historian Herodotus as early as the 5th century BC, "is the gift of the Nile." Nearly two-and-a-half millennia after his visit, and in spite of extensive technological and economic developments, it is still not an overstatement. Life and human activities in Egypt would be impossible without the waters of the Nile River.

Napoleon invaded Egypt in 1798. He shared the view of Herodotus on the importance of the Nile for the country's survival. He observed, "If I was to rule a country like Egypt, not even a single drop of water would be allowed to flow to the Mediterranean Sea." While Napoleon was not aware of the environmental and ecological roles the river waters play, his concept of using all the water available for human uses in a desert has some merit.

The situation is not very different at present. In fact, overall dependence of Egypt on the Nile waters has risen many times because its population has risen—from an estimated 3.8 million in 1800, around the time of Napoleon's invasion, to 72.7 million, according to the last census in 2006. The increase is more than 18-fold in a period of little over two centuries. The current projection is that the population will reach 124 million by 2050—a further increase of nearly 40 million people over the latest 2015 estimate.

For Egypt, which is the largest importer of wheat in the world, the major constraint to agricultural development is water, not land. Even after completion of the High Aswan Dam in 1970, the country's total arable land was still less than 4 % of its total land area. Currently, the rural population accounts for about 57 % of the total population, and their livelihoods primarily depend upon agriculture.

Like the entire developing world, Egypt is witnessing rapid urbanization. Because of the needs of a rapidly increasing urban population and consistently poor land-use planning, the country has been progressively losing productive agricultural land. This makes the country dependent on imported food. The area of Egypt's highly productive land (Class 1 soil) declined from 683.2 km^2 in 1992 to 618.5 km^2 in 2009. During the same period, the area of moderately productive (Class 2) land declined from 100.5 km^2 to 93.8 km^2 and marginally productive land (Class 3) from 209.1 km^2 to 198.3 km^2 (Shalaby et al. 2012). In 2006, it was estimated that an average Egyptian had to depend on 504 m^2 of arable land for food production. Fortunately, the decline in productive areas has been compensated by steadily increasing agricultural yields and improvements in management practices. With an ever-increasing population, a decreasing agricultural area, a serious water constraint, and rising expectations of the people for a better standard of living, the country has no choice but to improve its agricultural management practices and water-use efficiencies continually and significantly in the coming decades. This is an even more daunting task when the political turmoil of the recent years are considered. These factors will make it extremely difficult to improve the efficiencies of land and water management practices in the coming years.

The most challenging issues facing Egypt up to the year 2050 will be how to provide enough food, energy, housing, employment, a clean environment, and basic services, such as a water supply, wastewater treatment, education, health, transportation, and communication, to its current 84 million people, as well as to another estimated 40 million expected by that time. Furthermore, it should be noted that a very substantial middle class has emerged in recent decades. They are educated and affluent. Their number is likely to increase significantly in the coming decades. As in other parts of the world, the growing middle class will most certainly demand steady improvements in their standards of living and quality of life. If their expectations are not met, most probably there will be significant social unrest, which may have serious social, economic, and political repercussions for the country, including on land and water management practices.

The present and future expectations of Egyptians cannot be sustainably met without the efficient harnessing and use of the Nile waters. This is because water is a cross-cutting issue, essential for basic human needs such as drinking, agricultural production, energy generation, and employment creation through industrial, commercial, and agricultural developments, as well as for maintaining good health and a clean environment. Unquestionably, the Achilles heel of current and future development of the country is, and will continue to be, good water management processes and practices. Efficient and equitable water management is needed to contribute to steadily improving social and economic conditions. For an arid country like Egypt with only one primary source of water, the River Nile, the only option is water storage. Water can be stored in flood years and then used during drought years.

11.2 Water, Climate, and Development

The main source of water in the desert climate of Egypt is the River Nile. Rainfall is scanty. Only a small part of the country, the Mediterranean coastal strip, receives a comparatively higher average annual rainfall—100 to 200 mm—than the rest of the country. For example, annual rainfall south of Cairo is about 25–50 mm a year. In Central and Southern Egypt, several years may pass without any noteworthy rainfall. Throughout the country, rainfall is insufficient to grow most food crops in any year.

In the absence of significant, regular rainfall, Egypt, not surprisingly, does not have much renewable groundwater. Thus, human survival, agriculture, and the country's socioeconomic development depend on the flow of the River Nile. The river is shared by 10 countries, of which Egypt is the last downstream country. This geographical fact contributes to many complexities and uncertainties in terms of assuring water security in the future, which is essential for planning the country's future sustainable social and economic development. Fortunately for Egypt, in recent decades, two of its immediate upstream countries, Ethiopia and Sudan, did not use much water because of political and social turmoil, which constrained their economic and social development processes (Biswas 2011).

The hydropolitics behind the management of the Nile were complicated throughout the latter part of the 20th century. This has become even more complex in recent years (Biswas 1994; Biswas et al. 1997). The hydropolitics in the region have changed rapidly and significantly, especially in Ethiopia, which is currently constructing the Grand Ethiopian Renaissance Dam. Egypt is concerned that the construction and operation of this dam may reduce water availability for its steadily increasing future water needs (Abdelhady et al. 2015). Accordingly, the dispute between Egypt and Ethiopia on the construction of this new storage structure has become a contentious political issue. Egypt, not surprisingly, is concerned about its future water security and, thus, its economic development and political stability. Ethiopia also has similar concerns.

Irrespective of what happens in the future because of new storage structures that may be constructed in Ethiopia and Sudan, unless water management processes and practices in Egypt become significantly more efficient than they have been historically, the country will face major problems during the post-2020 period, both in terms of water quantity and quality. The time for incremental changes is now long over. The country needs significant and major structural changes in water management practices, which will be difficult to implement because of sociopolitical reasons, vested interests, and institutional inertia. However, the country has no alternative. It does not have the luxury of enough time to implement these long-needed structural changes, which have been mostly inconspicuous in the past by their absence.

Agriculture currently accounts for more than 80 % of total water use. Farming is by far the largest sector from where most Egyptians earn their livelihoods. Accordingly, the changes needed will be politically difficult to implement because a very large number of farmers are poor, illiterate, and barely eke out a living at present. Thus, making changes that may adversely affect their current lifestyles, at least over the short to medium terms, will be a very difficult process politically. Another major constraint facing Egypt is that even though water is a critical resource, data on its availability, demand, quality, and water-use efficiency by sectors and major users are not reliable and often could be contradictory. Without reliable, adequate data, proper planning and management of water will be a very difficult process, even under the best of circumstances.

11.3 Reservoirs and Climate Fluctuations

While the Nile has been a lifeline for Egypt for thousands of years, its annual flow has often fluctuated significantly from year to year, as well as within years. The Nile is one of the few rivers in both the developed and developing world for which reasonably accurate flow records have been available for several centuries. These

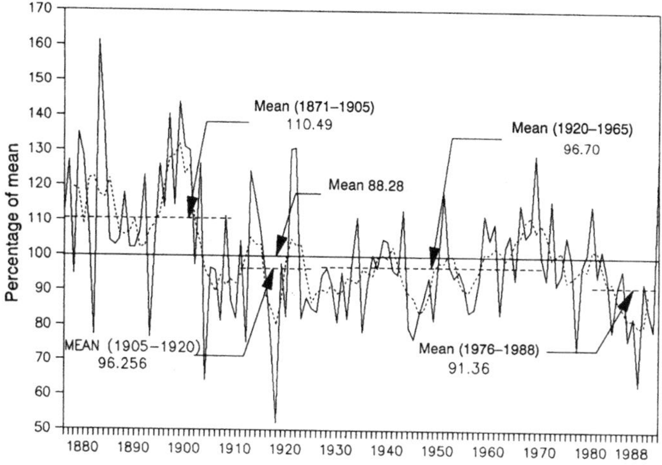

Fig. 11.1 Annualized Nile flow at Aswan (Abu-Zeid and Biswas 1992)

records show that the average yearly flow in the river has varied significantly, depending upon the periods over which average flow is estimated.

Figure 11.1 shows annualized Nile flow at Aswan as a percentage of the long-term mean flow over nearly 120 years. In hydro-meteorology, it is often assumed that the mean can be reasonably defined with around 30 years of data. Figure 11.1 shows that, depending on the period selected, the mean can be very different. For example, if a major storage structure had been designed in 1900 based on the previous 30 years of data or in 1950 based on the previous 50 years of data, the project design for the same location would have had significant differences. This means that the benefits and costs of a water infrastructure in the same location could be very different, depending on the period of data used by the planners and designers.

The problem is not limited to the Nile flow in Egypt. Similar problems would have been encountered at the White Nile at Mogren (Fig. 11.2) or the Blue Nile in Khartoum (Fig. 11.3) because of interannual and intra-annual river flow fluctuations. The flow records at Mogren can be easily divided into two distinct periods: 1912–1961 and 1962–1985. The mean annual flow during the period 1912–1961 was 25.6 billion m^3. For 1962–1985, it was 33.3 billion m^3, which is nearly one-third higher than during the previous 50-year period.

From the perspective of social and economic development in very arid areas of Egypt or Sudan, such annual variations in high and low river flow result in feast or famine without the construction of large storage reservoirs. Such reservoirs, by storing enormous quantities of water, smooth out downstream water availability and substantially reduce the adverse impacts of potential high floods or prolonged droughts.

Hydrologists and climatologists have often assumed that 30 years of continuously observed data can reliably define long-term rainfall and runoff patterns and

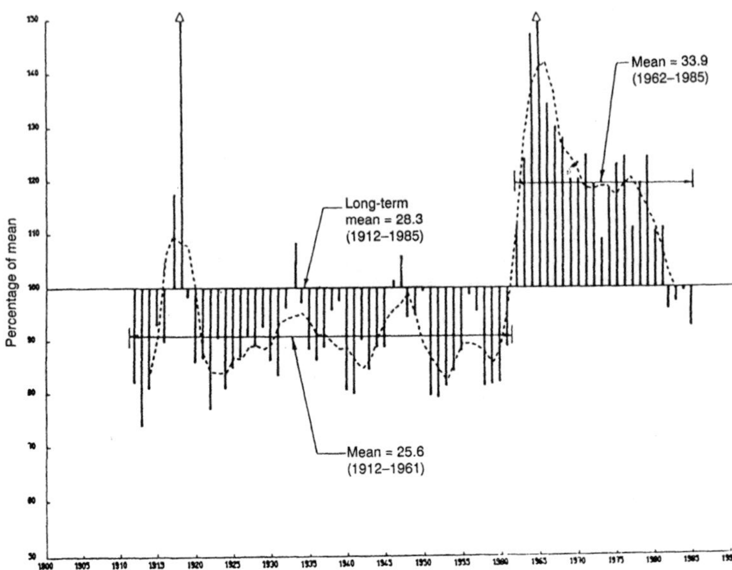

Fig. 11.2 Annual flow of White Nile at Mogren (Abu-Zeid and Biswas 1992)

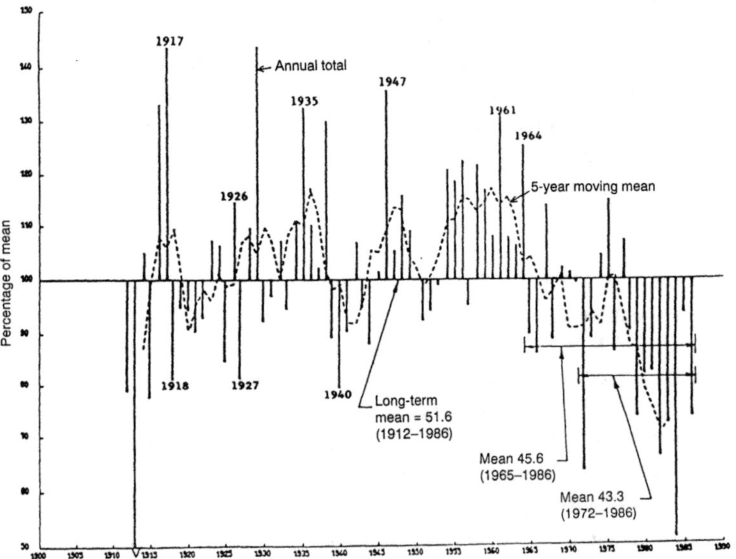

Fig. 11.3 Annual flow in Blue Nile at Khartoum (Abu-Zeid and Biswas 1992)

mean. The analyses of such data on the Nile, at three different places, clearly indicate that the interannual fluctuations in the Nile flow are often so different that this assumption is not valid, at least for this particular river.

11.4 High Aswan Dam and Lake Nasser: Hydro-political Developments

To stabilize the significant fluctuations in annual Nile flow, on which Egypt has been forever dependent for survival, Adrian Daninos, an Egyptian of Greek origin, first proposed in 1948 to construct a high dam at Aswan. The idea was that such a dam could control and manage the uneven flow regime of the river and, by doing so, would benefit the entire Egyptian population and contribute to economic growth. The proposed dam would also ensure that the Egyptians would have an assured supply of water, year after year, in perpetuity. The dam would further ensure that the periodic floods and droughts that have ravaged the country from prehistoric times could be successfully managed by creating a major storage reservoir. This reservoir is now known as Lake Nasser or the High Dam Lake.

In 1952, the Egyptian Government approved the proposed dam project and planning started. It was initially expected that the World Bank and the governments of the United States and the United Kingdom would finance the construction of the dam. However, the US Secretary of State, John Foster Dulles, unfairly and unilaterally decided in 1956 not to provide any funding for the construction of the dam for political reasons. Without the American funding, the British Government and the World Bank withdrew their offers to finance the project (Biswas and Tortajada 2012).

The dam was an important sociopolitical issue for President Nasser. Not only it was essential for Egypt to construct it to protect the country from regular floods and droughts, but also it was essential to smooth the Nile flow so that the country would have a reliable supply of water for agricultural production, hydropower generation, land reclamation, industrial and commercial development, and overall social and economic progress.

Psychologically, it was also important for Nasser to use the construction of what, at that time, was the largest dam in Africa as an important contribution to nation-building. This was especially relevant since, in 1952, King Farouk was dethroned and a new government of military officers took over the national leadership. Nasser felt that the construction of a large dam would give the newly independent nation confidence and pride. Successful construction of the dam would make the Egyptians proud of their country, instill confidence in the new political regime, and also contribute significantly its socioeconomic development. Thus, for both Nasser and Egypt, building the dam was more than a construction of a large infrastructure project: its symbolic and real contributions to Egypt's social and economic development pride and national unity should not be underestimated.

When western support for financing the dam was withdrawn in July 1956, a month after Nasser officially became the president, he promptly nationalized the Suez Canal. In his famous Alexandria speech of 26 July 1956, Nasser said, "The annual income of the Suez Canal Company was USD 100 million. Why not take it ourselves? We shall build the High Dam we desire." Accordingly, nationalization of the Suez Canal was a direct consequence of the withdrawal of the western offer to finance the construction of the High Aswan Dam. It resulted in the British, French, and Israeli invasion of Egypt. The war was an attempt to regain control of the Suez Canal and also to remove President Nasser from power. The United States publicly condemned the invasion and voted for the UN resolution to establish a peacekeeping force. No other dam in the entire history of the humankind has ever played such a major geopolitical role.

Nasser's view of the importance of the dam to the psyche of Egypt was outlined later in a speech in 1960: "We shall build the High Dam, but before building the High Dam, we have to first build the dam of dignity, the dam of integrity, the dam of liberty, and when we have built the dam of dignity, integrity and liberty, we shall have realized our hopes and we shall then surely build the High Dam."

The dam was eventually constructed with funding and technical assistance from the then Soviet Union. By all accounts, the dam has been remarkably successful. It has delivered much more than even its strongest supporters may have expected when it was first proposed in 1956. However, because of the then Cold War rivalry of the two superpowers and because the High Aswan Dam was the first successful foray by the Soviets into Africa, much false information was spread, often deliberately, by the western powers, intelligence authorities, media, and academics on the social and environmental costs of the dam. Regrettably, to a certain extent, this false information persists even today (Biswas and Tortajada 2012; Tortajada and Biswas 2017).

11.5 Role of Lake Nasser in Managing Climatic Fluctuations

When the High Aswan Dam was being planned, the overall concept was relatively simple and straightforward. Agriculture has never been possible in Egypt with only rainfall. The principal source of water is the Nile, which, historically, has had fluctuating interannual and intra-annual flows that have resulted in regular floods and droughts. The nature and extent of the roles of the dam in moderating the flow can be appreciated by the following facts. Before Lake Nasser became operational, the natural flow of the Nile fluctuated between 600 and 13,000 m^3/s, and the mean annual discharge to the Mediterranean Sea was 32 km^3/s (Abu-Zeid and El-Shibini 1997). After the construction of the dam, the discharge downstream varied from 700 to 2800 m^3/s. The variation between minimum and maximum discharge of the Nile was four-fold after the reservoir became operational, compared to nearly 22 times before its construction.

The differences in high and low water levels in the Nile, beyond Aswan, prior to the construction of the reservoir have been equally remarkable. These were around 7–8 m without the dam. After the reservoir was built, the difference has been around 2–3 m. Thus, downstream of Aswan, since the reservoir became operational, there has not been even a minor flood and, hence, no loss of life or damage to property.

In addition, the water flowing into the Mediterranean is now restricted to about 300 million m^3 in the winter season, only sufficient enough to provide the navigational requirements for the tourist boats. This is because during the winter, the irrigation water requirements in Egypt are at their lowest because the irrigation system is closed for annual maintenance. However, while irrigation water requirements are at their lowest during the winter, navigational water requirements for tourist boats during this period are at their highest. This is because tourism in Egypt during the winter is at its peak because numerous Europeans and North Americans like to take a break from their cold weather. The climate in Egypt during this period, especially in Aswan and Luxor, is very pleasant. In summer, the temperature in this region is very high and is not a pleasant time for tourism. Thus, the peak tourist arrivals in Egypt in this area are during November to January. Tourism has been a major source of income for Egypt, ranking only after the income from the tolls of the Suez Canal.

Without an adequate flow in the Nile, income from tourist boats would decline significantly. The tourist boats operating on the Nile are designed and constructed to require only a shallow depth of water. Thus, water released from the dam during the winter closure can be kept to a minimum, consistent with satisfying only the navigational water requirements of the boats.

Egypt's population in the 1950s was increasing rapidly. The majority of the households depended on agricultural activities for a living. Thus, a large reservoir was needed to store water in flood years so that it could be used during drought years. The High Aswan Dam provided the necessary storage to decrease the variations between high and low flows, eliminate flood damages, and ensure enough water was available during drought years for domestic needs, agriculture, and industrial uses.

The dam and reservoir system was completed in 1970. However, impoundment of Lake Nasser started in 1964. Figure 11.4 shows the variations in water level and the volume of water stored in Lake Nasser during 1964–2003. By storing a very large volume of water and significantly reducing the effects of climate fluctuations, Lake Nasser started playing an important role in Egypt's social and economic development. As President Nasser had anticipated, for an overwhelming majority of Egyptians, the High Aswan Dam became a source of national pride and joy. It has brought tremendous benefits to the country, both directly and indirectly, as well as political stability to the region.

The total cost of the dam, including subsidiary projects and the extension of power lines, was about EGP 450 million (Abul-Atta 1975, 1979; Biswas 2002). Water Minister Abul-Atta further estimated that the total cost was recovered in only 2 years, making it one of the most economically efficient large dams ever built anywhere in the world. The dam's annual return was estimated in 1975 at EGP 140 million in agricultural production, EGP 100 million in electricity generation, EGP 10 million from flood protection, and EGP 5 million as a consequence of improved navigation.

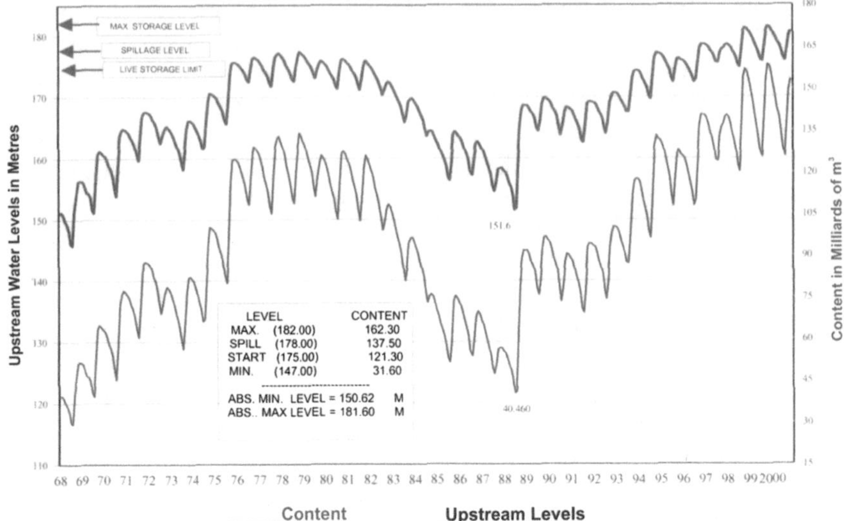

Fig. 11.4 Upstream water levels and volume of water in Lake Nasser (Tortajada and Biswas 2017)

11.5.1 Benefits of Lake Nasser

The overwhelming economic benefits from the construction of Lake Nasser are shown in Fig. 11.4. Egypt suffered from a prolonged drought between 1979 and 1986, followed by a very high summer flood in 1988. In the absence of Lake Nasser, the country would have undoubtedly faced very serious social, economic, and political turmoil because of the occurrences of these extreme hydrological events within a very short period of years. The two events had unprecedented adverse impacts on upstream neighbors like Ethiopia and Sudan. Because of the moderating effects of Lake Nasser, Egyptians did not feel any of the likely adverse impacts of the catastrophic prolonged drought, followed by a serious flood. Even with Lake Nasser, Egypt did come perilously close to feeling the adverse impacts of a prolonged drought. By early 1988, the volume of water available in Lake Nasser had declined to a perilously low level. Had the drought persisted for another two or three years, Egypt would have faced very serious economic hardships, and possibly social and political unrest. Lake Nasser saved the country from serious damages by attenuating the high and low flow of the Nile.

In addition, the dam generated hydroelectricity, which contributed to social, economic, and industrial development. During the 1950s, 1960s, and 1970s, Egypt had not discovered any oil or natural gas within its national jurisdiction. Thus, alternatives to generate electricity were limited. Electricity generated by the dam very significantly reduced the national balance of payment problems during the 1970s, when the country needed significant imports of food and energy for survival. Thus, the economic benefits accrued from Lake Nasser ensured that the Egyptian pound did not devalue as significantly as it might have without it. There is no question that without the reservoir, Egypt would have faced a serious balance of payments crisis.

The reservoir also provided a new water body for fish production, an important source of animal protein. Equally, the city of Aswan became an increasingly important tourist destination. The reservoir further ensured year-round navigation between Cairo and Aswan, which generated additional employments and revenues by boosting tourism and other associated economic and industrial activities.

11.5.2 Lake Nasser and Hydropower Generation

Water stored in Lake Nasser enabled hydropower generation. Originally, the power plant had 12 generating units, each with a capacity of 175 MW. Together they were capable of generating 10 billion kwh of electricity annually. The first two generators were brought online ahead of schedule to compensate for the destruction of thermal power plants during the 1967 Arab-Israeli war. Up to the early 1980s, electricity generated from Lake Nasser accounted for more than half that used by the country. Thus, not only did Lake Nasser ensure that the country did not feel the ravages of droughts and floods, but it also generated enough electricity to sustain more than half the country's domestic, industrial, and commercial needs.

In 1985, a total of 270 MW of additional capacity was added. Even as late as 2000, the hydropower contributed around one-fifth of the nation's electricity requirements. However, as time progressed and as Egypt's population and economic activities increased, the percentage of electricity generated by the reservoir waters started to decline because of increasing contributions from the newly

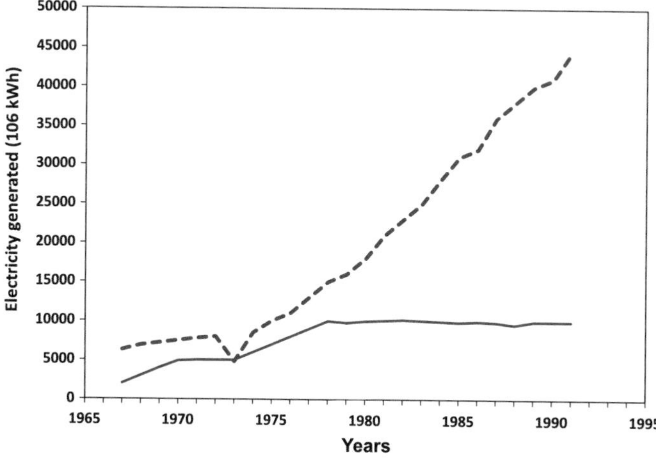

Fig. 11.5 Contribution of High Aswan Dam hydropower to Egypt's total electricity generation, 1967–1993 (Egyptian national statistics)

constructed thermal power plants. Figure 11.5 shows the contributions of Lake Nasser to Egypt's total electricity generation between 1967 and 1993.

It should be noted that when Lake Nasser became operational, Egypt had to import much of its fossil fuel. The hydropower generated from Lake Nasser significantly reduced the energy import bill of the country, and thus significantly reduced balance of payments problems. Even after oil was discovered and the country became self-sufficient and started to export oil, the availability of hydropower has reduced the amount of oil consumed nationally. This has meant that Egypt can export more of the oil it produced, and thus generate additional revenues. Without the reservoir, Egypt would have exported less oil, meaning significantly less revenue for the country.

During the early years of the operation of the High Aswan Dam hydropower plant, it accounted for over half the electricity generated in the country (Fig. 11.5). In 1978, hydropower accounted for 54.3 % of the country's electricity. Even though the total electricity generated by the dam has increased steadily over the decades because of better management and installation of increasingly more efficient machinery, its percentage share of national electricity production has steadily declined because more and more thermal power stations have come onstream.

11.5.3 Lake Nasser and Agricultural Development

The impacts of Lake Nasser on agricultural production have been very substantial for two important reasons. The first is land reclamation. When Lake Nasser became operational in 1970, Egypt's total arable land was estimated at 6 million feddan (1 feddan = 0.42 ha). By 1996, total cultivable land had increased by nearly 30 %, to 7.8 million feddan (Biswas 1995). This expansion was possible by using the water that was stored in Lake Nasser.

The second reason for steadily increasing agricultural production is because of steadily increasing cropping intensification. Before the construction of the reservoir, Egyptian farmers depended on irregular flows of the River Nile for agriculture. They did not have access to a reliable supply of water from Lake Nasser. Accordingly, agricultural yields in the farms were low. Farmers were lucky to get one good crop each year. Not surprisingly, the incomes of farmers were low. More than 90 % of Egyptian farmers are smallholders, with less than 3 feddan of land (Achthoven et al. 2004). Like most small farmers in developing countries, they are poor and their levels of literacy and risk tolerance are equally low. Somehow they manage to survive from these small plots of land.

The main advantage of a reliable supply of irrigation water from Lake Nasser has been that not only have crop yields increased, but cropping intensities have also steadily advanced. In recent decades, cropping intensities have reached around 180–200 %, especially in the old lands. The cropping intensities in the newly reclaimed land are somewhat lower, primarily because water still remains a constraint (Biswas 1993, 1995).

Lake Nasser ensured that Egyptian farmers had access to a reliable supply of irrigation water, year after year. By dampening the efforts of both high and low flows of the Nile through a vast reservoir and by enabling both vertical and horizontal expansion of agriculture, total agricultural production has increased very significantly. Agriculture accounted for 16.6 % of the gross domestic product (GDP) by 2000. It also contributed to 20 % of the country's export earnings. Agriculture further provided employment to more than 5 million workers, representing over 30 % of Egypt's entire labor force.

Lake Nasser not only stabilized the effects of climatic fluctuations that contributed to periodic high and low river flows, but it also ensured that a large number of Egyptian families have higher incomes, better standards of living, and a better quality of life. Our analyses indicate that had Lake Nasser not been constructed, Egyptians would have significantly lower standards of living at present. Without the reservoir, there is no doubt that there would have been serious social and political unrest, especially during the years when extreme hydrological events occurred (Tortajada and Biswas 2017).

11.6 Climate Change Implications

A major issue that the water profession has generally not faced up to now is how future potential climate change considerations can be seamlessly integrated with climate fluctuations factors that have occurred regularly in the past and will continue to occur in the future. Historically, climatic fluctuation has always been a fact of life. To a significant extent, climatic fluctuation considerations can now be adequately incorporated in planning, design, and operational processes of large reservoirs. We now have enough knowledge, technology, and management expertise to handle climatic fluctuation effectively. However, realistically, we are unable to plan, design, and operate new and existing reservoirs like Lake Nasser for both climatic fluctuation and climate change in an efficient, timely, and cost-effective manner. There are many reasons for this. Only the most important reasons for this inability will be discussed here in the context of Lake Nasser.

The first reason is the prevailing disconnect between water and climate change professionals. Climate change experts have been primarily concerned with averages, such as how average temperature may increase or how mean sea level may rise. In contrast, water professionals are interested not in averages but in extreme hydrological events, like prolonged droughts and high floods. Water resource systems are not planned or designed on the basis of average values, but on the basis of the occurrence of extreme events. This disconnect has had many major consequences.

Major storage reservoirs like Lake Nasser, which has a design life of 500 years, are invariably planned to withstand probable maximum floods. Estimating such floods has always been an art rather than a science. The process considers, among other factors, all river flow data that may be available. Then, based on analyses of these historical records, the largest 1-in-1000-year or even 1-in-10,000-year floods are estimated. Naturally, there are considerable uncertainties associated in estimating such large floods of very long return periods based only on short and limited periods of data, usually 20–60 years. In addition, river flows depend on many factors that change with time, such as land-use patterns, levels of urbanization and deforestation, rainfall patterns, groundwater recharge, and quantities of water abstracted upstream—all of which are impossible to predict with any degree of certainty over the long term.

In designing large reservoirs, probable maximum floods are generally estimated by considering the worst hydrological and meteorological conditions that could happen concurrently. These estimates are invariably conservative, and in many instances they are likely to consider both climatic fluctuations and some climate change issues, although how much it is impossible to say at present. The final figures selected often depend on the judgement and experience of the experts concerned.

The second reason is that the global circulation models (GCMs) that are currently available are of limited use for forecasting precipitation at river basin scales, especially over the catchment of a major river like the Nile. The uncertainties increase by several orders of magnitude when such uncertain estimates of precipitation are translated into river flows. Uncertainties increase even further when hydrologists use estimates of precipitation to forecast extreme events, such as high floods. The usefulness of such estimates at our present state of knowledge is dubious at best.

Let us consider the Nile flow at Aswan. Using currently available, well-regarded GCMs do not even give us the direction of river flow changes, let alone the magnitude of such changes. Consider three such GCMs. Two GCMs developed by the Goddard Institute of Space Studies (GISS) and Geophysical Fluid Dynamics Laboratory (GFDL), both well-respected institutions in the United States, give very different results. The GFDL model indicates a decline of 77 % in the River Nile flow. GISS indicates an 18 % increase. The GCM of the UK Meteorological Office, another well-regarded model, estimates a 12 % decrease in the river flow (Strzepek and Smith 1995).

In the face of such extreme uncertainties, not surprisingly, policymakers are reluctant to spend tens, if not hundreds, of millions of dollars to ameliorate the potential impacts of climate change. This is an especially difficult decision for a developing country like Egypt, which has limited financial resources but has many other pressing development priorities over the short, medium, and long terms that are certain, urgent, and real.

Be that as it may, whether climate change affects the Nile flow by an increase of 18 % or a catastrophic decline of 77 %, the storage option continues to remain viable. The country has to consider all forms and types of storage options and also to significantly improve its water use efficiency in all sectors as quickly as possible to assure its water security in the future. The country simply has no other choice.

There are similar uncertainties in other water-related issues as well. For example, take evaporation. The general thinking is that as the climate gets warmer, evaporation from Lake Nasser will increase and thus reduce the amount of water available. This consensus thinking is contradicted by a detailed analysis by Badawy (2009). He concluded that there will be a negligible (0.29 %) increase in evaporation over the entire lake by 2050. The explanation is that while temperature will rise, it will increase humidity and reduce wind speed—the main factors influencing evaporation losses.

With our present state of knowledge, unfortunately we cannot predict what the Nile flow regime may be during the post-2050 period with any certainty. Until it is possible to have actionable knowledge on which new, realistic, and cost-effective policies can be formulated, there is not much that can be recommended to the Egyptian government except to monitor what may happen in the future and to analyze the data as soon as they become available. This may indicate what could be reasonably accurate future trends on the basis of which appropriate policy measures could be considered and operational actions may be taken.

11.7 Lake Nasser and Climate Change

Lake Nasser has become even more important to Egypt than it has been previously because of likely future climate changes, even though their magnitudes are now highly uncertain. New and uncertain conditions have to be superimposed on the "normal" climatic fluctuations that have been witnessed from the Pharaonic times. The evolving uncertainties concerning future climate change have to be effectively superimposed on historical and future fluctuations in river flow attributable to other non-climate-related factors. This will further increase significantly the degrees of uncertainty and risk in terms of reservoir operation and management.

Lake Nasser has released more water in the years of low flows of the Nile (droughts) and stored more water during the years of high flows. Figure 11.6 shows the fluctuations in water level in Lake Nasser because of appropriate discharges from the lake to compensate for the effects of droughts and floods. It also shows the volume of water that remained stored in the lake after these discharges. This diagram alone clearly indicates the effectiveness of large storage reservoirs like Lake Nasser in ameliorating the changes in river flows resulting from climatic fluctuation. Climate change may exacerbate this fluctuation in terms of magnitude, frequency, occurrence, and duration. This means that in the future, storage reservoirs like Lake Nasser will have to play an even more important and critical role to concurrently counter the impacts of both climatic fluctuations and climate change factors. Management practices and operational procedures will have to be steadily and progressively improved as and when the situation warrants.

Accordingly, an effective alternative will be to consider, more and more, all types of water storage practices, big and small, surface and underground, to deal with the two major scourges of floods and droughts in the future. The operational and management practices of water storage structures may have to change over time, depending on the changes observed. The changes in practices have to be complemented with rapidly increasing efficiencies in all types of water uses.

Based on preliminary analyses carried out by the Third World Centre for Water Management, our view is that for a large basin like the Nile, covering 10 countries

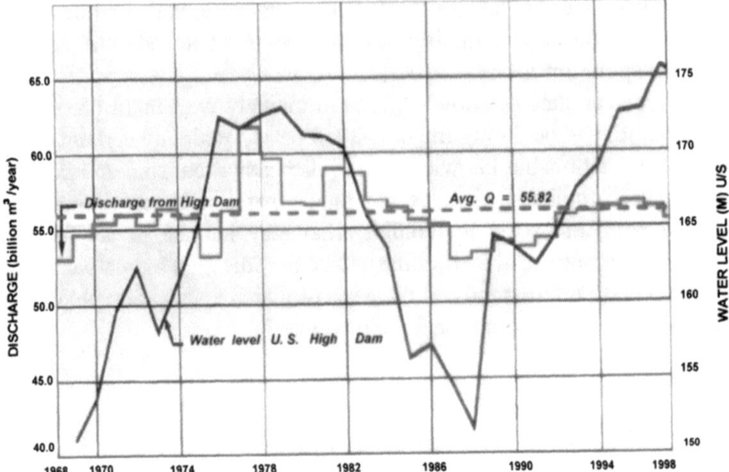

Fig. 11.6 Water levels in Lake Nasser and discharges from High Aswan Dam, 1968–1998 (Tortajada and Biswas 2017)

and spanning some 3,112,370 km^2, the overall impact of future climate change will probably be neutral over the entire river basin. Some parts of the basin may become drier, but other areas may receive more precipitation. An increase in evapotranspiration may contribute to more demand for water because of an increase in demand for supplementary irrigation water. Simultaneously, the water supply may become increasingly uncertain and more unpredictable than has been witnessed in the past.

A country like Egypt, with a desert climate, high temperatures, and an increasing middle class, may further have to meet a surging demand for energy for air conditioning in the warmer months. Because electricity cannot be produced without water, the demand for water for electricity generation may intensify steadily in the coming decades. Thus, total demand for water for agriculture, electricity, and other human and ecosystem needs is likely to increase further, unless the current efficiencies can be improved significantly.

11.8 Concluding Remarks

Lake Nasser has been very beneficial for Egyptians. An objective analysis by Strzepek et al. (2008) indicated that the economic benefits of Lake Nasser added EGP 4.9 billion to the country's GDP in 1997, or about 2 %. Thus, the social, economic, and psychological benefits of Lake Nasser for Egypt have been very substantial.

All three countries in the lower Nile Basin—Egypt, Sudan and Ethiopia—have to negotiate the use of the Nile waters with their neighbors. Populations and economic activities will increase steadily over the next five decades, necessitating more and more water. All Nile Basin countries have to significantly improve water use efficiency in the coming years. There is simply no other alternative. Otherwise, water will increasingly become a major constraint to further development and advances in the standard of living of the Egyptian people. It may also become a serious political issue between the three countries.

To cope with climatic fluctuations and expected climate change, all Nile Basin countries will have to collaborate even more actively and consistently than they have ever done historically. They also must concurrently maximize water-use efficiency by all users and in all sectors. For Egypt, Sudan, and Ethiopia, increasing water-use efficiency must become a national priority. The three countries must consider all means available to improve efficiency, including economic instruments, education, technology, social and cultural behavior change, and institutional strengthening.

In the Nile Basin, all water used in Burundi and Rwanda, and more than half the water used in Uganda, is produced internally. Accordingly, these countries may have more control of their sources of water and, thus, of planning and managing water. However, for both Egypt and Sudan, most of the water they use originates outside their national boundaries. This will make water management progressively more and more difficult and complex for these two tail-end countries of the Nile Basin. Furthermore, both of these countries must consider all cost-effective means available to increase water storage. Simultaneously, they must make water management practices, both in terms of quantity and quality, significantly more efficient than they have ever been in recorded history.

Water demand will increase further because Egypt's population is expected to grow by 40 million by 2050. Egypt does not have other rivers or other suitable locations where another major storage reservoir could be built. Thus, Egypt's most feasible alternative to meeting water demand in the future, and also to meeting the twin challenges of climate change and climatic fluctuation, is to make water management practices and processes progressively more efficient within the coming two to three decades. Water-use efficiency in domestic, industrial, and agricultural sectors can be very significantly improved with existing knowledge, technology, and management practices. Egypt and Sudan must also find new ways of managing agricultural water demand, such as by increasing the use of drought-resistant and pest-tolerant crops, crops that can grow in marginal and saline water, and by keeping abreast of scientific and safety issues concerning genetically modified crops. For domestic and industrial consumption, Egypt must also consider desalination in an aggressive manner.

The current scientific consensus is that extreme hydrological events, such as droughts and floods, are likely to be more intense and frequent in the future. According to the figures quoted by Karl et al. (2009) of the US Global Change Research Programme, for the United States at least, the amount of rain falling during the most intense 1 % of the storms during the past 50 years has increased by almost 20 %. This does not mean that the Nile flow into Lake Nasser has had, or will have, similar changes. Reliable data, as well as the level and quality of research on rainfall and river flow in Egypt and upstream countries, leave much to be desired.

Irrespective of what has happened in recent years and what may likely happen in the coming decades, Egypt has to be prepared to significantly improve management of Lake Nasser, its only major water storage infrastructure. Ensuring water security in Egypt in the coming decades means that the country will have to run ever faster and faster simply to remain in the same place.

References

Abdelhady D, Aggestam K, Andersson D et al (2015) The Nile and the grand Ethiopian renaissance dam: is there a meeting point between nationalism and hydrosolidarity? J Contemp Water Res Edu 155(1):73–82

Abul-Atta AA (1975) Egypt and the Nile after the construction of the high Aswan Dam. Ministry of Irrigation and Land Reclamation, Cairo

Abul-Atta AA (1979) After the Aswan. Mazingira 11:21–27

Abu-Zeid MA, Biswas AK (1992) Some major implications of climatic fluctuations on water management. In: Abu-Zeid MA, Biswas AK (eds) Climatic fluctuations and water management. Butterworth-Heinemann, Oxford, pp 227–238

Abu-Zeid MA, El-Shibini FZ (1997) Egypt's Aswan high dam. Int J Water Resour D 13(2):209–218

Achthoven TV, Merabet Z, Shalaby KS et al (2004) Balancing productivity and environmental pressure in Egypt: toward an interdisciplinary and integrated approach to agricultural drainage. Agriculture and rural development working paper 13. World Bank, Washington, DC

Badawy HA (2009) Effect of expected climate changes on evaporation losses from Aswan high dam reservoir. In: 13th International water technology conference, Hurghada, Egypt

Biswas AK (1993) Land use for sustainable agricultural development in Egypt. Ambio 22(8):556–560

Biswas AK (1994) International waters of the middle east: from Euphrates-Tigris to Nile. Oxford University Press, Oxford

Biswas AK (1995) Environmental sustainability of Egyptian agriculture: problems and perspective. Ambio 24(1):16–20

Biswas AK (2002) Aswan dam revisited: the benefits of a much maligned dam. Devel Cooperation 6:25–27

Biswas AK (2011) Cooperation or conflict in transboundary water management. Hydrolog Sci J 56 (4):662–670

Biswas AK, Tortajada C (2012) Impacts of the High Aswan Dam. In: Tortajada C, Altinbilek D, Biswas AK (eds) Impacts of large dams: a global assessment. Springer, Berlin, pp 379–396

Biswas AK, Kolars J, Murakami M et al (1997) Core and periphery: a comprehensive approach to Middle Eastern water. Oxford University Press, Oxford

Karl TR, Melillo JM, Peterson TC (eds) (2009) Global climate change impacts in the United States. Cambridge University Press, New York

Shalaby A, Ali R, Gad A (2012) Urban sprawl impact assessment on the agricultural land of Egypt using remote sensing and GIS. J Land Use Sci 7(3):261–273

Strzepek KM, Smith JB (1995) As climate changes: international impacts and implications. Cambridge University Press, Cambridge

Strzepek KM, Yohe GW, Tol RSJ et al (2008) The value of the high Aswan dam to the Egyptian economy. Ecol Econ 66(1):117–126

Tortajada C, Biswas AK (2017) Hydropolitics of the Aswan high dam. Springer, Berlin (forthcoming)

Chapter 12
Greater Security with Less Water: Sterkfontein Dam's Contribution to Systemic Resilience

Mike Muller

Abstract The Sterkfontein Dam, South Africa's third largest dam by volume, is an important component of the Vaal River system, which supports the water security of a socially and economically important inland region of the country. The design, construction, and four decades of operation of the dam provides a useful illustration of the role of storage in general and dams in particular; the contribution of system management approaches to water security; the place of interbasin transfers in such systems management approaches; and how the water-energy nexus may function in practice. There have been substantial changes in the wider socioeconomic context over the past 50 years, particularly in relation to the national energy system, and the region experienced a severe drought in 2016. The operation of the Sterkfontein Dam provides some insights into how one component of a system can help to manage drought impacts, how this operation may have to be adapted to address the emerging challenges posed by changing contexts, and some limitations and risks that may arise. At a larger scale, the case also provides an unusual example of substantial innovation during implementation and raises questions about the supportive institutional context that enabled these innovations to be made. Finally, the case also demonstrates how practice can influence policy, both in subsequent South African water policy development as well as the global polices developed at the 1977 United Nations (UN) Water Conference in Mar del Plata, Argentina.

Keywords Water resource management · Water supply · Climate change · Adaptation · South Africa · Sustainable development

M. Muller (✉)
Wits School of Governance, University of the Witwatersand, Johannesburg, South Africa
e-mail: mike.muller@wits.ac.za

C. Tortajada (ed.), *Increasing Resilience to Climate Variability and Change*, Water Resources Development and Management,
DOI 10.1007/978-981-10-1914-2_12

12.1 Introduction and Background

12.1.1 *Water Requirements of the Vaal River System: A "Problem-Shed" not a Watershed*

The Vaal river system supports a region that includes three of South Africa's six metropolitan areas and a significant proportion of the South African economy. Gauteng, the core province, accounts for 33 % of the gross domestic product (GDP) and 24 % of the population (Stats 2014, 2015). Although data is not collected by catchment or water supply region, a conservative estimate is that at least 40 % of the national GDP and 30 % of the population are supported by the system.

While the importance of mining has declined, the evolution of the region's economic development was determined by mineral deposits, particularly gold and coal. The area supplied by the Vaal system lies on both sides of the continental divide between the Orange River (which flows west to the Atlantic) and the Limpopo River (which flows east to the Indian Ocean). As a consequence, the Vaal system serves as what some authors call a "problem-shed" rather than a naturally defined watershed (Fig. 12.1).

Situated high in the catchment of the two main rivers, the water requirements of the rapidly expanding urban and industrial settlements soon exceeded the capacity of the local sources in normal seasons, leaving them vulnerable to the vagaries of a

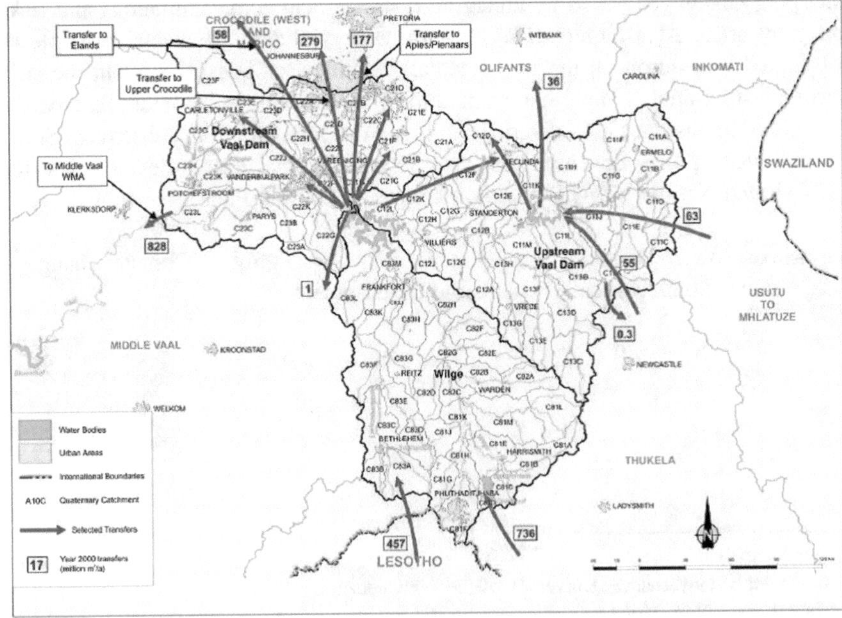

Fig. 12.1 Upper Vaal water management area (Reproduced from DWAF 2003)

highly variable climate. Over the first century of urban and industrial development, from 1870 to 1970, the resources of local rivers within the Vaal catchment were developed and extensively interlinked in a relatively complex system (Tempelhoff 2003).

While the planning of water resource development is built up from catchment level perspectives, the distribution systems follow patterns of urban and industrial development that do not follow catchment boundaries. Most of the municipal and industrial users are served by a regional utility, Rand Water, which was established in 1907. Rand Water's jurisdiction covers three major catchments (Vaal, Crocodile West/Marico, and Olifants) and includes 18 municipalities, including three metros, and all or part of five of South Africa's nine provinces. It currently supplies an average of over 4 million m^3 of water a day, all derived from the Upper Vaal system.

12.1.2 Water Resources of the Region

The primary challenge for water supply in the region is the extreme variability of the sources. One measure of this is the variability of the overall Orange River system, of which the Vaal is a tributary. The river's mean annual runoff is around 12,000 Mm^3 per annum; during the drought year of 1982, only 1000 Mm^3 reached the river mouth; in the flood year of 1988, that rose to 26,000 Mm^3. Variability at the tributary level is even greater (McKenzie et al. 1998).

The other key challenge for water resource managers posed by the Upper Vaal water management area is that, compared with other water management areas, its population and economic activity are in inverse relation to its mean annual runoff (Fig. 12.2). This relationship is even more acute in neighboring water management areas, notably the Crocodile West system, much of whose needs are met from the Vaal system and its wastewater discharges (Fig. 12.3).

By the 1960s, it was already becoming clear that even if the full potential of the Vaal River was developed, it would not be able to meet the region's growing needs. This was a strategic challenge because, in addition to mining and industrial development, the region had become the energy hub of the country with the majority of its electricity generated from coal deposits in the Vaal and neighboring Olifants river basins. New sources were required and their development was a high political priority. As a result, consideration turned to neighboring river systems.

12.1.3 Institutional Context

In the 1960s, institutional responsibilities for the water schemes in the Vaal system were clear and uncontested. Legislation in 1956 (South Africa 1956) had given the

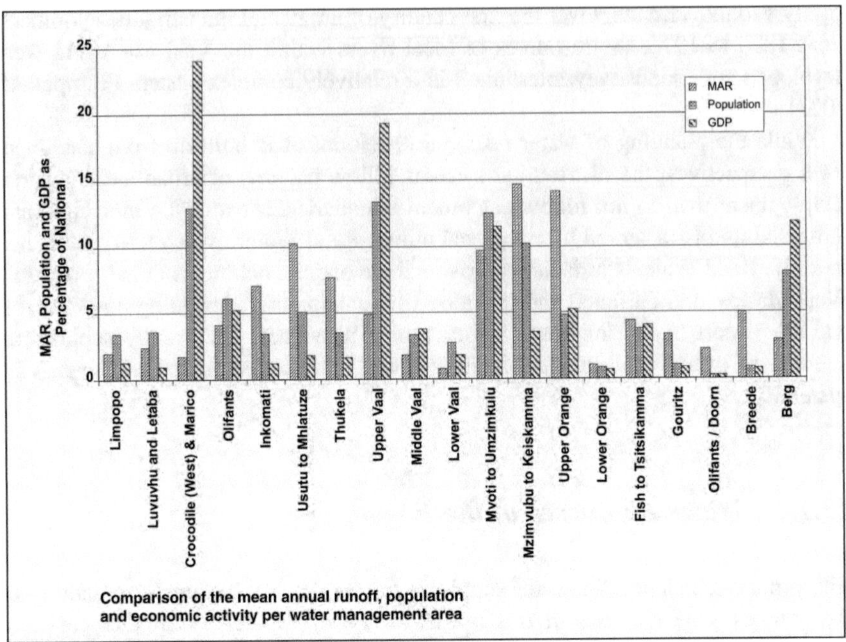

Fig. 12.2 Mean annual runoff (MAR), population and gross domestic product for South Africa's 19 water management areas (Reproduced from DWAF 2003)

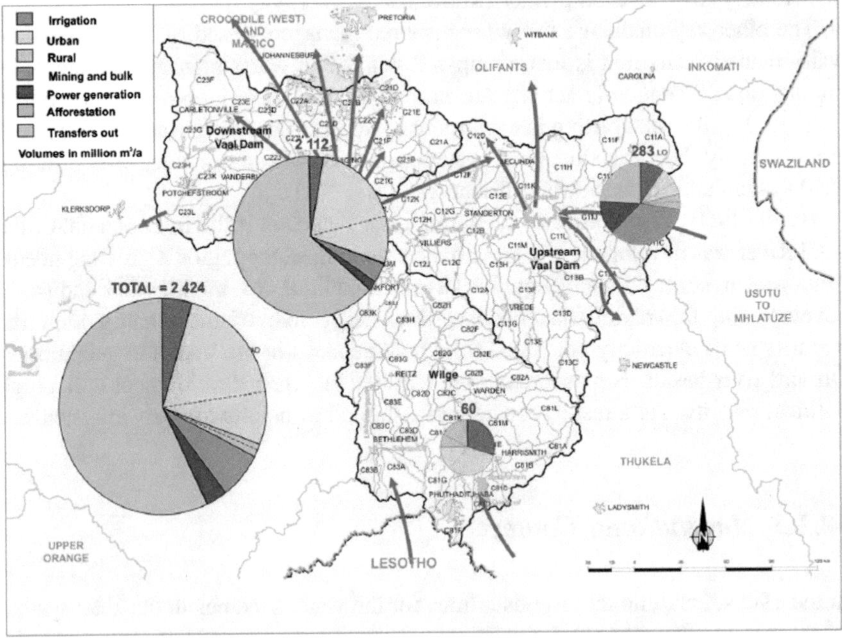

Fig. 12.3 Upper Vaal water management area sectoral water requirement, including transfers to other water management areas (Reproduced from DWAF 2003)

national Department of Water Affairs (DWA)[1] overall responsibility for all activities from resource planning, project identification, development, and subsequent operation. The legislation also introduced some economic discipline into project planning, requiring formal white papers to be presented to Parliament for information, confirmation, and, where appropriate, allocation of budgetary funds for all large projects.

In parallel, electricity supply utility ESKOM had the mandate to plan, invest, and operate the infrastructure required to provide a cheap and reliable electricity supply to the growing society and economy on an economic basis. The customers for the water supply were primarily large industrial users (including ESKOM but also SASOL, the state-owned oil-from-coal producer as well as large-scale mining operations) and municipal users, many of whom were served through the treatment and transmission system of the Rand Water Board.

The existence of a simple and robust set of public institutions enabled decisions to be taken and implemented relatively swiftly. The substantial economic base of the principal users also assured bulk water suppliers of the financial viability of their operation. In the case of large industrial and urban supplies, systems were planned and designed on a full cost recovery basis.

12.1.4 The Sterkfontein Dam in the Vaal System

The Sterkfontein Dam is a key component of the overall Vaal system. The volume of water stored in its reservoir (2650 Mm^3) is the third largest in South Africa. At the time of its completion in 1982, the dam itself was recorded as one of the ten largest earth embankment dams in the world (19.8 million m^3). A peculiarity of the dam is that it has no spillway because the stream on which it is built is too small to generate significant flood flows, with a catchment area of just 191 km^2 (Elges 1982).

This draws attention to some of the dam's other characteristics, which are relevant to the focus of this case study. It is an important component of a larger interbasin transfer scheme—in effect and intent, an off-channel storage dam filled from sources beyond its natural catchment. It was built not with the primary purpose of supplying water to local users but rather to increase the yield and improve the assurance of supply of a much larger system of dams and links within the catchment of the Vaal River. In modern terminology, its purpose was explicitly to increase the assurance and enhance the resilience of the overall system.

Water is transferred to the dam by pumping from the upper reaches of the Tugela (now known as Thukela), the third largest river in South Africa. When the national power utility became involved as one of the principal beneficiary users of the

[1]The department's name has changed from "Water Affairs" to "Water Affairs and Forestry" to "Water and Sanitation." What was the "Electricity Supply Commission (ESCOM)" is now, formally, ESKOM.

scheme (the majority of their coal-fired power stations are located in the Vaal and neighboring Olifants river basins, largely dependent on water supplies from the Vaal system) other opportunities were identified. As a consequence, the scheme was eventually developed as a combined pumping/pumped storage scheme.

The dam and associated works were first conceived of in the 1960s to meet the growing needs of the Vaal system. Most of the infrastructure was planned and built in the 1970s. The overall scheme, which included the construction of the pumped storage components, including an internal balancing dam, was only completed in the early 1980s. During this period, there were substantial changes in approach and emphasis in water resources management, which were reflected in changes to the overall conception of the scheme and the dam that forms part of it.

This was long before climate change-related concerns were raised and the importance of resilience emphasized. However, many of the approaches that are today considered to deal with the potential impacts of climate change were already introduced to cope with extremes of climate variability to which the region is vulnerable. So, when analyzed against today's criteria of resilience, sustainability, and contribution to inclusive development, the dam and associated works still offer important lessons for future water resource development in general and storage in particular.

From a water resource management perspective, it is also notable that the Sterkfontein Dam and the scheme of which it was part were a product of new approaches to the analysis and design of water resource systems that emerged from the Harvard Water Program (Maass et al. 1962 and see Reuss 2003). As such, the scheme provided a practical illustration, at scale, of the power of those approaches and thus influenced the direction of subsequent water resource policy and management. Although environmental concerns were only just beginning to be recognized, this was also one of the first water projects in South Africa—and internationally—for which an environmental impact analysis was undertaken (Roberts and Erasmus 1982).

12.2 The Evolution of the Tugela-Vaal Scheme Concept

12.2.1 Choosing from Limited Options

The Sterkfontein Dam emerged as an element of a larger scheme for the augmentation of the Vaal system at the time of a far wider review of water policy and strategy by the national government. This review, undertaken by a Commission of Enquiry beginning in 1966 and concluding in 1970, was prompted in part by concern about the long-term viability of development strategies that had been focused on major infrastructure development to expand irrigation in a country in which water was already scarce.

In its comprehensive set of conclusions and recommendations (South Africa 1970a), the Commission found that demand-side interventions were required to complement supply-side investments. The reuse of wastewater was seen both as a means to provide a supply stream to the periphery of the system as well as to support irrigation. The agricultural sector was characterized as having much lower value use than urban and industrial applications, which would have to have first call on available water. Water conservation, water pricing, and trading as well as administrative reallocation were all identified as important instruments to deal with future challenges, supported by the 1956 Water Act, which had given national government greater control over what had previously been a largely private resource, with access disputes resolved in specialized courts.

Encouraged by the Commission's findings and recommendations, the government placed restrictions on agriculture's use of water in the Vaal basin. As a matter of policy, it was decided that dry cooling would have to be used in new inland power stations despite the lower efficiencies and higher costs that would be incurred. If nuclear power was to be developed, it would have to be at coastal locations where seawater could be used for cooling. Nevertheless, even after the application of these and related measures, supply augmentation was still required to meet the needs of the inland areas in general and the Vaal system in particular (Stephenson 1971).

Options for the Vaal system were limited. Rivers flowing north and northwest towards the Limpopo were already heavily committed. There was more capacity in rivers flowing to the east; some small transfers were feasible from the Usutu and Komati (which were subsequently developed to supply the power stations in neighboring areas). Other options included the Caledon River, another tributary of the Orange or the Orange River itself, from where it enters South African territory from Lesotho, 500 km from Johannesburg. The largest proximate source was the Tugela, at the southern tip of the Vaal catchment. However, like the Komati and Usutu, any transfer from the Tugela would require extensive—and expensive—pumping across the escarpment which forms their watershed with the Vaal.

By the middle of the 1960s, the focus of the discussion had narrowed to two options: the Tugela River and an ambitious proposal to divert the Orange River close to its source in the highlands of Lesotho (until 1966, the British protectorate known as Basutoland) through the so-called Oxbow scheme. Although the Oxbow scheme would likely be able to supply water at a lower unit cost than the Tugela, because it did not involve pumping, it was felt to be technically and institutionally easier and, importantly, quicker to implement a scheme on the Tugela (Stephenson 1971). This was not least because of the administrative and diplomatic complications that would be involved in building a water scheme in a newly independent neighboring country.

12.2.2 The Tugela Decision and the Storage Option Lead to Sterkfontein

Given the growing urgency, a formal decision was taken in a 1967 white paper presented to Parliament, proposing to construct the Spioenkop Dam on the Tugela

River as the source for a transfer scheme. The site chosen was far enough downstream from the watershed to be able to provide a reliable supply of the transfer volumes that were then considered to be required without impacting upon downstream users. The Tugela was still relatively undeveloped and there was limited potential for future downstream development (van Robbroeck 2015).

The initial intention was to build a system that would pump directly from the Spioenkop Dam, over the escarpment into a holding dam that fed into the Wilge River and then the Vaal Dam. Construction of the Spioenkop Dam began in 1968. However, it was immediately evident that it would be very expensive to pump the large volumes required to provide the required assurance of supply during drought years, over 40 km and a lift of over 700 m. Reflecting on the initial project concept, the Commission of Enquiry noted:

> Originally, it was proposed to increase the assured yield of the Vaal River to 1064 million gallons per day (mgd) by pumping 180mgd from the Tugela River. As it is not possible to predict exactly the onset of a drought, it would be necessary to provide a continuous flow of 180mgd … by pumping from the Tugela at all times except when the Vaal river dams were overflowing.

Further disadvantages of this approach identified by the Commission were that:

- Keeping the Vaal system dams full would result in high evaporation losses and good rains might lead to a loss of expensive pumped water if they overflowed.
- Very large works would be required to lift the water almost 750 m.
- Between 60 and 100 MW of electrical power would be required.
- The use of the full 180 mgd (147) in the Tugela basin itself would be lost.

In response to these challenges, a major change of approach was proposed, even as construction of Spioenkop began. It was found through simulations of the system under the worst drought conditions on record that, if sufficient additional storage was built in the Vaal catchment itself, this could augment supply during drought periods. The storage could be filled gradually, requiring lower transfer flows. This would enable these flows to be diverted higher up in the catchment, reducing the pumping head. Capital and operational costs could thus be substantially reduced and there would be less impact downstream in the donor catchment. Evaporation losses could also be reduced if an appropriate site could be found.

With construction of the Spioenkop Dam now well under way, this major conceptual change could have been politically embarrassing if it suggested that the original decision had been wrong. In that event, the 1969 white paper reporting on the project presented the change as a phasing arrangement, stating simply:

> The construction of the Spioenkop Dam on the Tugela River … is under way and progressing well. After completion, the Spioenkop Dam will have a net assured yield of 249 mgd, of which 150 mgd could be made available for use in the Vaal River catchment. *It is proposed to utilize* 60 mgd *of this amount now* [1969], and the balance at a later stage in the development of the Tugela Vaal Project. (South Africa 1969).

Next, it was necessary to identify the high-level storage required. In a 1970 report (South Africa 1970b), the subject of the second phase was broached and it was proposed to build a

> dam on the Elands River near the town of Harrismith, to "receive water from the Spioenkop Dam on the Tugela River, with the purpose of storing water to meet shortages in the supply from the existing dams on the Vaal River".

It was explained that these shortages might arise during times of severe drought. The construction of the dam would constitute the second phase of the Tugela-Vaal Project and the two phases would, together, increase the net assured yield to the Vaal River system from 884 to 1064 mgd (723–871 Mm^3/d).

Once again, the change was approved but a further iteration was still needed. It transpired that the site identified for the dam had been zoned as part of the then government's plans for racial segregation. As the official then responsible explained:

> When the contents of the draft White Paper became known at the Department of Bantu Administration and Development, they protested that the dam would submerge much of the arable land in the then proposed Qua-Qua Homeland. Consequently, we had to urgently look for an alternative storage site for the pumped water from the Tugela. I located a suitable site at Sterkfontein on the Nuwejaarspruit, close to the watershed between Vaal and Tugela (van Robbroeck 2015).

In this case, there was at least a clear (political) reason for a change in the project concept. This opened the way to formalize two further innovations. First, the reduction in the average flow rate needed to provide the required level of assurance meant that water for transfer could be diverted higher up the catchment, saving money by reducing pumping costs (but rendering the pipeline from Spioenkop unnecessary). The second, more fundamental, was to introduce a power dimension to the scheme.

12.2.3 Drakensberg Pumped Storage: The Final Iteration

As already noted, the initial Tugela-Vaal project was clearly going to require substantial power to pump water over the escarpment, an issue that had to be addressed with power utility ESKOM. However, there were other items for discussion. Although South Africa has only limited hydropower potential due to the relatively small average flow in its rivers, there was precedent for the development of hydropower in other river projects and there were good operational relations between the DWA and ESKOM. Therefore, consideration had been given to the development of hydropower from Spioenkop Dam and the lower river where there was perhaps 600 MW potential available (Stephenson 1971).

In this process, the potential for locating a pumped storage system in tandem with the basic pumping arrangement was identified. This was made possible by the relocation of the storage dam and pumping station closer to the escarpment. Since the pumping infrastructure was only to be used occasionally, during times of drought, this prospect was intuitively attractive because it would allow the infrastructure to be used more intensively (van Robbroeck et al. 1970). Because the national power utility was building a new fleet of large coal-fired stations, this idea had merit for them, too. Coal-fired power stations can usefully be complemented by a peaking power plant that reduces the need to ramp generation up and down to follow demand fluctuations.

Once again, project implementation was running ahead of these conceptual innovations:

> The idea came too late for implementation of the first phase of the Tugela-Vaal. When the planning for the second phase was on the go, I went to see the then Secretary, Jacques Kriel, with the request that he approach Escom for investigating a joint scheme. I think that he thought that the proposal was too grand, and ignored it (van Robbroeck 2015).

In the event, there was a further serendipitous development. An engineer with knowledge of the proposals left the DWA to work for ESKOM, where he discussed the pumped storage opportunity with his managers:

> They saw the merit and then got in touch with Water Affairs. This time it was taken seriously and the rest is history. At a meeting on 2 April 1974 … it was agreed to implement a joint project (van Robbroeck 2015).

The cost of pumped storage is significantly influenced by the cost of providing the water storage required (up to 60 % of the total capital cost of a system). The storage volume is determined by the potential application of the system—a system operated to manage weekly peaks, with storage refilled only over weekends, may require 50 times more storage than one that only meets daily peaks (Norman 1982). Other conditions being equal, some of these costs could be shared if a pumped storage scheme was linked to a larger water storage project.

Pumped storage has many advantages for the managers of a national power grid. In a South African context, a coal-fired unit took several hours to bring into service and its load changed only slowly, at 2 % of its total rating per minute. A pumped storage system could be brought from zero to full load in 30 s; the change from maximum pumping to full generation is made in less than 10 min: "From the point of view of rapid response, pumped storage plant has clear technical advantages over coal-fired plant for peaking duty". Including a pure pumping function extended the flexibility because, "in an emergency, this pumping can be curtailed and then made up later when more plant is available" (Norman 1982).

However, the introduction of a pumped storage component required one final set of changes to the original scheme design and the detailed design of Sterkfontein. To provide the storage needed to ensure that the proposed transfer volumes could be achieved, it was decided to construct the Woodstock Dam above the diversion barrage at Driel (South Africa 1974). It also had to allow the pumped storage system to function even when the main Sterkfontein reservoir was almost empty, which might occur under severe drought conditions. This required a subreservoir to be built within the main reservoir (Fig. 12.4).

The resulting Driekloof Dam was classified as a large dam in its own right. Although its capacity, only 35 million m^3, is less than 2 % of the volume stored by the main reservoir, the dam wall is 46 m high and 500 m long. A particular design

Fig. 12.4 The Driekloof dam-within-a-dam (Photograph by author)

challenge was that it had to withstand rapid weekly fluctuations of up to 22 m and was subject to pressure from both sides. It also needed a spillway that would allow water to flow into the main reservoir when the scheme operated in transfer pumping mode (Elges 1982).

A final, critical benefit of integrating the pumped storage function was that the transfer capacity would no longer be a constraint. Previously, the department had warned that:

> … should the Sterkfontein Dam be drained during a critical drought, it would take approximately seven years before it could be pumped to its full capacity. With the increased pumping capacity of 20 m^3/s [from 11 m^3/s] the Sterkfontein Dam could be pumped to its full capacity within four years, if there is water available in the Tugela River, thereby lessening the risks attached to a drought (South Africa 1984).

The limit to how much water can be transferred is now what is available on the Tugela side, given the far higher pumping capacity installed for energy generation.

12.2.4 The Completed Configuration

The inclusion of the pumped storage scheme required some alterations on the Tugela side, but these were simply adaptations of the configuration originally proposed. The revised project specification was finally approved by DWA and ESKOM in April 1974 (van Robbroeck 2015). In summary, over the 8 years from 1966 to 1974, the overall scheme concept had evolved as follows.

The construction of the Spioenkop Dam was started in 1968. Water was to be pumped over the escarpment into a holding dam in the Wilge tributary of the Vaal. This would supplement the Vaal's natural flow to meet the peak needs during a drought while providing sufficient capacity to sustain existing uses (and flows) in the donor catchment.

In 1969, it was decided to increase the storage in the recipient catchment. This allowed water to be transferred at a slower rate over a longer time period. It also meant that water could be captured higher in the donor catchment, reducing costs of both the infrastructure and pumping.

In 1970, it was recognized that, because the infrastructure would only be used occasionally at times of low flow in the Vaal system, it could be used for other purposes when it was not required for pumping. When access to the initial Java storage dam site was denied, an even better site was found at Sterkfontein, which facilitated the integration of a pumped storage function and an increase in water storage capacity.

In 1974, it was agreed to develop the next stage of transfer as a joint pumped storage system. This required construction of two additional dams (Woodstock and Driekloof) and substantial alterations to the intake works.

This evolution in approach was captured in a series of white papers tabled to Parliament as the planning process evolved and the detailed structure of the project was clarified (South Africa 1974, 1980, 1984; and supplementary reports). The evolution of the concept was not always clearly communicated but had to be presented when budget allocations were required.

12.3 Implementation and Operation

The construction of the scheme over a period of almost two decades, with initial plans significantly adapted to reflect the changes in project concept, is described below. However, even as it reached completion, the scheme was already contributing to the management of climate extremes in its supply area. Operational arrangements were elaborated as experience was gained—a process that is ongoing as operators manage the impacts of the El Niño drought of 2015/16 in an electricity supply environment that has changed radically since the scheme was commissioned.

12.3.1 Construction

While the design of the overall scheme was still evolving, the construction of many elements had already begun. The Spioenkop Dam was completed in 1972. Three weirs were built higher up the river to divert small tributaries into the Driel barrage on the main stem from which water was pumped into a 37 km canal linking to the Jagersrust forebay of the main high-level pumping station. The first phase of the scheme was completed in 1974 with the construction of the first stage of the Sterkfontein Dam, a 69 m high embankment, creating a reservoir with a capacity of 1200 Mm^3. As the transfer design changed, a decision was made in 1974 to build the Woodstock Dam to assure the increased transfer flows that the pumped storage system would enable.

The initial transfer capacity was 130 million m^3/year, at which rate the Sterkfontein reservoir would take nearly a decade to fill. Once full, however, the assurance it provided allowed other dams in the system to be drawn down faster than before and increased the overall system yield by 180 mgd (147 Mm^3/d), almost three times the actual physical transfer rate. Meanwhile, the Spioenkop Dam now served simply to reregularize the flows in the river to continue to assure supplies to downstream users.

This initial arrangement was superseded once the pumped storage system was completed. The duplication of the initial pumping works was no longer required and all water transfer was undertaken using ESKOM's systems. The pumped storage system (including the internal dam) was completed in 1982, at the same time as the second phase of Sterkfontein Dam, which saw it raised to its final height and storage capacity of 2600 Mm^3. At its conclusion, the overall system yield had been increased by 1545 Mm^3/year to 2345 Mm^3/year, an increase of 800 Mm^3/year made possible by a transfer of just 347 Mm^3/year. Much of this was due to a 34 % increase in the yield that could safely be drawn from the Vaal River itself, by virtue of the supply assurance provided by Sterkfontein (van Robbroeck 1982).

12.3.2 Operation

The operation of the water transfer element of the scheme was initially undertaken in terms of a framework agreement with ESKOM, which had given DWA responsibility for the construction and operation of the dams and associated civil works while ESKOM was responsible for the construction and operation of the core pumped storage system, from intakes to discharges.

Operating costs were also determined on a practical basis. ESKOM's operating model at that stage was to use "tied" coal mines, which had cost-based contracts for ESKOM's supply. Because there were limited additional operating costs incurred by either party, the agreement was that DWA would simply pay ESKOM's marginal costs—essentially, the cost of coal used to generate the power to pump its share of the water. Even at full capacity, this was a relatively small proportion of that used in the power storage operations.

Operating arrangements evolved over time but there have always been seasonal evaluations of the state of the system which guide operational decisions. A formal review of operating procedures (DWAF 2006) described how:

> ... the operation of Vaal Dam is driven mainly by the downstream releases and the water abstracted by Rand Water. Releases from Sterkfontein Dam are usually only made if Vaal

> Dam's storage is depleted to the minimum operating level, or if Sterkfontein Dam is full and there is still water available in Woodstock Dam for transfer via the Thukela Vaal Transfer Scheme … The normal operating rule, with the objective of maximising yield in the system, is to continue the transfer until Vaal and Bloemhof Dams are full. However, during wet hydrological conditions when the dam levels are relatively high or in the case where excess supply capability in the system is present, the transfer volume is reduced to save pumping costs. These deviations are only implemented if it is proven, through scenario analysis, that the long term assurance of supply will not be jeopardised.

12.3.3 Operational Performance and Responses

While the theoretical benefits of a systems approach to the planning of the Tugela-Vaal transfer are clear, they can only be realized if systems are operated appropriately, as one of the DWA officials at the time reminded his colleagues (Alexander 2010). The operating procedures and their results are thus as important as the procedures used in planning.

However, the true test of the operating rules is the system's performance when it is under stress. This is more difficult to assess for the Sterkfontein Dam than it would be in the case of a stand-alone dam serving a single, clearly identified set of uses and users. The challenges are accentuated when the objective of the system of which the dam is part is to provide assurance in circumstances that are expected to arise only once in 20–50 years. A further test is whether operational approaches are adapted in the light of experience and changing circumstances.

Three incidents over the life of the Sterkfontein Dam have served to provide a test of its contribution to the overall functioning of the system:

- The severe drought that affected many parts of South Africa, ending in 1983;
- A series of dry years between 1992 and 1995;
- The El Niño-related drought beginning in the 2014/15 season that continues to affect the region.

12.3.4 1978–1983 Drought and Subsystem Crisis

Shortly after the completion of the second phase of the dam, the central regions of South Africa experienced a significant El Niño-related drought, characterized as the most serious since the 6-year event of the 1930s. At this stage, the Sterkfontein Dam had not yet been filled to its full capacity; more significantly, its contribution was limited because the main impact of the drought was felt in another subcatchment where the Grootdraai Dam had been built to supply many of the regions' power stations. Because there were no interlinkages between the Grootdraai subsystem and the Wilge River subsystem that was supported by Sterkfontein, the situation became critical. As explained by one of the officials responsible at the time (van Robbroeck 2015):

> Since 1978, river flow had been below average and was exceptionally low in the 81/82 and 82/83 seasons. It was projected that all the dams supplying the Highveld would run dry by September 1983. This threatened the whole country's electricity supply! … The only feasible (sic) was a temporary scheme consisting of seven weirs in the Vaal River between the low water level in the Vaal Dam reservoir and Grootdraai. A pumping station downstream of each of these weirs would lift the water, resulting in the river running upstream over a distance of 208 km … All these works had to be and were completed within four months. Ironically, after operating for a while, the temporary part of the scheme could be abandoned due to heavy rains filling the dams, which even washed away some of the weirs.

What the 1982 event demonstrated was that, in the absence of internal linkages, providing supply assurance for the entire system would not necessarily extend assurance to all its subsystems.

12.3.5 1992–1996 Droughts and System Strengthening

The 1992/96 drought was not as severe as that of the early 1980s, in the Vaal system at least. However, it provided a reminder that the Grootdraai system was still vulnerable, although new links had been built to supply the subsystem from the Usutu and Komati. However, it served to affirm the strategic function of Sterkfontein once again. In a review of the performance of the overall system shortly after the drought, it was noted that in the 2002/2003 season:

> A conditional surplus of about 300 million m3/a exists in the Vaal River System. The available surplus is qualified as "conditional" since it is only available if all the transfers are fully operational. In practice the volume of water conveyed through the Thukela-Vaal Transfer scheme will be determined annually, effectively operating the system such that the water demands are in balance with the supply. The quantity transferred (and the pumping costs) will thus increase over time to match the growth in the water requirements (DWAF 2003).

One reason for the continued vulnerability of the Grootdraai subsystem was that SASOL and some of ESKOM's large power stations' rivers depended on a single canal link with very limited onsite storage. This meant that the flows could not be interrupted for canal maintenance, which put the integrity of the canal and hence the energy operations at risk. To address this, it was decided in 2004 to build a 115 km link (TCTA 2016) that now allows Vaal Dam water, assured by Sterkfontein, to be supplied into the Grootdraai system:

> Eskom indicated that in order to minimise the total risk of water supply, the pipeline option to transfer water from Vaal Dam would be the preferred scheme to augment the Eastern Subsystem. The Vaal Dam pipeline option has the advantage that water can be accessed from both the Thukela River and Lesotho Highlands rivers, which decreases the vulnerability with respect to localised droughts in the catchments of the Eastern Sub-system (DWAF 2003).

Construction of the Vaal River Eastern Subsystem Augmentation Project (VRESAP) reinforced the assurance of supply to the Grootdraai subsystem. However, increasing assurance in one subcomponent may reduce assurance in other components unless this is addressed, as it was in this case.

12.3.6 2016 Drought

The impact of the further development of the system and the evolving context in which it operates is being tested during the current drought event. This began in some parts of the Vaal system with a relatively dry 2014/15 season and continued into a full-fledged drought, linked to a strong El Niño event, in 2015/16. This event has drawn public attention to the function of the Sterkfontein Dam when substantial releases were made from the dam at the end of 2015. As part of the government's response to the drought, the incumbent Minister explained that:

> As part of the normal operating rules, the department does a 3-year view in May of each year of how the system is going to react. Based on what we saw during the May runs this current year, it was decided to release some water from Sterkfontein Dam to Vaal Dam to create storage in the Sterkfontein Dam, which will then be filled with water pumped from the Tugela River.
>
> For years now, we were experiencing above normal rainfall and runoff, which means that we only had to pump enough water to replenish that what was lost through evaporation in Sterkfontein Dam. But, because of the system storage being lower than before, and to avoid having to impose possible restrictions in the near future, we decided on this additional pumping. Gauteng receives most of its water from the Vaal River System and should therefore be fine for the coming couple of years. We are also monitoring the situation on a monthly basis (Mokonyane 2015).

It is notable that while a number of smaller urban areas have suffered severe water restrictions, those in the Vaal system have remained relatively secure and no resource-related restrictions have yet been imposed. Only if the drought continues into the 2016/17 season is it likely that restrictions will have to be introduced in the Vaal system (DWS 2015).

12.4 Performance in a Changing Context

Some of the social and economic assumptions made when the scheme and its individual components were designed and built are no longer valid. This is not surprising given that the entire political and economic structure on which they were based has changed radically. In addition, global economic developments have had significant impact on South Africa. There has also been a substantial evolution in understanding of water-related environmental issues and radical shifts in social preferences with respect to the protection of the environment and mitigation of climate change.

These contextual changes have an impact on the operation of the project and its effective achievement of its objectives and may indeed introduce changes in those objectives. Because it is inevitable that the broad economic and social context in which the scheme operates will change over the decadal life of major infrastructure, it is of interest to assess the extent to which the scheme has the flexibility to adapt as needed.

12.4.1 Social: Increased Water Demand and New Water Uses

In 1970, the water requirements of the black majority of the population were of limited significance. Urban residents had only limited supplies—perhaps 15 % of the consumption per capita of the affluent minority whose suburban lifestyles had high consumptive water demands. It was policy at that time to restrict the access of black people to the cities and expected growth rates were thus relatively low.

By 1990, this had already changed as economic and political pressures forced the white government to allow black residence in urban areas. After a democratic government was established in 1994, there was a drive to improve water services, including waterborne sanitation, in many urban areas, changing the profile of and trends in water consumption and substantially increasing wastewater flows, which posed further management challenges in relation to system water quality.

However, social factors did impinge on the project. The Sterkfontein site was only adopted because the first choice site was rejected for political reasons: it would have reduced the already inadequate arable land available in a "homeland" to which African communities in the region were to be restricted. Contestation about forced population removals is an important political issue in modern South Africa that goes far wider than relatively small-scale relocation for specific projects. It affected millions of people across the country while Sterkfontein and related projects, undertaken for a public purpose that is still relevant today, affected at most a few thousand.

A greater concern for the Tugela-Vaal scheme is that past social policies have left a legacy of large, poor, semi-urban communities effectively stranded in areas of limited economic opportunity. These communities were not directly served by the system; water supply for the 330,000 people now living in the Maluti-a-phofung local municipality, where Sterkfontein is located, came initially from small local dams. However, as services were expanded and service levels improved, water demand began to overtake the supply capacity and the Sterkfontein reservoir was an obvious source of augmentation. This eventuality had been noted during the original development, but it was assumed that the requirements would be insignificant in the overall scale of the project.

12.4.2 Environment and Amenity

While social issues are not a major concern in this case, environmental issues are far more important today than they were in the 1960s and 1970s. The Sterkfontein Dam remains an important example of the utility and challenges of interbasin transfers, which are still controversial in some circles. Impacts on the Vaal side have been limited, not least because the stream on which the dam was built made little contribution to overall environmental flows and was of limited environmental significance. The reservoir is now the site of a small number of relatively low-impact tourist developments that have generated small but significant additional economic opportunities in this otherwise pastoral community.

The Sterkfontein Dam and Tugela-Vaal project were also in the forefront of environmental planning in the 1960s and 1970s. The environmental impacts were greater on the Tugela side, an area of great natural beauty that was already a well-developed tourism and recreation destination. This was recognized in the early stages of the scheme with the implementation of a very early example of an environmental impact assessment and environmental management plan. Public health concerns that an interbasin transfer might transmit the vectors of the parasitic disease bilharzia (schistosomiasis) from the Tugela Basin where it was endemic to the Vaal have proved to be unfounded (Pretorius et al. 1976).

Sterkfontein's initial operating rules sought to maximize system yield at the desired level of assurance. Accordingly, since evaporation losses from the Vaal Dam were substantially higher than at Sterkfontein, releases would only be made if at the start of the season the Vaal Dam levels had dropped towards 15 %. However, other considerations emerged. Amenity values became important as the Vaal Dam was developed for aquatic sports and leisure activities where major fluctuations in reservoir levels had direct economic impacts. At a more systemic level, water quality concerns have required releases to be made to provide additional dilution to reduce salinity levels.

Political preferences and technical challenges have changed substantially since the scheme's inception. A further challenge is that the 1998 National Water Act (South Africa 1998) requires the establishment of environmental flows for all water resources. Although the framework allows for less onerous requirements in heavily used watercourses, it is still intended that minimum flows will be maintained; these have not yet been implemented in many reaches of the Vaal system.

12.4.3 An Economy in Transformation

In the formal economy, the importance of mining has declined dramatically since the 1960s. Although the Vaal region remained a significant producer of coal—both for local energy production and for export—gold mining shrank to low levels and was not compensated for by the emergence of other commodities, such as platinum and chrome for which demand was smaller.

Mining continued to have impacts of a different sort as mines closed. Poorly planned closures led to the decant of polluted water from old workings, with significant impact on the water quality of the system. Although the mines are not the predominant source of contamination—estimated to account for just 13 % of the salt load in the Vaal river system—the mine water problem and the impact of deteriorating water quality on downstream users helped to draw attention to the need to manage water quality as well as quantity. Vaal Dam releases were required in dry periods to maintain total dissolved solids (TDS) levels below critical limits (Herold et al. 2006). This impacted the levels of assurance provided by the system and has required new operating rules, affecting the way in which Sterkfontein's storage is deployed.

12.4.4 An Energy Landscape Changing with Climate

The energy supply landscape has also changed dramatically since Phase 2 of the scheme was completed in 1982. Initially, the scheme used electricity that was cheaply available, almost as a byproduct of ESKOM's relatively inflexible fleet of large coal-fired power stations. By 2005, some of those energy benefits had been eroded. A lack of policy coherence in the energy sector had seen responsibility for decision-making become diffused and confused. This undermined planning and decision-making. Almost unnoticed, South Africa's national grid slipped from overcapacity to chronic undercapacity, aggravated by poor maintenance of a now ageing coal fleet.

By 2010, there was no longer any cheaply available off-peak power; instead, peak supplies were being met by extremely expensive diesel and gas generators while the rest of the fleet just managed to cope with off-peak demand. Despite this, demand had to be curtailed, sometimes including widespread, if structured, load-shedding. Although pumped storage helped the utility to meet weekday peaks by pumping over weekends, these developments undermined the logic of using off-peak power to pump water and maintain the levels in Sterkfontein.

This was of little concern during a decade of good rainfall. However, Sterkfontein had been built to prepare for a drought and the current hydrological drought is putting considerable strain on the Vaal system. At the time of writing (March 2016), a substantial release had been made from Sterkfontein—both to maintain rapidly declining levels in downstream reservoirs and to "make space" for further transfers during 2016. This is in line with standard operating procedures. However, it is notable that the refilling of the reservoir is proceeding more slowly than planned as ESKOM husbands its limited capacity. Although new generating capacity, from both renewable and coal sources, is being built, it would appear that the era of surplus off-peak generating capacity is over.

Climate change has not yet affected Sterkfontein's operation directly or the Vaal system more generally and there is as yet no firm indication of significant impacts, although a warming climate will increase evaporation rates and may reduce river runoff. Reviews of the future impact of climate change have been inconclusive (Lutz et al. 2013). While there is reasonable confidence that rainfall will be reduced in the western part of the country, there is no strong evidence for rainfall changes along the eastern escarpment, which is the main catchment for the Tugela-Vaal transfer and the Vaal system itself (Groves et al. 2015). For this reason, it is considered that current approaches to manage annual variability will be adequate to manage and adapt to the consequences of climate change. However, global response to climate change has put the role of coal into question. The primary impact is thus through changes in the arrangements for and costs of generating the electricity needed for the pumped transfer.

12.4.5 Technical Responses

In response to the emerging challenges, there has been a review of the approaches to system management and new guidelines have been produced to guide resource management in both normal and drought conditions. These guidelines are applicable countrywide but use the Vaal system as an example. They are based on the systems modelling approaches first used in the Vaal and now incorporate considerations such as quality management and the maintenance of environmental flows while continuing to monitor and manage dam and groundwater levels, abstractions, flows (including return flows), and general compliance (especially during droughts).

The guidelines emphasize the dynamic nature of the process and the need to continually adjust approaches to accommodate new demands on the system. They continue to emphasize that:

> The relevance of guidelines will only be confirmed through their implementation. Water managers are therefore invited to use them with a critical mind and to provide the necessary feedback that will assist in improving future versions (DWAF 2006).

A fundamental question is thus whether the operating discipline that has been established over the past four decades will continue into the future and allow sufficient adaptation to cope with changes in the broader environment.

12.5 Discussion

The review of the role and performance of Sterkfontein Dam, as a component of the Tugela-Vaal transfer and the larger Vaal system, over four decades, necessarily covers a range of perspectives, from social and economic to political and institutional in addition to the more obvious technical and environmental. It is helpful to consider the interactions and intersections between these dimensions in order to understand how initial innovations were introduced, the extent to which those are still relevant in a different context, and the potential for further innovation and adaptation.

12.5.1 Has the Tugela-Vaal Scheme Provided Value for Money?

The history of the project, with its many changes and adaptations, might reasonably lead to the conclusion that it was economically inefficient. Certainly, if the final concept had been agreed before construction began, some of the infrastructure would have been built later or not at all. Against this, however, it must be recognized that there was a growing demand for water that had to be responded to, as the 1982 drought demonstrated. A failure to maintain supplies to the Vaal Dam, which might have resulted from delays, would have had economic impacts far greater than any additional project costs incurred by apparently premature decision-making. This was also the logic that deferred the choice of the Oxbow diversion of the Orange River—and, in the event, that scheme was also fundamentally redesigned when finally implemented.

It must also be recognized that the systems analysis approaches were new and not integrated into planning when the project began. Similarly, the expansion of the coal-fired power station fleet, which underpinned the pumped transfer element of the project, had not been decided upon. Therefore, there was a timing mismatch and if there had been no flexibility in the conceptual design, it would not have been possible to take advantage of cheap power for pumping. Given the wider economic costs of supply failure, it can thus be concluded that the changes in conceptual design were reasonable—and appropriate—adaptive changes to changing circumstances.

The operational agreements were also remarkably pragmatic. The water-for-coal agreement might appear naïve today. However, the underlying logic was sound. At the time, the electricity utility had contracts with "tied suppliers" with set volumes and payment schedules. It was thus advantageous to have a mechanism to transform surplus off-peak power of very limited value into much more commercially valuable peaking power; as the operator of a range of generators with different cost characteristics, ESKOM was able to make a rigorous evaluation of these benefits. The water authorities, by sharing the use of what had been planned to be single-use infrastructure, achieved an energy cost substantially less than the commercial tariffs that a pure pumping scheme would have had to pay.

Some decades later, it is not just academically interesting to make a more rigorous analysis of the terms of this agreement and determine whether it was indeed as mutually beneficial as it appears at this level of review. As has been explained above, the changing energy context has changed the balance of economic benefits. The question now is the larger one of how the economics of pumped storage and transfers will operate in a world in which the integration of unpredictable and variable renewable energy sources is becoming the new priority.

12.5.2 *The Political Conditions that Enabled Public Sector Innovation*

In recent decades, the role of the public sector has been widely challenged since it is considered to be less efficient than market-driven arrangements. The evolution of the Tugela-Vaal process could certainly have been portrayed as an example of disastrously flawed public management had the product not been an elegant and effective contribution to the achievement of its strategic objectives. Therefore, it is interesting to consider the institutional characteristics that enabled this outcome.

The literature on developmental states posits the requirements for success as the existence of a competent and committed bureaucracy (or, in this case, technocracy) embedded in a state that is guided politically by clear strategic objectives, with the power to pursue them (Evans 2008). The South African state at the time had many of these characteristics. Although grossly skewed and deliberately excluding the black majority of the population, it had its own developmental logic. Faced with an increasingly hostile external world, by virtue of its racial policies, the South African state had to manage strategic sectors of the economy effectively. It had thus mobilized its community behind what was portrayed as a strategy for national survival, formally defined as a "total strategy" (an evolution well documented in Nattrass and Ardington 1990).

The Tugela-Vaal scheme reflected these drivers and characteristics. Water and energy supplies were clearly of critical strategic importance for the inland urban/industrial/mining hub centered in the Vaal river basin. The imminence of international sanctions had focused political attention on the need to develop generation capacity (which depended on the importation of capital equipment) and the coal-fired power stations would require an adequate and assured source of water.

It was possible to promote this agenda because the technical cadre of white professionals 'embedded' in the public service were seen to have common interests with the dominant (white) political class. There was thus a degree of trust between politicians, administrators, and technicians and an openness to innovation. This relationship extended beyond the individual Department of Water Affairs to its relations with the power utility.

It is almost quaint in today's world of accountant- and economist-driven investment funding processes to read that, in the Tugela-Vaal project:

> Agreement was reached with Eskom that the DWA would build the two dams involved, in exchange for which, the water needed for supplementing the Vaal, would be lifted off-peak and free of charge, except for the cost of the coal burned. The agreement was drafted by T Stoffberg of Escom and myself. Both were of the opinion that the water user and the electricity user were one and the same and that exact division of the benefits was not necessary (van Robbroeck 2015).

The development of the professional cadre was deliberate and supported by targeted recruitment of technical specialists from European countries. There was an extensive program, led from a ministerial level, to allow some of the best and brightest candidates to study at leading institutes across the world. Aside from PhD study at universities like MIT and Colorado, this included attachments to the US Bureau of Reclamation and Army Corps of Engineers as well as cooperation with European technical institutes on technical design issues. The need to develop high-level human resources was explicitly recognized by the 1970 Commission of Enquiry. However, the pioneers of this cohort had already brought to South Africa the systems management concepts that underlay the final incarnation of the Tugela-Vaal transfer and associated Drakensberg pumped storage scheme.

Although there was a formal process for the Parliamentary political approval of projects, the senior administrators felt able to support substantial changes proposed by their technicians as implementation proceeded. For this reason, the coincidence of timing between the Tugela-Vaal scheme and emergence of systems analysis approaches to water resource design did not block innovation, even though its adoption must have cost significant political and reputational capital.

It is also interesting to note how practice influenced policy and vice versa. Some developmental state efficiencies had already been introduced in the 1956 Water Act, which required formal white papers, including economic analysis, to justify projects before they were initiated. These were subsequently supplemented and strengthened by requirements to undertake cost-benefit analysis and environmental impact assessments and to use systems analysis approaches to the planning of water schemes. The documentary and contextual evidence suggests that there was significant interaction between the 1970 Commission of Enquiry and the Tugela-Vaal transfer process.

However, the cooperation at an operational level illustrates the value of collaborative rather than competitive approaches, which are inherently, although not always practically, easier to achieve in the public rather than the private domain. As Norman (1982) concluded, from the perspective of the electricity utility:

> Bearing in mind the need to reach agreement on many aspects of the planning, design and operation of Drakensberg as well as on a fair apportionment of capital and operating costs between WSCM and DWA it is felt that the Drakensberg Project is not only a significant engineering achievement but also an outstanding example of cooperation between all parties involved.

12.5.3 Technical Perspectives: Systems with Storage as Supply Assurance

The technical key to the successful development of the Tugela-Vaal scheme was the introduction of systems analysis approaches that focused on assured system yield rather than on a simple summation of the volumes of water supplied into a system. This enabled the fundamental challenge of managing an extremely variable hydrology to be addressed in a cost-effective manner in relatively complex physical systems comprising a number of sources, storages, and links. The decision to build a very large storage scheme almost entirely dependent on transfers was bold, although the technical logic guiding it was strong. The contribution of the scheme in general and Sterkfontein in particular to alleviating the impact of the 1982 drought in the Vaal supply area strengthened its reputation as a successful innovation. This was enhanced when it was found that, in practice, transfer volumes were significantly lower than anticipated.

The development of a cooperative pumped storage/transfer scheme was both more obvious in concept and more of a coincidence in adoption. It still impacted on broader local practice. As acknowledged by ESKOM, "the Drakensberg scheme paved the way for Eskom's second pumped storage project at Palmiet in the Cape" (ESKOM 2016). This 400-MW station provides greater flexibility to the supply of electricity in the Western Cape region, where generation is anchored by the Koeberg nuclear power station, which has similar characteristics to large coal stations. The Palmiet scheme, planned in 1982 and completed in 1988, also provides interbasin transfers albeit on a smaller scale than the Drakensberg scheme, at an average 22.5 Mm^3/year. This was not done with the third large scheme (Ingula), approximately 60 km north of Sterkfontein, on the same escarpment as the Tugela-Vaal system. One reason for this was that the water sector had decided that the next tranche of supply augmentation would come from Lesotho. Proposals to use a proposed fourth pumped storage scheme to lift water required for a large semi-rural population in the Limpopo province were terminated when ESKOM decided that more pumped storage was not required at present.

The final development, further enhancing the achievement of system reliability, was the linkage, described above, of the Vaal Dam to the Grootdraai subsystem. This enabled direct support from Sterkfontein into that subsystem, which reduced the risks posed by reliance on a single-source infrastructure as well as increasing overall assurance—and flexibility of allocation should there be a severe event.

12.5.4 The Current Conjuncture: Challenges of Continuity and Trust in Storage Management

The operation of a system of water storage is sensitive to political pressures. Stored water is valuable, particularly when there is a shortage. As a consequence, its management cannot be regarded as a purely technocratic exercise. This has been demonstrated during South Africa's current drought when, in a number of locations, municipal water supplies have remained unrestricted until their source dams ran dry. In many cases, this was the result of a political reluctance to implement restrictions, despite technical advice to the contrary. Similar situations have occurred with hydropower generation in neighboring countries; the Kariba Dam, between Zimbabwe and Zambia on the Zambezi river, has fallen to extremely low levels, severely constraining electricity generation, not least because officials disregarded earlier attempts to reduce power generation to avoid deeper cuts. An even more egregious example of political engagement in the operation of storage occurred in Brazil during Sao Paulo's recent drought (Muller 2015).

The operation of Sterkfontein Dam has not yet seen such impacts, although there are preliminary signs that similar behavior may occur, not least through the identification of the dam as a source of water for local urban populations that could have found alternative sources. Elsewhere in the Vaal system, releases were made from Lesotho dams to supply a number of towns that face critical shortages, which were again aggravated by a failure to conserve limited resources.

One consequence of the changing context in which the Sterkfontein Dam is operated as part of the Vaal system is that system operation rules will have to be adapted and the overall approach may need more radical change. As new demands are placed upon it (from within and outside of the system), the contribution of a fixed volume of storage to levels of assurance will be reduced.

Furthermore, as new data on climate variability, particularly extreme events, is incorporated in systems models, the assured yield of the system changes. This may raise concern amongst policymakers about the reliability of the data and methods used, although a robust indicator of reliability problems would be if there was no change in estimates after an extreme event.

Because policy decisions reflect the perceptions of policymakers, there is a need for structured technical engagement and input at this level to ensure that policymakers have confidence in the systems when there are political pressures to ignore operating rules. In the case of Sterkfontein and the Vaal system generally, a particular challenge is that the generation of technicians who developed the system has retired.

There has been limited institutional continuity, due to a failure to sustain the high-level technical cadre built up after 1970 as well as the changes in the wider public service. This break in continuity could pose risks to the effective operation of the system. If the current drought is prolonged, the very visible availability of stored water in Sterkfontein will raise questions about its deployment to meet the needs of local communities. Allowing water from Sterkfontein to be discharged into the system would respond to an understandable desire to see "national infrastructure" supporting local services and mitigating the impact of drought, even though its effect might be the reverse.

12.6 Conclusions and Lessons

The Sterkfontein Dam has played a central role in assuring the water security of the region served by the Vaal system and thus supported its economy and people. It has shown how storage can play a strategic role, independent of the source of the water to be stored. The location of Sterkfontein's storage in the system has allowed new priorities to be addressed.

At a time when national energy generation capacity was limited, the associated pumped storage facility provided flexibility that allowed maintenance to be undertaken and supplies to be maintained on a predictable basis as well as reducing expenditure on high-cost peaking power alternatives, such as diesel generation. As the share of variable and relatively unpredictable wind- and solar-based renewables in the system grows, pumped hydro will increasingly be used to support their integration into the system due to its rapid response times.

Other lessons can be drawn from the planning, construction, and operation of the Sterkfontein Dam, which has been demonstrated to have achieved its objectives of supporting water security for a strategically important region of South Africa over four decades.

A multipurpose systems analysis approach should be taken in planning water resource development in water-constrained regions with high and growing demands. This has been a policy doctrine in South Africa (1970a), later reinforced at the 1977 United Nations Water Conference in Mar del Plata, Argentina (UN 1977). The case of Sterkfontein has demonstrated that the involvement of key water users in planning and implementation can often generate innovation and achieve better results.

Within such systems, the strategic storage often needed to provide assurance of supply may be provided separately from the main sources of streamflow. At Sterkfontein, this was both cost-effective and helped to reduce environmental impacts, not least through the reduction in evaporative losses.

Storages that are part of interlinkages between different river basins may strengthen resilience because there is often a lower probability of below average flows occurring simultaneously in both basins.

The long-term nature of water resource infrastructure investment means that there is a likelihood of contextual changes during the life of a scheme, which need to be provided for in their design and operational arrangements. While appropriately located and managed storage can make a valuable contribution to system resilience, there is a risk that the assurance may prove to be illusory if operating rules are not regularly reviewed and adapted to reflect changes in the wider physical, social, and economic context.

Where extensive pumping is required in a water resource system, energy planning considerations, including the overall profile of and trends in energy system supply and demand, should be taken into account. The water supply assurance for coal-fired stations, underpinned by Sterkfontein, has been its major contribution to the reliability and resilience of power supply, although pumped storage also assists the management of grid stability in multigenerator power grids—a function that is becoming more important as intermittent renewables are introduced into the system.

The example of Sterkfontein suggests that, while the storage requirements of pumped storage will usually be orders of magnitude less than that for increasing water supply assurance, the pumping capacity for generation is likely to be far greater than required for water supply purposes. On the other hand, the flexible demand offered by pumped transfers to storage may further contribute to the integration of renewable energy generation into grid systems during periods of surplus production that have been observed in other systems. Such multipurpose uses can increase assurance, resilience, and efficiencies, adding value to the individual functions performed.

Beyond the normal environmental concerns to be addressed, the implications of the operational characteristics of strategic storage within the particular system need to be considered. These may be significantly different from other storages, with longer (decadal) periods at full capacity coupled with short (annual) periods during which the reservoir may be drained relatively rapidly to very low levels, which will have an impact on recreational and related uses.

Finally, it should be noted that there were substantial elements of luck involved in the evolution of the Tugela-Vaal scheme, of which Sterkfontein's storage is a critical component. This in large measure was due to the phased implementation process, which allowed options and opportunities to be better explored and understood. However, without the political requirement to move the upper reservoir, the pumped storage option might never have arisen; had a staff member not moved from the water department to the power utility, the opportunity for cooperation in pumped storage may not have been discovered in time. This in turn may simply prove that good luck is often the product of hard preparatory work and institutions that are sufficiently responsive to take advantage of opportunities that may arise.

References

Alexander WJR (2010) Analytical methods for water resource development and management. http://www.droughtsandfloods.com/Chapter%2013%20Development%20of%20operating%20rules.pdf

DWAF (2003) ISP: Vaal river system overarching report no: P RSA C000/00/0103. Department of Water Affairs and Forestry, Pretoria

DWAF (2006) Guidelines for water supply systems operation and management plans during normal and drought conditions, Shand and Furumele, DWAF Report No. RSA C000/00/2305 (Appendix C: Vaal Pilot Study C2.2). Department of Water Affairs and Forestry, Pretoria. https://www.dwaf.gov.za/documents/policies/droughtguidevol2oct06anc.pdf

DWS (2015) Vaal river system: annual operating analysis 2015/2016—meeting record. Department of Water and Sanitation, Pretoria

Elges HFWK (1982) The dams of the Tugela-Vaal project. Civ Eng South Africa Drakensberg Pump Storage Scheme 24(8):375–377

ESKOM (2016) Drakensberg pumped storage scheme: technical information. ESKOM, Johannesburg. http://www.eskom.co.za/OurCompany/MediaRoom/Documents/DrakensbergTechnicalinfo.pdf. Accessed Mar 2016

Evans P (2008) In search of the 21st century developmental state. The centre for global political economy, University of Sussex Working Paper 4. Brighton, Sussex

Groves D, Mao Z, Liden R, Strzepek KM, Lempert R, Brown C, Taner MÜ, Bloom E (2015) Adaptation to climate change in project design. In: Cervigni R, Liden R, Neumann JE, Strzepek KM (eds) Enhancing the climate resilience of Africa's infrastructure. World Bank, Washington, DC, p 131

Herold CE, Le Roux PJ, Nyabeze WR, Gerber A (2006) WQ2000 Salinity model: enhancement, technology transfer and implementation of user support for the Vaal system. Umfula Wempilo Consulting, Pretoria, South Africa

Lutz J, Volkholz J, Gerstengarbe FW (2013) Climate projections for southern Africa using complementary methods. Int J Clim Chang Str 5(2):130–151

Maass A, Hufschmidt MA, Dorfman R, Thomas HA Jr, Marglin SA, Fair GM (1962) Design of water-resource systems: new techniques for relating economic objectives, engineering analysis, and governmental planning. Harvard University Press, Cambridge, MA

McKenzie RS, Stoffberg FA, Little PR (1998) An overview of the orange river replanning study. Civ Eng Johannesburg 6:21–24

Mokonyane N (2015) Minister Nomvula Mokonyane: status of drought. Department of Water and Sanitation, Pretoria. http://www.gov.za/speeches/minister-nomvula-mokonyane-status-drought-1-nov-2015–0000. Accessed Mar 2016

Muller M (2015) Political blockages cannot be allowed to let taps run dry. Business Day, 23 January, Johannesburg

Nattrass N, Ardington E (1990) The political economy of South Africa. Oxford University Press, Oxford

Norman HB (1982) Power system aspects of pumped storage plant. Trans SAIEE 73:150

Pretorius SJ, Oberholzer G, Van Eeden JA (1976) The Tugela-Vaal state water scheme as a bilharzia risk. S Afr Med J 50(25):968–972

Reuss M (2003) Is it time to resurrect the Harvard water program? J Water Resour Plan Manag 129 (5):357–360

Roberts CPR, Erasmus JJ (1982) Environmental considerations of the Drakensberg pumped storage scheme. Civ Eng South Africa Drakensberg Pump Storage Scheme 24(8):361–363

South Africa (1956) National water act, act 54 of 1956. Government Printer, Pretoria

South Africa (1969) Report on the proposed Tugela-Vaal government waterwork. Doc. No. W. P. U-'69. Department of Water Affairs, Pretoria

South Africa (1970a) Report of the commission of enquiry into water matters, R.P.34/1970. Pretoria

South Africa (1970b) Supplementary report on the proposed Tugela-Vaal government water project. Doc. No. W.P. W-'70. Department of Water Affairs, Pretoria

South Africa (1974) Report on the proposed second phase of the Tugela-Vaal government water project with the Drakensberg pumped storage development. Doc. No. W.P. R-74. Department of Water Affairs, Pretoria

South Africa (1980) Fifth supplementary report on the Tugela-Vaal government water project (increased cost of second phase, and proposed raising of Sterkfontein Dam). Doc. No. W. P. H-80. Department of Water Affairs, Pretoria

South Africa (1984) Seventh supplementary report on the Tugela-Vaal government water project (extension of phase II). Doc. No. W.P. J-84. Department of Environment Affairs, Pretoria

South Africa (1998) National water act, act 36 of 1998. Government Printer, Pretoria

Stats SA (2014) Gross domestic product annual estimates 2004–2013, statistical release P0441. Statistics South Africa, Pretoria

Stats SA (2015) Mid-year population estimates 2015, statistical release P0302. Statistics South Africa, Pretoria

Stephenson D (1971) Integrated development of the Orange-Vaal and Tugela basins. S Afr J Sci 67:457–463

TCTA (2016) VRESAP. Trans-Caledon Tunnel Authority, Pretoria. http://www.tcta.co.za/#!vresap/yid36. Accessed Mar 2016

Tempelhoff JWN (2003) The substance of ubiquity: rand water: 1903–2003. Kleio Publishers, Vanderbijlpark

UN (1977) Report of the United Nations water conference, United Nations, New York. www.ircwash.org/sites/default/files/71UN77-161.6.pdf. Accessed Dec 2014

van Robbroeck TPC (1982) The Drakensberg project: water and power for South Africa. Civ Eng South Africa Drakensberg Pump Storage Scheme 24(8):343–345

van Robbroeck TPC (2015) Water engineering: memories of a career in water engineering and management. http://lieberheim.blogspot.co.za/p/blog-page_70.html. Accessed Feb 2016

van Robbroeck TPC, Pullen RA, Graber BW, van Schalkwyk A (1970) Proceedings of convention: water for the future. Water year 1970, Pretoria

Index

C. Tortajada (ed.), *Increasing Resilience to Climate Variability and Change*,
Water Resources Development and Management,
DOI 10.1007/978-981-10-1914-2

X

Y

Printed in the United States
By Bookmasters